Field Guide to

WISCONSIN GRASSES

Field Guide to

WISCONSIN GRASSES

EMMET J. JUDZIEWICZ
Department of Biology and Museum of Natural History, University of Wisconsin–Stevens Point

ROBERT W. FRECKMANN
Department of Biology, University of Wisconsin–Stevens Point

LYNN G. CLARK
Department of Ecology, Evolution, and Organismal Biology, Iowa State University

MEREL R. BLACK
Department of Biology, University of Wisconsin–Stevens Point

THE UNIVERSITY OF WISCONSIN PRESS

The University of Wisconsin Press
1930 Monroe Street, 3rd Floor
Madison, Wisconsin 53711-2059
uwpress.wisc.edu

3 Henrietta Street, Covent Garden
London WC2E 8LU, United Kingdom
eurospanbookstore.com

Printed in the United States of America

Library of Congress Cataloging-in-Publication Data
Judziewicz, Emmet J., author.
Field guide to Wisconsin grasses / Emmet J. Judziewicz, Robert W. Freckmann, Lynn G. Clark, and Merel R. Black.
pages cm
Includes bibliographical references and index.
ISBN 978-0-299-30134-7 (pbk. : alk. paper) — ISBN 978-0-299-30133-0 (e-book)
1. Grasses—Wisconsin—Identification. I. Freckmann, Robert W., author.
II. Clark, Lynn G., author. III. Black, Merel R., author. IV. Title.
QK495.G74J83 2014
584′.909775—dc23
2014007281

Contents

Acknowledgments vii

Morphology: Grass Structure 3

Agrostology: The Study of Grasses 23

Grasses in Wisconsin Plant Communities 33

Keys to the Grass Genera of Wisconsin 61

Field Guide to Wisconsin Grasses 91

Glossary 321
References 327
Illustration Credits 331
Taxonomic Index 335

Acknowledgments

We are especially grateful to the staff of the Wisconsin State Herbarium in Madison, especially to Hugh H. Iltis, Theodore S. Cochrane, Mark A. Wetter, and Ken Cameron for their help over the years for providing access to the herbarium and the statewide specimen database. We are also grateful to the estates of Elsie H. Froeschner (1913–2006) and Anna Gardner (1958–2006) for permission to use their wonderful line drawings and photographs of Iowa grasses. The Wisconsin-Iowa grass connection is a close and long-standing one: Both Clark and Freckmann received their PhD's in grass systematics at Iowa State University under Dr. Richard W. Pohl (1916–93), a Milwaukee native who received his undergraduate degree at Marquette University; Froeschner's drawings were first used in Pohl's "The Grasses of Iowa" (1966); Clark and Judziewicz collaborate on research on tropical bamboos (which are also grasses).

We thank the curators of herbaria who welcomed us: Tom Lammers and Neil Harriman (UW–Oshkosh), Tim Gerber (UW–La Crosse), Gary Fewless (UW–Green Bay), and those who contributed photographs: William S. Alverson, Derek Anderson, Matthew Bushman, Joan Elias, Eric J. Epstein, Gary Fewless, Steve Garske, Hugh Iltis, Kitty Kohout, Jim Meeker, Arthur Meeks, Thomas A. Meyer, Scott Milburn, Christopher Noll, Robert C. Roe, Suzanne Sanders, and Paul Skawinski; and Barney Lipscomb for the photograph of the painting of Lloyd Shinners.

For assistance in the project we also thank Hugh Iltis, Carol Kropidlowski, Kevin Doyle, Larry Leitner, Anita Cholewa, Rebecca Gregory, Mary Barkworth, and the 2013 UW–Stevens Point agrostology class. Finally, we thank Gwen Walker and Sheila Leary at the University of Wisconsin Press for their continued interest in our natural history book endeavors.

Judziewicz thanks the University of Wisconsin–Stevens Point and Dean Chris Cirmo for the Justus F. and Barbara J. Paul Faculty Award, which enabled much of this work to be done during his sabbatical, and his mother, Lucinda M. Judziewicz (1921–2011), who attended the awards ceremony but passed away shortly thereafter. Freckmann thanks his wife, Sally. Clark thanks the late Dr. Richard W. Pohl for sharing his incredible knowledge of grasses; the late Anna Gardner for her beautiful close-up photography, which made the website Grasses of Iowa possible; and the Fred Maytag Foundation for providing the major funding for the development of that website. Black dedicates her portion of the book to her grandson, Taiji, and thanks her husband, Joel, for his support through the years.

About the Illustrations

The illustrations were drawn by Elsie H. Froeschner for Richard W. Pohl's treatment of the grasses of Iowa (1966); many were redrawn by Anna Gardner. A few are from Hitchcock and Chase (1951). About two-thirds of the grass photographs were taken by Gardner from Iowa specimens and the remaining one-third by other, mainly Wisconsin photographers. A complete list of illustration credits precedes the index at the end of the book.

Sources

Keys and species descriptions and distributions are from Wisconsin specimens examined at the following herbaria: University of Wisconsin–Stevens Point, University of Wisconsin–Madison, University of Wisconsin–Oshkosh, University of Wisconsin–Green Bay, University of Wisconsin–La Crosse, and Milwaukee Public Museum, as well as data in published treatments, especially Flora of North America Editorial Committee (2003, 2007), Barkworth et al. (2007), and *Field Manual of the Michigan Flora* (Voss and Reznicek 2012).

Status

Some species have a special status, which is listed beneath the species epithet. A species' status may change over time. The following is a list of possible statuses and their definitions in this guide.

Endangered: Refers to species that are endangered in the state of Wisconsin. Describes any species whose continued existence as a viable component of this state's wild animals or wild plants is determined by the state Department of Natural Resources to be in jeopardy on the basis of scientific evidence.

Threatened: Refers to species that are threatened in the state of Wisconsin. Describes any species that appears likely, within the foreseeable future, on the basis of scientific evidence, to become endangered in the state of Wisconsin.

Special Concern: A Wisconsin designation that describes a species about which some problem of abundance or distribution is suspected but not yet proven. The main purpose of this category is to focus attention on certain species before they become threatened or endangered in the state.

Field Guide to

WISCONSIN GRASSES

Morphology

Grass Structure

What Is a Grass?

A grass is an herbaceous plant in the family called the Poaceae. Grasses are differentiated from all other plants by their possession of a compound inflorescence of discrete, small inflorescences called spikelets, usually composed of two glumes and one to many florets, each floret usually composed of a lemma and palea enfolding a tiny flower that matures into a fruit called a grain or caryopsis.

There are 668 genera and 11,160 species of grasses in the world (Stevens 2001 onward). Grasses are the most economically important plant family because they are the source of the four most-eaten cereal grains in the world (rice, wheat, corn, and sorghum); they are critical as pasturage (bluegrass and oats) upon which cattle, sheep, and goats graze; they are a major source of sugar (sugarcane) and energy (sugarcane, corn, and switchgrass biofuels); and they are fermented and/or distilled into alcoholic beverages such as beer and whiskey. Eighty-seven genera (13 percent of the world's total) and 232 species (2 percent of the world's total) occur without cultivation in Wisconsin.

How Grasses Differ from Grass-Like Plants such as Sedges and Rushes

Grasses (family Poaceae), sedges (family Cyperaceae), and rushes (family Juncaceae) are graminoid or grass-like plants often confused with each other. All three are common in Wisconsin and can be found in a wide variety of habitats, although grasses tend to dominate in upland sites, while sedges and rushes are found in wetlands.

Members of the Poaceae family are common worldwide, although more species are found in tropical and warm temperate areas than in colder regions. Grasses are usually more abundant in dry and open habitats, including deserts.

The Cyperaceae family includes about 98 genera and 5,430 species (Stevens 2001 onward). About 235 species occur in Wisconsin. Sedges are also distributed worldwide but are commoner in moist to wet regions, rarely in deserts.

The Juncaceae family includes 7 genera and about 430 species (Stevens 2001 onward); 28 species occur in Wisconsin. Most rush species grow in the colder northern regions in wet areas.

There is no single vegetative character that can be used to tell grasses, sedges, and rushes apart. Instead, combinations of characters are most useful. In practical terms, however, if you hold a grass-like plant at arm's length and see these features, then you can be 99 percent certain that you are holding a grass:

the leaves are evenly distributed along the culm (stem) and not clustered near the base;
the leaf blades are flat or only slightly folded along the midrib;
the leaf blade margins are smooth, not harsh, to the touch when stroked downward from the tip toward the base; and
the inflorescence has a single main axis.

Some grasses, however, do not follow the first three "rules."

The following paragraphs differentiate the three families in more detail.

Culms (stems)

Comparison of sedge (Cyperaceae) culms (*two large stems on left*: top is a species of bulrush, *Schoenoplectus*, bottom a species of woolgrass, *Scirpus*) with small hollow grass (Poaceae) culm (*right*).

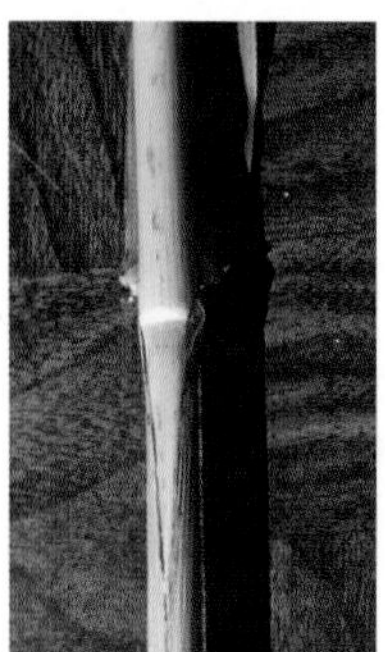

Grass showing open, unfused leaf sheath embracing stem (culm) at a node (the whitish swelling of the culm where a leaf is attached).

Grass nodes are conspicuous and often swollen, with growth tissue capable of turning up a lodged (bent-down) stem. Sedge and rush nodes are usually inconspicuous and not swollen.

Grass internodes are usually hollow, although warm-season grasses (especially those storing food during good growing conditions for later flowering) are often solid. Sedge and rush internodes are usually solid, although emergent aquatic species have pulpy centers, which often break down with age.

Grass stems are almost always terete (i.e., round in cross section, like the body of a snake) and are only rarely flattened. Sedge stems are usually triangular in cross section, rarely terete, flattened, or four-sided. Rush stems are usually terete, rarely flattened.

Leaves

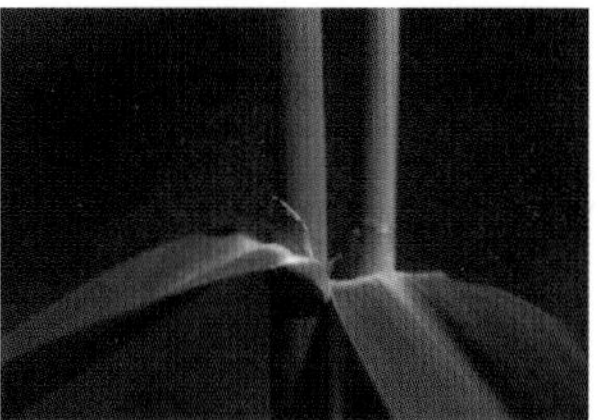

Grass leaf showing membranous ligules and more or less flat leaf blades.

Sedge (*Carex*) leaf showing fused (closed) leaf sheath and high-arching, rim-like ligule fused to the base of the blade.

Sedge (*Carex*) leaf showing characteristic pleated (W-shaped) blade.

Comparison of sedge (*left*, *Carex*) with grass (*right*, *Echinochloa*) culms; in sedges the leaves are clustered toward the base and/or tip of the culm, while in grasses the leaves are produced all along the culm.

Two-ranked aspect of grass (*Setaria*) leaves, the leaves diverging at the nine o'clock and three o'clock positions.

Three-ranked aspect of sedge (*Carex muskingumensis*) leaves, the leaves diverging at the ten o'clock, two o'clock, and six o'clock positions.

Grass leaves are usually fairly evenly spaced along the stem, but they are sometimes clustered in a basal rosette and very rarely clustered at the base of the inflorescence. Sedges and rushes usually have the leaves clustered at or toward the base of the stem and often also at the base of the inflorescence.

Grass sheaths are usually open (overlapping but not fused) or sometimes closed (fused). Sedge sheaths are almost always closed, and rush sheaths are usually closed.

Grass sheaths have a distinct collar; sedges and rushes do not.

Grass sheaths usually have a ligule composed of hairs, a membrane, or a membrane tipped with hairs. Most sedges lack a ligule; some have a membranous ligule. Nearly all rushes lack a ligule.

Grass, sedge, and rush leaves are alternate (i.e., only one leaf arises at a node). Grass blades are almost always developed and are two-ranked (i.e., they lie in a single vertical plane on alternating sides of the stem), except in basal rosettes, where they tend to be unranked ("unranked" means with the leaves seeming to be produced in all directions).

Sedge blades are usually three-ranked (i.e., they lie in three vertical planes arising from the apices of the triangular stem), although they tend to become unranked when clustered together. Rush blades can be unranked, three-ranked, or sometimes two-ranked. Many sedge and some rush leaves do not develop blades.

Grass blades are usually flat, sometimes folded on the midrib, and sometimes involute (i.e., rolled lengthwise). The margins are usually smooth, sometimes retrorsely serrated (i.e., with serrations pointed downward), rarely antrorsely serrated (i.e., with serrations pointed upward). Sedge blades are often partially folded on the midrib with an additional fold near the margins so that they are W-shaped in cross section. The margins are roughened with serrations pointed upward. Rush blades are flat, or involute, or rolled into a hollow tube with distinct cross partitions. The margins are smooth.

Inflorescences

Comparison of the inflorescences of two grasses (Poaceae), a sedge (Cyperaceae), and a rush (Juncaceae). The two grasses (*left*) have a main central axis, while the sedge and rush have several branches, no one of which is dominant.

Sedge (*Scirpus pendulus*) inflorescence, showing the lack of a central axis. Also note the fused leaf sheath (*lower left*).

Rush (*Juncus pelocarpus*) inflorescence, showing the lack of a central axis.

The most common grass inflorescence is a panicle with a main axis and the branches arising singly or in whorls at several levels. Some grasses have a single two-sided spike; some others have one-sided spikes in a pinnate or palmate arrangement. Sedges and rushes have panicles with radiating main branches and no central axis. Some sedges have a solitary spikelet, while others have a spike, a raceme, or a condensed panicle of spikelets. In many species in all three families, spikelets or flowers are clustered into small dense glomerules.

Flowers

The flowers of almost all grasses, sedges, and rushes are wind-pollinated, small, and without showy sepals and petals. Most are bisexual (the major exception is the sedge *Carex*), but some grasses and sedges are monoecious or dioecious. Rushes are bisexual or rarely dioecious.

Most grasses lack sepals or petals, although the two or three lodicules may represent modified petals. Sedges usually have scales, hairs, or bristles representing the sepals and no petals. Rushes have six green or brown dry or hardened sepals and petals.

Most grasses and sedges have three stamens, rarely one to six, with pollen shed singly. Rushes have six or three stamens, with pollen shed in tetrads (groups of four).

Grasses usually have two feathery stigmas (three in bamboos, one in corn). Most sedges have two or three slender stigmas, and rushes have three slender stigmas.

Spikelets

Oat (*Avena sativa*) spikelet, showing prominent basal glumes.

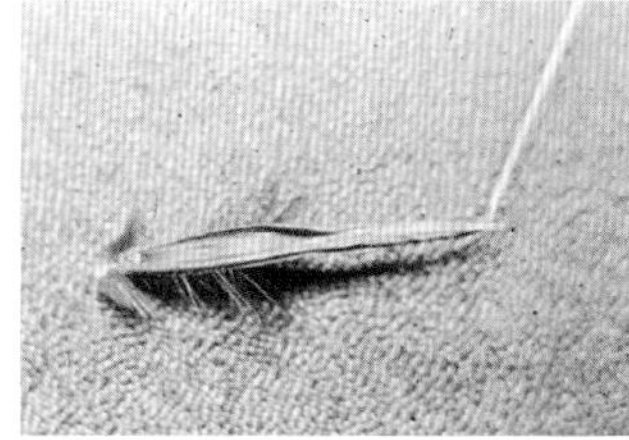

Grass floret, showing a dense beard of basal hairs (*left*) and a long awn (*right*).

Sedge (*Cyperus diandrus*), showing about fifteen spikelets, each with undifferentiated spikelet scales.

Grass and sedges have spikelets; rushes do not.

Grass spikelets are flattened, with the parts two-ranked. Sedge spikelets are usually cylindrical or ovoid, with the parts spirally arranged, although some are flattened and two-ranked, as in three-way sedge (*Dulichium*).

Grass spikelets usually have two glumes (empty bracts) at the base. Sedge spikelets do not have empty glumes; each bract subtends (embraces and partly conceals) a flower.

Grass flowers or fruits are enclosed between two scales, the lemma and the palea. Sedge flowers or fruits are usually subtended by a single scale and are usually partly visible.

Grass spikelets usually have one to twelve florets. Sedges usually have a large number of florets, often dozens or even hundreds per spikelet.

Grass spikelets have a continuous axis (the rachilla), which often breaks at the base of each floret. Sedge spikelets usually have a continuous axis, which does not break up at maturity.

Fruits and seeds

Most grasses have a grain (caryopsis) fruit; it is usually dispersed within the floret or the entire spikelet. Sedges have an achene (the seed is separated from the fruit coat by a small air space); it is usually dispersed with the sepal bristles, within a perigynium (*Carex*), or alone. Rushes have a three-seeded to usually many-seeded capsule; the tiny seeds are dispersed separately.

Most grass seeds are relatively large (compared to seeds of sedges and rushes) and have a fleshy to hard endosperm; a few have a liquid endosperm. Sedge seeds have a flour-like or firm endosperm. Rush seeds have a fleshy endosperm.

The Parts of a Grass Plant

We recommend Clark and Pohl (1996) to those who want to learn more about grass morphology. The following discussion is freely adapted from Freckmann (1977).

For students and many amateur botanists (and even some professional botanists), running an unknown grass specimen through a dichotomous key seems to be an overwhelming task. There are widespread beliefs that grasses are difficult to learn, that they hybridize, and that they all look alike! Although there is an element of truth in these beliefs, the same things may be said about composites, sedges, mustards, and other plant families. The real obstacles, though, usually lie in the beginner's lack of understanding of grass structure and terminology and the difficulty of getting started in the key.

Annual versus perennial habit: bunchgrasses, rhizomes, and stolons

The clump-forming annual habit of bearded sprangle-top (*Diplachne fusca* subsp. *fascicularis*): the entire plant has weak roots and is easily picked up.

The perennial habit of prairie dropseed (*Sporobolus heterolepis*): the dense clump or tussock resprouts after a fire, with a porcupine-like aspect.

The strongly rhizomatous habit of reed canary grass (*Phalaris arundinacea*).

The base and underground portion of a grass exhibit a few variations. Many grasses have rhizomes: horizontally spreading underground stems, which bear small yellowish scale-like leaves and clumps of roots at regular intervals. Good examples are Kentucky bluegrass (*Poa pratensis*), quackgrass (*Elymus repens*), and reed canary grass (*Phalaris arundinacea*). At the opposite extreme are the so-called bunchgrasses, which, unlike rhizomatous species, spread very little laterally each year. An example of a bunchgrass is prairie dropseed (*Sporobolus heterolepis*), which dies down to near ground level each winter and regenerates from buds and clusters of small leaves telescoped together and crowded on a crown of stems and roots usually just below ground level. Still other grasses are annuals, arising from seed during the growing season and dying back completely by the end of the season.

One of the most difficult decisions to be made in keying out a grass is the choice as to whether it is an annual or a perennial bunchgrass. This decision is usually easier to make early in the growing season, since the perennial can be expected to be well developed with several stems and branches before many annuals are much more than delicate seedlings. But by the latter part of the growing season, the decision rests on a fine and sometimes subjective distinction: the perennial has a tougher, thicker culm base and often bears more buds and branches clustered closer together at this base than does the annual. Finally, very late in the season, the annual will probably have most of its leaves dead or dying, leaving almost nothing green or fleshy near the base, indicating that all of its food reserves are going into fruits for next year's population, whereas the perennial is greenish or "fleshy" at the base. A simple test that often works is that an annual plant can be easily pulled from the ground and has a minimal root system.

Culms (stems)

Compared with the spikelet characteristics, the structures of grass stems and leaves are fairly simple. The grass stems (called culms) are usually herbaceous, although sometimes hardened. The nodes (the points where leaves are attached) are conspicuous and often swollen, with meristematic (actively growing and dividing) tissue capable of turning up a lodged or fallen stem. The portions of the culms between the nodes are called internodes. These are usually hollow, although in the panicoid grasses such as corn and big bluestem (those storing food during good growing conditions for later flowering), they are often solid. Grass culms are almost always terete (i.e., round in cross section) or rarely flattened, as in Canada bluegrass (*Poa compressa*) and goose grass (*Eleusine indica*). A majority of our species are so-called cool-season grasses (they have the C_3 photosynthetic pathway) and have hollow culms. Common examples are bluegrasses (*Poa*) and bromes (*Bromus*). A minority are warm-season grasses (they have the C_4 photosynthetic pathway) and have solid culms; examples are switchgrass (*Panicum virgatum*), big bluestem (*Andropogon gerardii*), and corn (*Zea mays*).

Leaves

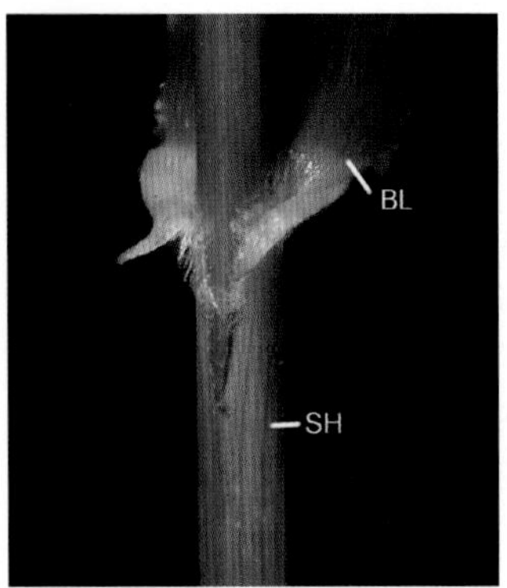

Ear-leaved brome (*Bromus latiglumis*) showing auricles and closed (fused) leaf sheath.

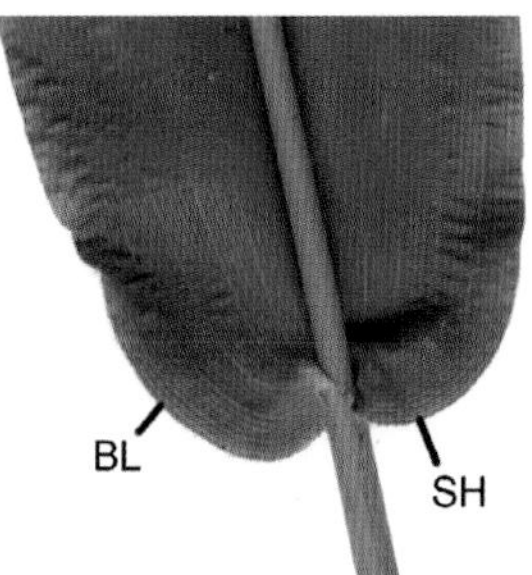

Cordate (heart-shaped) leaf blade base and open leaf sheath in broad-leaved panic grass (*Dichanthelium latifolium*), with open (unfused) leaf sheath.

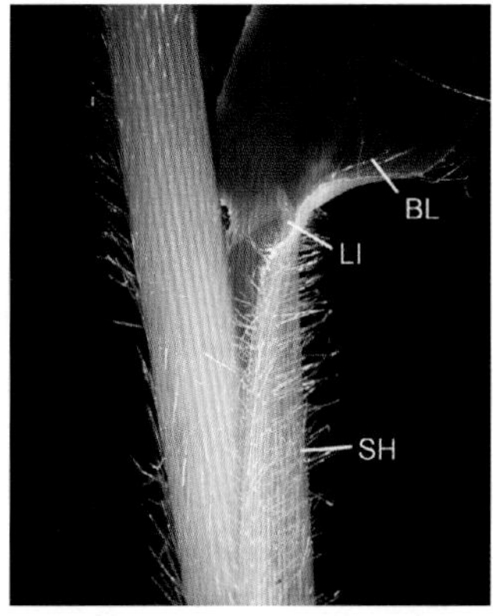

Hairy ligule and open leaf sheath of witch grass (*Panicum capillare*).

Grass leaves are usually fairly evenly spaced along the culm but sometimes clustered in a basal rosette (as in prairie dropseed, *Sporobolus heterolepis*). The basal part of each leaf (the sheath) is usually open, with overlapping but unfused margins. In about a half dozen genera, the margins are fused for about a third to nearly their full length. These are keyed out in Key 3. The summit of the leaf sheath has a ligule composed of either a membrane (the majority of our genera), hairs, or a membrane tipped by hairs. In a few grasses, such as ear-leaved brome (*Bromus latiglumis*) and the meadow fescues (*Lolium*), the summit of the leaf sheath bears a pair of ear-like auricles that project from either side and sometimes partly wrap around the culm.

Grass leaf blades are almost always developed and are two-ranked (i.e., they lie in a single vertical plane on alternating sides of the culm), unless crowded into a basal rosette, where they tend to be unranked. The blades are usually flat, sometimes folded on the midrib, and sometimes inrolled (involute, i.e., rolled lengthwise). The margins are usually smooth, sometimes retrorsely serrated (i.e., with serrations pointed downward), or rarely antrorsely serrated (i.e., with serrations pointed upward).

Inflorescences

The most common grass inflorescence is a panicle of spikelets (discussed below) with a main axis and the branches arising singly or in whorls at several levels. Some grasses have a single two-sided spike; some others have one-sided spikes in a pinnate (like the veins in a feather) or palmate (finger-like) arrangement. In some genera the spikelets are crowded into small dense clusters.

Spikelets

The flowers of grasses are aggregated into discrete little "packages" called spikelets. Grass spikelets are usually flattened, with the parts two-ranked (i.e., arranged back and forth on both sides of a central axis). Each spikelet has two empty bracts called glumes at the base; exceptions are cut-grass (*Leersia*) and wild rice (*Zizania*), which do not have glumes. Above the glumes, individual grass flowers are enclosed between two scales, the lemma and the palea. The lemma, palea, and enclosed flower together are called a floret, a basic and very important component of grass identification. The number of florets per spikelet can range from just one to twelve or more. These florets are held on a zigzag axis (the rachilla).

Grass identification relies heavily on the characters of the grass spikelet. Spikelets are usually present with characters sufficiently developed for use with a key for several weeks. Nevertheless, collecting a specimen too early in the season can be misleading; for example, bluegrasses (*Poa* species) and other genera may have two or more florets (see below) "telescoped" together, with only the lowermost floret apparent. Many a beginner has called a very young Kentucky bluegrass (*Poa pratensis*) a red top (*Agrostis gigantea*) because he or she failed to realize that several very immature florets were "nested" within one apparent floret and assumed that the plant was a species with only one floret.

Some keys require one to determine the way in which the spikelet falls apart (disarticulates) at seed maturity; however, only mature specimens show this character, and they may shatter quickly upon examination. Other keys require measurement of the lengths of the stamens, which are very short-lived and rarely present in mature spikelets.

We have constructed our keys, we hope, in such a manner as to avoid some of these pitfalls, emphasizing characters that can be easily seen (at least with a hand lens) and measured. But to learn the grasses, one must cultivate patient powers of observation, looking for tiny awns and how they are attached to the spikelets, counting fine nerves on lemmas, and looking for sterile prolongations beyond the florets.

To reiterate, because of the heavy use of spikelets in grass keys, the structure of spikelets must be understood. A good way to begin this process is to examine the spikelets of the cereal grain oats (*Avena sativa*). Oats have large, reasonably typical spikelets with well-differentiated parts. The inflorescence bears about five to a dozen spikelets and may rebranch once or twice, meaning that it may be defined as a panicle, one of the commonest types of grass inflorescences. It is important to keep in mind when examining species with dense inflorescences, or with spikelets clumped together, that no branching occurs within a spikelet; therefore, one has not yet reached a single discrete spikelet if one encounters branching. But in oats, each spikelet is well set off on its own wiry stalk or pedicel.

Two papery scales, each about 20 mm long, mark the base of each spikelet. The lower of the two is the lower or first glume; the one attached slightly above it and opposite it is the upper or second glume. With the exception of cut-grass (*Leersia*) and wild rice (*Zizania*), all Wisconsin grass spikelets have these paired glumes at their base. The glumes may be quite unequal in length and even in shape, and they range from much shorter than the spikelet to slightly longer than it. Bending back either glume reveals no flower or seed inside of it and the scale of the floret above it—glumes, by definition, are always "empty."

An oat spikelet (*Avena sativa*) showing large glumes (*far left and right*) and two florets, the lower floret (*right*) with an awn.

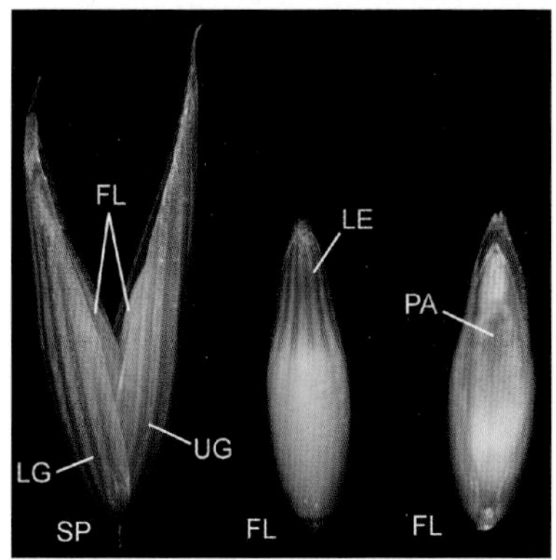

Parts of an oat spikelet.

The remainder of the oat spikelet consists of two or three distinct units called florets joined in a slightly zigzag line by short stalks called rachilla joints. The most conspicuous part of each floret is a tough scale, about 10–15 mm long on the lower floret, called the lemma. A tough wiry bristle-like structure (the awn), also about 10–15 mm long, often extends upward and outward from near the middle of the back of the lemma. The lemma is roughly tubular and surrounds a smaller, mostly hidden scale called the palea, which acts like a lid on the floret. The appearance of the floret changes little throughout the existence of the spikelet, except for some increase in its plumpness or roundness.

A flower lies within the lemma and palea at first. The flower consists of an ovary, two styles with two feather-like stigmas, and three stamens. Since the lemma and palea are firmly closed, the anthers and the stigmas could not protrude during pollination if it were not for the pair of tiny lodicules at the base of the floret. In most grasses the lodicules swell with water immediately prior to pollination and push apart the lemma and palea, permitting the anthers to shed pollen into the air and the feathery stigmas to "strain" the air currents for pollen. However, this mechanism is poorly developed in oats, since they, like a minority of grasses, are primarily self-pollinated.

After pollination, the stamens and lodicules quickly shrivel and die. The ovary of the pistil, however, enlarges as its single seed develops. Later nearly the entire interior of the floret is filled by this one-seeded grain, increasing the roundness and turgidity of the floret. Cultivated oats, as a result of human selection, show little tendency to break up at this stage, since this would interfere with the harvest. Wild oat spikelets, in contrast, break up cleanly at the base of each floret. Each floret, containing a one-seeded grain, is usually dispersed alone, leaving only the glumes attached to the pedicel of the parent plant.

The breakup of a spikelet at the beginning of seed dispersal is called disarticulation in many keys—we have tried to downplay this character in our keys, but it is still useful to be able to tell whether a spikelet disarticulates "above or below the glumes" at maturity. The choice is usually between those spikelets that break between each of the florets, leaving the glumes attached to the parent plant, and those spikelets that do not break up but fall off the parent plant completely as a single unit. A decision can be made easily with a mature fruiting specimen: if some pedicels end only in glumes, the specimen is an example of the former type, and if some of the pedicels are bare, it is of the second type. A less than mature specimen can present problems. If the spikelet has two or more florets, each bearing a fruit, it is almost certain that the spikelet breaks up between the florets (an example is bluegrass, *Poa* species). If the spikelet includes only one floret bearing a fruit, the spikelet may be of either type. If it is a "dorsally compressed" (compressed along the "spine" and "belly," as is the human body) spikelet, it usually does not break up; the whole spikelet is dispersed as a unit. If the floret has hairs at the base or a long awn that might aid in wind or animal dispersal, the spikelet is probably dispersed alone, leaving the glumes behind. If the glumes have awns, or are of unusual shape, or would seem to aid in dispersing the spikelet, it is likely that the whole spikelet leaves the parent plant.

Before setting aside the oat spikelet, note that it is somewhat flattened or two-sided rather than completely rounded in cross section. The glumes and, to a lesser extent, the lemmas are folded back from their midrib so that the flat parts of the spikelet lie at the edge of the glumes, not on their midribs. Almost all spikelets are somewhat flattened, and those flattened in this manner are usually described as being "laterally compressed," much like the body of a fish such as a bass or bluegill.

After finishing with oats, try to examine a spikelet of brome grass (*Bromus*), bluegrass (*Poa*), love grass (*Eragrostis*), or manna grass (*Glyceria*). They are laterally compressed (like a fish) and consist of three or more florets above a single pair of glumes. Each spikelet is well set off on its own pedicel. They all differ from oats, however, in that their glumes are relatively small, so small that they are both usually shorter than the lowermost floret.

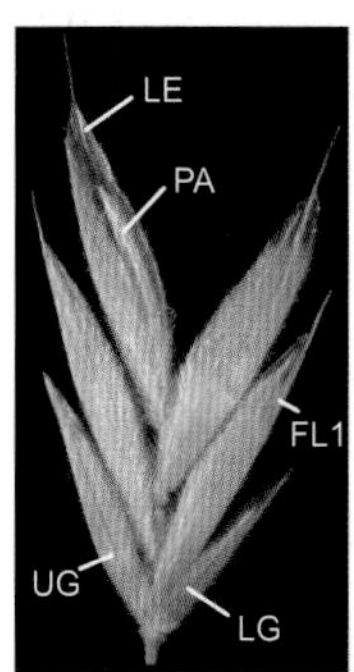

Inflorescence of quackgrass (*Elymus repens*) showing two-ranked spikelets (alternating on each side of the axis).

Spikelet of prairie brome (*Bromus kalmii*).

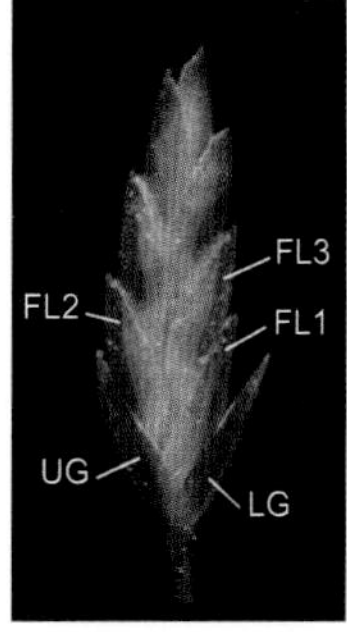

Spikelet of purple love grass (*Eragrostis spectabilis*).

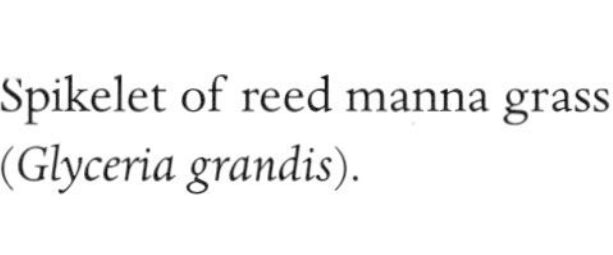

Spikelet of reed manna grass (*Glyceria grandis*).

Quackgrass (*Elymus repens*) should be the next grass that one observes. It has a spikelet essentially like those discussed above, but the spikelets lack individual stalks or pedicels and are attached directly to the main axis of the inflorescence. One spikelet is attached at a particular level, and the inflorescence axis is partially recessed to accommodate it. The next spikelet is attached above it on the opposite side, and the inflorescence axis continues upward.

Next, a single-flowered spikelet should be examined. Red top or tickle grass (*Agrostis*) is a good example, but unfortunately the spikelets are quite small—here is where a good hand lens or dissecting scope is useful. Timothy (*Phleum pratense*), which is so common in fields, is a bit more difficult to interpret: The inflorescence is a dense, pencil-like cylinder, with the spikelets very crowded together. The glumes are wide and U-shaped, and they almost conceal the floret inside. Canada bluejoint (*Calamagrostis canadensis*) is an abundant tall slender grass in marshes and ditches and on shores and banks. The base of the floret is surrounded by long hairs—the floret also has a short delicate awn produced from the base that is easy to miss among the dense hairs. Reed canary grass (*Phalaris arundinacea*) occupies the same habitats as bluejoint and differs superficially in having rather wide leaves and a more contracted panicle. Most beginners assume that reed canary grass spikelets also consist of a single floret surrounded by long hairs. They do not, however. The long hairs do not encircle the floret but arise in two distinct tufts from small scales at the two edges of the florets. These tiny scales actually represent vestigial florets, and thus reed canary grass is usually keyed out under the "two or more" florets per spikelet section in most keys.

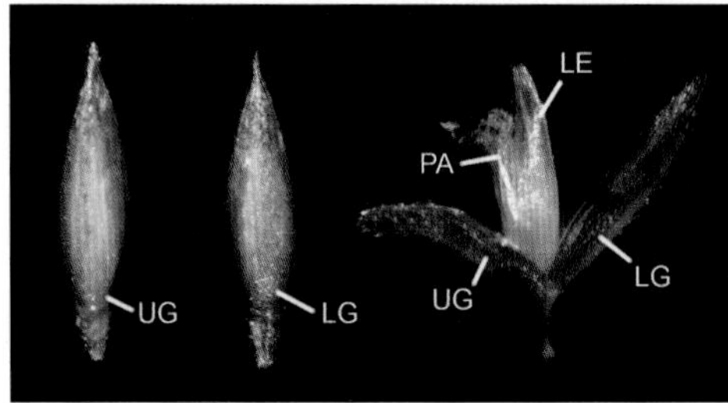

Redtop (*Agrostis gigantea*).

Timothy (*Phleum pratense*).

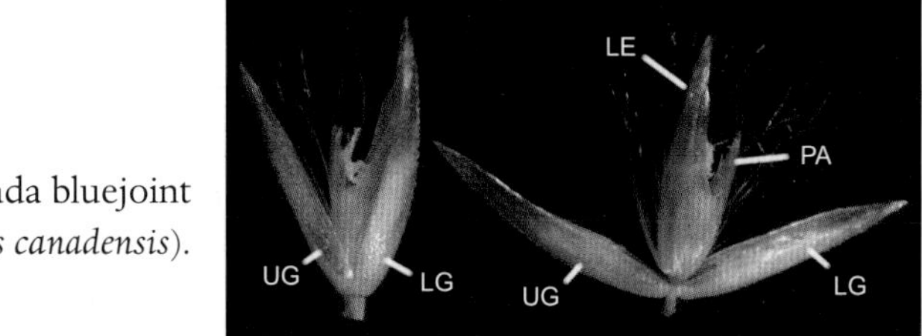

Canada bluejoint (*Calamagrostis canadensis*).

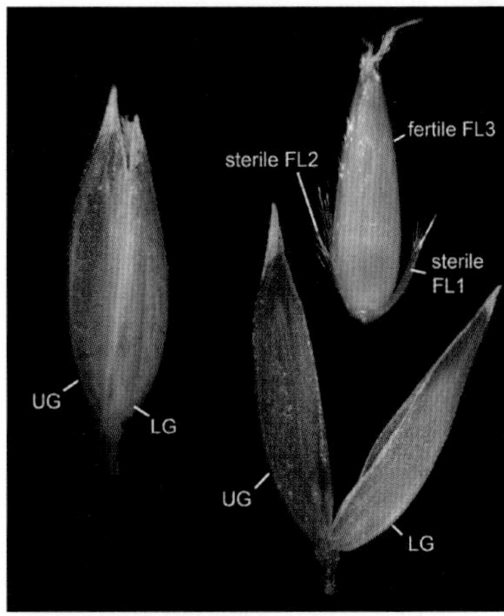

Reed canary grass (*Phalaris arundinacea*).

The last major spikelet variation that should be studied is the panicoid spikelet, as commonly represented by foxtails (*Setaria*), crabgrass (*Digitaria*), switchgrass (*Panicum virgatum*), and barnyard grass (*Echinochloa*). These spikelets are plump and teardrop- or beetle-shaped, with the more flattened sides centered on the midrib of the glumes and lemmas—they are therefore either dorsally compressed (like the human body) or terete (round in cross section, like the body of a snake). The spikelets of most of these species appear to have three (rather than two) glumes: a short lower glume; an upper glume about as long as the spikelet and borne on the opposite side of the spikelet from the lower glume; and a third glume-like structure also about as long as the spikelet and located inside the lower glume. This "third glume" is actually the lemma of the lower floret. It may be "empty," as in a true glume, or there may be a membranous palea or even a male flower within it, but, in any case, it is called the lemma of the sterile lower floret. The lemma and palea of the upper floret are of a texture different from that of the glumes and lower lemma: they are usually tough, thick, and nongreen and often (but not always) smooth, shiny, and white.

Spikelet of yellow foxtail (*Setaria pumila*), showing subtending bristles that do *not* fall attached to the spikelet.

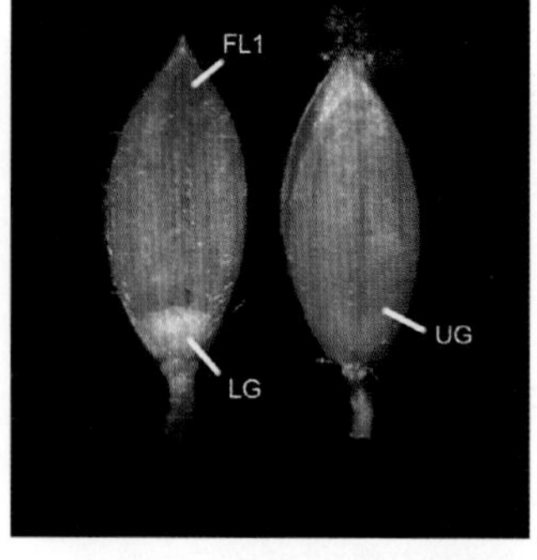

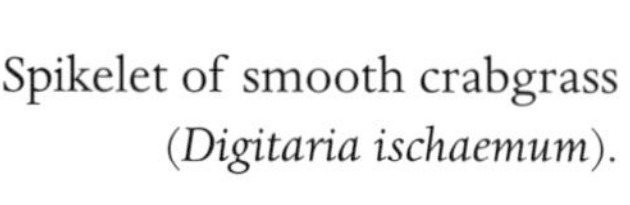

Spikelet of smooth crabgrass (*Digitaria ischaemum*).

Spikelet of switchgrass (*Panicum virgatum*).

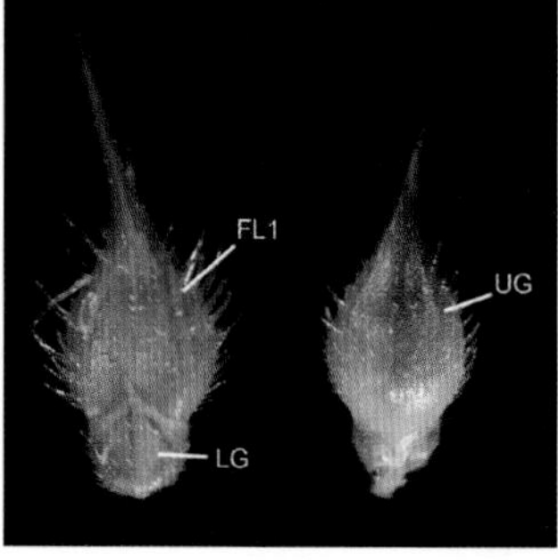

Spikelet of American barnyard grass (*Echinochloa muricata*).

Flowers

The flowers of all of our grasses are tiny, short-lived, and wind-pollinated; they lack showy sepals and petals. Most grass genera have perfect flowers—that is, both the stamens (male parts) and pistil (female part) are found in the same flower. Exceptions include corn (*Zea*) and wild rice (*Zizania*), where the sexes separate in differently shaped spikelets on the same plant; this is called the monoecious condition. In corn the males are found above the females, while in wild rice the females are found above the males.

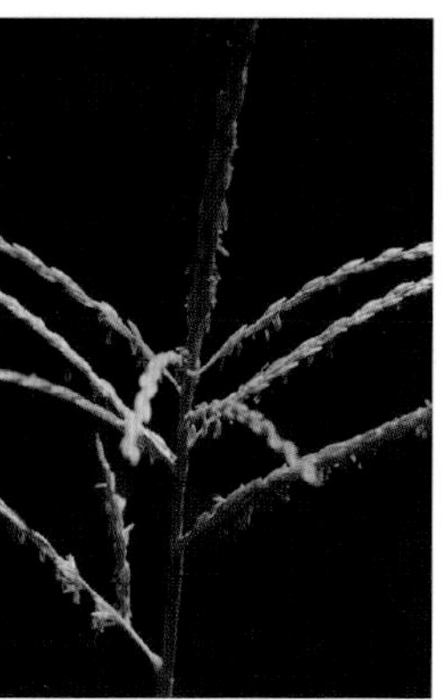

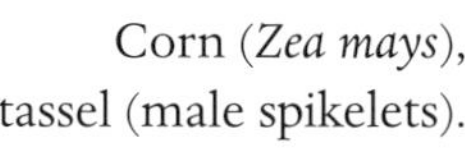
Corn (*Zea mays*), tassel (male spikelets).

Corn (*Zea mays*), ear (female spikelets).

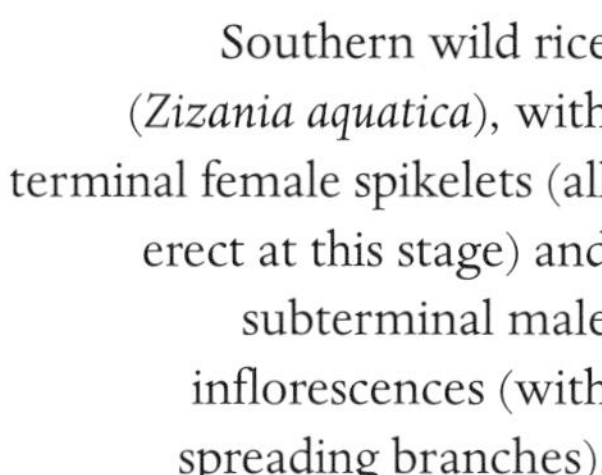
Southern wild rice (*Zizania aquatica*), with terminal female spikelets (all erect at this stage) and subterminal male inflorescences (with spreading branches).

Other exceptions include a few genera in which the male and female spikelets are found on separate plants (the dioecious condition): examples are the rare weeds buffalo grass (*Bouteloua dactyloides*) and salt grass (*Distichlis spicata*). Grass flowers lack green sepals and colorful petals, but they do have two or three tiny short-lived structures called lodicules that may represent modified petals. Most of our grasses have three stamens or rarely just one; the stigmas are feathery in appearance, and there are just two (only one in corn).

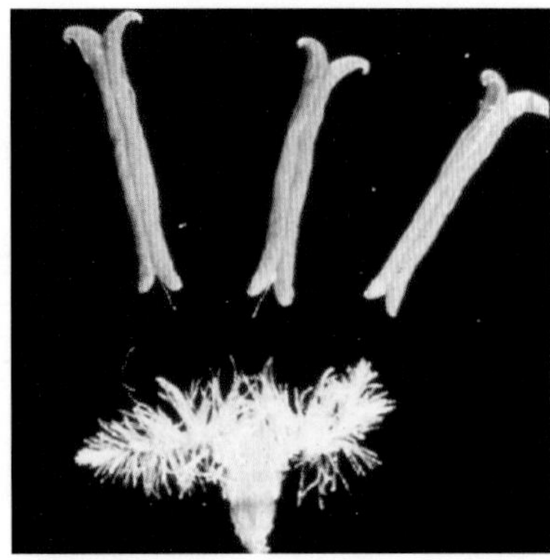

Grass flower showing anthers (*top three*) and pistil with feathery stigmas (*bottom*).

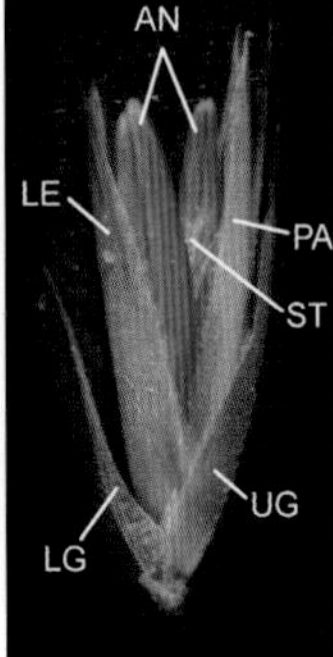

Spikelet of side-oats grama (*Bouteloua curtipendula*) showing orange-red anthers.

Fruits and seeds

All grasses have a grain (caryopsis) type of fruit, familiar to anyone who has cooked rice or eaten corn on the cob. The grain is usually dispersed within the floret or as part of the entire spikelet. The actual grass seed is within the grain and is usually completely fused to the fruit coat. Exceptions are dropseed (*Sporobolus*), in which the seed, when moistened, pops out of the fruit coat.

Agrostology

The Study of Grasses

How Are Grasses Named and Classified?

The scientific names used in this book follow the current generic concepts of "A Worldwide Phylogenetic Classification of the Poaceae (Gramineae)" by Soreng et al. (2014). For species epithets, we have followed the accepted names in the Tropicos Plant Database (2014). If either Fassett's *Grasses of Wisconsin* (1951) or Barkworth et al.'s *Manual of the Grasses for North America* (2007) uses a different name, we cross reference that name under the species entry and in the index. We generally ignore subspecies and varieties in this book.

The common names follow a variety of sources but ultimately represent our personal preference. Unlike, say, ornithologists, we botanists have the freedom to pick and choose any common name that we like for a plant. Since many common and distinctive grasses have names that are either awkward ("long-awned shorthusk" for the genus *Brachyelytrum*; "black-seeded rice grass" for *Patis racemosa*), misleading ("wild oats" for *Chasmanthium latifolium*), or simply nondescriptive (there are forest, wood, and woodland bluegrasses!), we have indulged in inventing a few new ones of our own, devising as we deem appropriate. For example, our "red dot panic grass" (*Dichanthelium oligosanthes* var. *scribnerianum*) has often been called "Scribner's panic grass," and our "elbow grass" (*Panicum dichotomiflorum*) sports a host of bland monikers. It is up to you, the users of this book and others like it, to ultimately determine whether these names "stick" or not. It is our view that common names should be just that: colloquial appellations that people use every day and that are continuously voted on in the court of public opinion. Common names should be open to amendment or improvement if someone comes up with a newer, cleverer, more mnemonically helpful epithet—much as folk songs are amended and improved down through the generations.

A detailed discussion of the classification of grasses is beyond the scope of this field guide. Briefly, however, modern classifications integrate molecular, morphological, anatomical, and genetic data to produce a classification that reflects the likely evolutionary descent of the family. Below is the most recent classification of the family available (Soreng et al. 2014). The classification is hierarchical. A name that ends in *-oideae* (such as Ehrhartoideae) represents a *subfamily*, the taxonomic rank below family. The next lowest rank is the *tribe*, whose name ends in *-eae*. Below each tribe are *genera* (singular, *genus*). An asterisk (*) indicates that the genus denoted has no native species in Wisconsin; if there is no asterisk, then at least one species is native here. The numbers in the first set of parentheses following the genus name represent first the total number of species growing without cultivation in Wisconsin and then the number of exotic species included in that total. The number in the second set of parentheses is the page number in the field guide section of the book where you can find this genus.

Subfamily Aristidoideae
- Tribe Aristideae
  - *Aristida* (7, 1) (104)

Subfamily Arundinoideae
- Tribe Arundineae
  - *Molinia** (1, 1) (236)
  - *Phragmites* (1, 0) (263)

Subfamily Chloridoideae
- Tribe Chlorideae
  - *Bouteloua* (4, 2) (115)
  - *Chloris** (1, 1) (142)
  - *Cynodon** (1, 1) (147)
  - *Diplachne** (1, 1) (174)
  - *Distichlis** (1, 1) (175)
  - *Eleusine** (1, 1) (180)
  - *Muhlenbergia* (12, 1) (238)
  - *Tridens** (1, 1) (310)
  - *Triplasis* (1, 0) (311)
- Tribe Eragrostideae
  - *Eragrostis* (10, 6) (194)
- Tribe Zoysieae
  - *Calamovilfa* (1, 0) (138)
  - *Crypsis** (1, 1) (146)
  - *Spartina* (1, 0) (299)
  - *Sporobolus* (6, 6) (302)

Subfamily Danthonoideae
- Tribe Danthonieae
  - *Danthonia* (2, 2) (150)

Subfamily Ehrhartoideae
- Tribe Oryzeae
  - *Leersia* (3, 0) (224)
  - *Zizania* (2, 0) (317)

Subfamily Panicoideae
- Tribe Chasmantheae
  - *Chasmanthium** (1, 1) (141)
- Tribe Paniceae
  - *Cenchrus* (1, 0) (140)
  - *Coleataenia** (1, 1) (144)
  - *Dichanthelium* (16, 1) (155)
  - *Digitaria* (3, 2) (171)
  - *Echinochloa* (4, 2) (176)
  - *Eriochloa** (1, 1) (201)
  - *Panicum* (6, 3) (251)
  - *Paspalum* (1, 0) (257)
  - *Setaria** (6, 6) (289)

Tribe Sacchareae
Andropogon (1, 0) (100)
Miscanthus⋆ (1, 1) (235)
Schizachyrium (1, 0) (285)
Sorghastrum (1, 0) (295)
Sorghum⋆ (3, 2) (296)
Zea⋆ (1, 1) (315)
Subfamily Pooideae
Tribe Brachyelytreae
Brachyelytrum (2, 0) (120)
Tribe Bromeae
Bromus (13, 8) (123)
Tribe Diarrheneae
Diarrhena (1, 0) (154)
Tribe Hordeae
Agropyron⋆ (1, 1) (91)
Elymus (12, 1) (181)
Hordeum⋆ (3, 3) (220)
Leymus⋆ (1, 1) (227)
Pascopyrum⋆ (1, 1) (256)
Secale⋆ (1, 1) (288)
Triticum⋆ (1, 1) (314)
Tribe Meliceae
Glyceria (6, 1) (208)
Melica (2, 0) (231)
Schizachne (1, 0) (284)
Tribe Poeae
Agrostis (5, 2) (92)
Alopecurus (4, 1) (96)
Ammophila (1, 0) (99)
Anthoxanthum (2, 1) (102)
Apera⋆ (1, 1) (104)
Arrhenatherum⋆ (1, 1) (109)
Avena⋆ (2, 1) (110)
Avenella (1, 0) (111)
Avenula⋆ (1, 1) (112)
Beckmannia⋆ (1, 1) (114)
Briza⋆ (1, 1) (122)
Calamagrostis (3, 1) (133)
Catabrosa (1, 1) (139)
Cinna (2, 0) (143)
Cynosurus⋆ (1, 1) (148)
Dactylis⋆ (1, 1) (149)
Deschampsia (1, 0) (152)
Festuca (8, 4) (203)
Graphephorum (1, 0) (215)
Holcus⋆ (1, 1) (219)
Koeleria (1, 0) (222)
Lolium⋆ (3, 3) (228)
Milium (1, 0) (234)
Phalaris⋆ (2, 2) (260)
Phleum⋆ (1, 1) (262)
Poa (15, 6) (267)
Polypogon⋆ (2, 2) (280)
Puccinellia⋆ (2, 2) (281)
Sclerochloa⋆ (1, 1) (287)
Sphenopholis (2, 0) (300)
Torreyochloa (2, 0) (308)
Trisetum (1, 0) (313)
Ventenata⋆ (1, 1) (315)
Tribe Stipeae
Hesperostipa (2, 1) (216)
Nassella⋆ (1, 1) (249)
Oryzopsis (1, 0) (250)
Patis (1, 0) (258)
Piptatheropsis (2, 0) (266)

The History of Agrostology in Wisconsin

Increase A. Lapham

Agrostology is the study of grasses. The first grass collections made in Wisconsin—or at least the earliest dated collection in the 31,892 Wisconsin grass specimens now in the current database—were made by pioneering Wisconsin conservationist Increase A. Lapham (1811–1875) (Janik 2007). He published the first treatise on Wisconsin grasses, listing 76 species for the state (Lapham 1853). Lapham collected 24,000 Wisconsin plant specimens beginning in 1836, but unfortunately nearly all were destroyed in an 1884 fire that burnt down Science Hall in Madison.

Norman C. Fassett

During the late nineteenth and early twentieth centuries, collectors such as Charles Goessl, L. S. Cheney, and John J. Davis made important grass collections statewide, with professional collector Charles Goessl in particular providing valuable baselines for many weedy exotics. Harvard-educated Massachusetts native Norman C. Fassett arrived as the director of the University of Wisconsin herbarium in Madison in 1925. He began an active statewide inventory of vascular plants, concentrating on aquatic macrophytes but not omitting grasses; he was most active before 1940.

The brilliant Lloyd H. Shinners arrived in 1938 as Fassett's graduate student and proceeded to begin a statewide survey of Wisconsin grasses, culminating in his 1943 PhD dissertation at the University of Wisconsin (Ginsburg 2002). He and his mentor, Fassett, eventually had a falling out, and Shinners's dissertation was in large part recycled by Fassett to produce his 1951 book, *Grasses of Wisconsin*, which, by mutual agreement, makes no mention of Shinners or his papers on Wisconsin grasses (Shinners 1940, 1941, 1943, 1944). Shinners, born in Blue Sky, Alberta, Canada, but a resident of Milwaukee from a young age, did not neglect the weedy grasses of his adopted hometown of Milwaukee, especially those found along railroads. His colleague Grant Cottam claimed that Shinners's small stature made him the perfect "railroad botanist," easily able to step precisely from rail tie to rail tie without overstepping rails, as a taller person would do (personal communication to Judziewicz in 1981). Shinners went on to a productive career at Southern Methodist University in Dallas. His legacy is the foundation of an herbarium that is the largest in the entire southern United States and is currently part of the Botanical Research Institute of Texas.

Lloyd H. Shinners

Fassett died from a brain tumor at age fifty-four. He was succeeded by Hugh H. Iltis, who became director of the Madison herbarium in 1955, a position he held until 1995. Born of German heritage in what is now the Czech Republic, he and his family fled the Nazis just before World War II. Iltis, a leading environmentalist and authority on the evolution of corn, led an active Wisconsin plant-collecting program, especially during the first decade of his directorship. He particularly concentrated on the grasses of the Driftless Area west of Madison and trained numerous students who went on to make significant contributions to our knowledge of Wisconsin's grass flora. Among these are James H. Zimmerman, Theodore Cochrane (especially Rock County), Frank Crosswhite, Bruce Hansen, Michael Nee (especially Richland County), and Emmet Judziewicz (especially Oconto County and islands in the Great Lakes) (see table 1).

Hugh H. Iltis

Table 1. Wisconsin Grass Collectors

Name (birth–death; affiliation)	*Number of grasses collected*	*Years active*	*Main collection areas*
Robert W. Freckmann (1939–; UWSP)	1,804	1961–present	Statewide, especially central Wisconsin
Lloyd H. Shinners (1918–71; WIS)	1,760	1937–48	Statewide
Hugh H. Iltis (1925–; WIS)	1,453	1955–present	Statewide, especially the Driftless Area
Gary Fewless (1948–; UWGB)	1,369	1978–present	Northeastern Wisconsin, especially Brown and Marinette Counties
Norman C. Fassett (1900–1954; WIS)	1,308	1924–51	Statewide
Charles Goessl (1866–1941; MIL)	1,290	1903–40	Statewide
Emmet J. Judziewicz (1953–; WIS, UWSP)	1,115	1975–present	Statewide, especially Apostle Islands and Door and Oconto Counties
Theodore S. Cochrane (1942–; WIS)	947	1966–present	Statewide, especially Rock County
Michael Nee (1947–; WIS)	940	1965–present	Statewide, especially Richland County
George C. Becker (1917–2002; UWSP)	655	1964–1965	Central Wisconsin
Alvin Bogdansky (1933–; UWSP)	641	1975–present	Statewide, especially Wood County
Frank C. Seymour (1895–1985; WIS)	596	1948–55	Lincoln County
Neil A. Harriman (1938–; OSH)	ca. 500	1964–present	Statewide, especially Winnebago and Outagamie Counties
Lellen S. Cheney (1858–1938; WIS)	472	1890–1927	Statewide, especially Wisconsin River valley, Lake Superior region

Note on affiliations: MIL = Milwaukee Public Museum; OSH = University of Wisconsin–Oshkosh; UWGB = University of Wisconsin–Green Bay; UWL = University of Wisconsin–La Crosse; UWSP = University of Wisconsin–Stevens Point; WIS = University of Wisconsin–Madison.

Milwaukee native Robert W. Freckmann earned his PhD at Iowa State University in 1967 under agrostologist Richard W. Pohl, another Milwaukee native. Freckmann then taught at the University of Wisconsin–Stevens Point beginning in 1968, collecting grass statewide but concentrating on the central counties and eventually becoming Wisconsin's most prolific grass and vascular plant collector. Beginning in the 1970s, he mentored Wisconsin Rapids paper mill worker Alvin Bogdansky, who also became a prolific collector of the state's grass flora, as well as numerous other students.

Robert W. Freckmann at Avoca Prairie, Iowa County, in 2007

Gary Fewless, a native of Goodman in Marinette County, founded and curates the University of Wisconsin–Green Bay herbarium, where, since 1978, he has built up a herbarium rich in the grasses of northeastern Wisconsin, especially Brown, Marinette, Oconto, and Door Counties. St. Louis–born Neil A. Harriman became curator of the University of Wisconsin–Oshkosh herbarium in 1964 and has since been a prolific collector of the state's flora; he and colleague Katherine D. Rill concentrated on the area around Lake Winnebago.

Gary Fewless

Wisconsin Grass Biodiversity

Fassett (1951) treated 66 genera and 190 species of Wisconsin grasses, of which 67 (36 percent) were exotic. The present treatment treats 87 genera and 232 species growing in the state without cultivation; of these, 106 species (46 percent) are exotic. These numbers represent an increase of 42 species, or 22 percent, in the past sixty-two years.

Wisconsin's grass flora of 232 species is richer than the flora of states to the west but poorer than that of states (and one Canadian province) to the south

Table 2. Grass Species in Nearby States and Province

	Grass species (total)	*Species in common with Wisconsin*	*Percentage in common with Wisconsin*
Illinois	309	211	68
Iowa	202	151	75
Michigan	264	215	81
Minnesota	196	164	84
Ontario	273	188	69

Sources: Illinois (Swink and Wilhelm 1994; USDA Plants Database 2014); Iowa (Grasses of Iowa 2013); Michigan (Voss and Reznicek 2012); Minnesota (Cholewa 2011); Ontario (USDA Plants Database 2014).

and east (see table 2). Even small tracts in Wisconsin can harbor surprising grass diversity:

Whiting Triangle, Portage County, 12 ha, 67 species (Freckmann 1978)
Butternut Pines, Oconto County, 16 ha, 51 species (Judziewicz 2004a)
Kaukamo Spruces, Bayfield County, 8 ha, 33 species (Judziewicz 2004b)

If we exclude newly recognized segregate species (e.g., northern shorthusk, *Brachyelytrum aristosum*), only three new native grass species have been discovered in Wisconsin since 1950: *Melica smithii*, Smith's melic grass (1964); *Muhlenbergia richardsonis*, mat muhly (1989); and *Danthonia compressa*, flattened oat grass (2006).

On the other hand, twenty-seven new exotic grasses have been discovered during the same period, an average of almost one every other year. The ones marked with an asterisk (*) either have become seriously invasive or have the potential for becoming so:

Setaria faberi,* nodding foxtail (1953)
Agropyron cristatum, crested wheatgrass (1956)
Anthoxanthum odoratum,* sweet vernal grass (1956)
Bouteloua dactyloides, buffalo grass (1956)
Distichlis spicata, salt grass (1956)
Leymus arenarius,* lyme grass (1959)
Miscanthus sacchariflorus,* Amur silver grass (1959)
Tridens flavus, purple-top (1961)
Setaria verticilliformis, barbed bristle grass (1964)
Eriochloa villosa,* wooly cup-grass (1966)

Holcus lanatus, velvet grass (1966)
Bromus squarrosus,* corn brome (1973)
Eragrostis trichodes, sand love grass (1974)
Glyceria maxima,* tall manna grass (1975)
Apera interrupta, silky bent grass (1981)
Ventenata dubia, ventenata grass (1981)
Hordeum pusillum, little barley (1983)
Chasmanthium latifolium, wild oats (1984)
Poa bulbosa,* bulblet bluegrass (1984)
Molinia caerulea,* purple moor grass (1986)
Sclerochloa dura, fairgrounds grass (2001)
Dichanthelium clandestinum, deer-tongue grass (2006)
Avenula pubescens, downy alpine oat grass (2008)
Poa arida, plains bluegrass (2008)
Polypogon monspeliensis, rabbit's-foot grass (2009)
Cynodon dactylon, Bermuda grass (2011)
Chloris verticillata, windmill grass (2012)

The number of exotic grasses collected in the state (by decade), after rising steadily and peaking from 1960 to 1990, has apparently declined slightly since 1990. It is not certain whether this represents a true decline or whether the diminished collecting intensity of the last twenty years means that fewer exotic grasses are being discovered or reverified. The numbers of exotic grass species collected by decade are:

1880s, 25
1890s, 31
1900s, 41
1910s, 49
1920s, 52
1930s, 66
1940s, 67
1950s, 71
1960s, 74
1970s, 76
1980s, 75
1990s, 69
2000s, 71

Finally, the grasses in the following list occur within about 100 km of the Wisconsin border and may eventually be found in the state. Species denoted by an asterisk are exotic.

Agrostis capillaris,* Rhode Island bent (possible statewide)
Andropogon hallii,* silver bluestem (possible in the west)

*Andropogon virginicus,** broomsedge bluestem (southeast)
Digitaria filiformis, slender crabgrass (south)
Eragrostis reptans, creeping love grass (south)
*Eriochloa contracta,** prairie cup grass (south)
*Eriocoma hymenoides,** Indian rice grass (west)
Melica mutica, twinflower melic grass (south)
Muhlenbergia bushii, nodding muhly (southwest)
Poa chapmaniana, Chapman's bluegrass (south)
Poa interior, interior bluegrass (northwest)
Sphenopholis nitida, shiny wedge grass (south)
*Thinopyrum ponticum,** tall wheatgrass (south)

Grasses in Wisconsin Plant Communities

Grasses occur in a wide variety of natural communities in Wisconsin, and this chapter aims to briefly describe the major plant communities with which grasses are associated, drawing freely on the descriptions presented in Black and Judziewicz (2009). For further information on other plants and plant communities, we recommend these books and resources: Curtis (1971), Cochrane and Iltis (2000), Hoffman (2002), Cochrane, Elliot, and Lipke (2006), Hipp (2008), Waller and Rooney (2008), Skawinski (2011), Voss and Reznicek (2012), and *Plants of Wisconsin* (2014). Species marked with an asterisk (*) are considered exotic.

Prairies

Prairies occupied about 5–7 percent of Wisconsin's presettlement landscape, and, along with members of the sunflower family (Asteraceae), grasses dominated prairie landscapes. The following is a very brief overview; for a fuller discussion of Wisconsin prairies, see Cochrane and Iltis (2000) and Cochrane, Elliot, and Lipke (2006). The natural disturbance regime in prairies featured frequent fires, both natural (lightning) and human-caused (Native Americans burning for hunting success and other reasons). After settlement, prairies were degraded in several ways. Settlers first allowed their cows to graze widely, reducing plant diversity. They then plowed and destroyed the prairie sod to plant crops; if the area was poorly drained, settlers ditched and tiled it, altering the hydrology and leading to the decline of many species. Finally, they suppressed fires, with the result that many prairies, especially those too steep to graze, were invaded by woody vegetation, especially shrubs and eastern red cedar (*Juniperus virginiana*).

Dry prairies

In presettlement times, dry prairies covered about 2 percent of southern and western Wisconsin, most frequently on steep south- or west-facing sandstone and dolostone bluffs above major rivers such as the Wisconsin, Mississippi, Chippewa, and St. Croix but also as "sand prairies" on flat alluvial terraces along these same rivers. Gravelly dry hill prairies are also found in the southern Kettle Moraine in Waukesha and Walworth Counties. Yet, since settlement, dry prairies have fared better than either mesic or wet prairies (Cochrane and Iltis 2000). Common wildflowers in this community include asters (*Symphyotrichum*), goldenrods (*Solidago*), flowering spurge (*Euphorbia corollata*), common spiderwort (*Tradescantia ohioensis*), bush-clovers (*Lespedeza*), pasque flower (*Anemone patens*), bird's-foot violet (*Viola pedata*), prairie coreopsis (*Coreopsis palmata*), prairie-clovers (*Dalea*), lead-plant (*Amorpha canescens*), and blazing-stars (*Liatris*).

Andropogon gerardii, big bluestem
Aristida dichotoma, Curtiss's three-awned grass
Aristida longespica, Kearney's three-awned grass
Bouteloua curtipendula, side-oats grama
Bouteloua hirsuta, hairy grama
Dichanthelium perlongum, long-stalked panic grass
Dichanthelium wilcoxianum, Wilcox's panic grass
Elymus canadensis, Canada wild rye
Festuca octoflora, six-weeks fescue
Hesperostipa spartea, needle grass
Koeleria macrantha, June grass
Melica nitens, tall melic grass
Schizachyrium scoparium, little bluestem
Sorghastrum nutans, Indian grass
Sporobolus compositus, meadow dropseed
Sporobolus heterolepis, prairie dropseed

Dry prairie, Cassville Bluffs State Natural Area, Grant County, August 19, 2012.

Dry prairie, Spring Green, Sauk County.

A small set of calciphilic (lime-loving) muhly grasses are found on partly forested dry dolostone bluffs with a strong prairie element along major rivers in western Wisconsin:

Muhlenbergia cuspidata, plains muhly
Muhlenbergia sobolifera, creeping muhly
Muhlenbergia sylvatica, forest muhly

Sand barrens

Sand barrens, or blowouts, are most common on the sandy terraces along the Mississippi and lower Wisconsin Rivers but sometimes also occur on abandoned farms in central Wisconsin. With fire suppression and the failure of agriculture on these poor soils, this community, never very large, is fast disappearing because of succession to woody plants. Where still found, these communities do have a suite of interesting grasses. Common woody associates are jack pine (*Pinus banksiana*), false heather (*Hudsonia tomentosa*), and common juniper (*Juniperus communis*).

Aristida basiramea, forked-tip three-awned grass
Aristida purpurascens, arrow-feathered three-awned grass
Aristida tuberculosa, dune three-awned grass
Calamovilfa longifolia, sand reed
Cenchrus echinatus, sandbur
Dichanthelium commonsianum var. *euchlamydeum*, Shinners' panic grass
Dichanthelium leibergii, Leiberg's panic grass
Dichanthelium meridionale, slender panic grass
Digitaria cognata, fall witch grass
Eragrostis spectabilis, purple love grass

Koeleria macrantha, June grass
Muhlenbergia schreberi, nimble-will
Muhlenbergia tenuiflora, slender muhly
Panicum virgatum, switchgrass
Paspalum setaceum, bead grass
Schizachyrium scoparium, little bluestem
Sorghastrum nutans, Indian grass
Sporobolus cryptandrus, sand dropseed
Tridens flavus,* purple-top
Triplasis purpurea, purple sand grass

Sand barrens with jack pine and clumps of false heather (*Hudsonia tomentosa*); habitat for purple sand grass (*Triplasis purpurea*), Millston, Jackson County, July 2, 1997.

Dune three-awned grass (*Aristida tuberculosa*) in sand barrens, Cruson Slough, Sauk County, August 15, 2012.

Moist (mesic) prairies

Mesic prairies covered about 2–3 percent of southern Wisconsin in presettlement times, mostly on flat to gently rolling, rich-soiled, well-drained sites that were quickly grazed, then plowed and converted to agricultural uses. As a result, this is one of the rarest natural community types in Wisconsin, with perhaps only a few hundred acres still in existence, often in long narrow strips along railroads (Cochrane and Iltis 2000; Cochrane, Elliot, and Lipke 2006). This is the classic "tall grass prairie" and is the Wisconsin natural community most dominated by grasses; it shares many wildflower associates with those listed under dry prairies and wet prairies.

Andropogon gerardii, big bluestem
Dichanthelium leibergii, Leiberg's panic grass
Dichanthelium oligosanthes var. *scribnerianum*, red-dot panic grass
Dichanthelium perlongum, long-stalked panic grass
Elymus canadensis, Canada wild rye
Hesperostipa spartea, needle grass
Muhlenbergia racemosa, upland wild-timothy
Panicum virgatum, switchgrass
Schizachyrium scoparium, little bluestem
Sorghastrum nutans, Indian grass
Sporobolus heterolepis, prairie dropseed

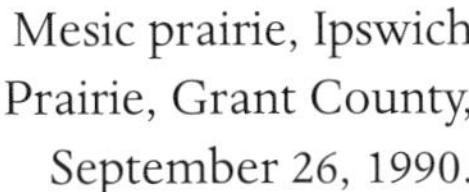

Mesic prairie, Ipswich Prairie, Grant County, September 26, 1990.

Wet prairie, Waupaca County.

Wet prairies

Wet prairies covered from 1–2 percent of southern Wisconsin in presettlement times on rich but poorly drained soils that were ultimately drained by ditching and converted to agriculture. Like mesic prairies, present-day wet prairie remnants are small, rare, and scattered (Cochrane and Iltis 2000; Cochrane, Elliot, and Lipke 2006). Wet prairies intergraded with sedge meadows and calcareous fens; if sedges are present, the dominant species are likely to be *Carex stricta*, *C. lacustris*, *C. pellita*, and *C. trichocarpa*. Common wildflowers include prairie dock and compass-plant (*Silphium*), mountain-mint (*Pycnanthemum virginianum*), prairie phlox (*Phlox pilosa*), black-eyed Susan (*Rudbeckia hirta*), culver's-root (*Veronicastrum virginicum*), shooting-star (*Primula meadia*), prairie blazing-star (*Liatris pycnostachya*), and golden alexanders (*Zizia*), as well as various asters (*Symphyotrichum*) and goldenrods (*Solidago*).

Andropogon gerardii, big bluestem
Anthoxanthum hirtum, sweet grass
Bouteloua curtipendula, side-oats grama
Calamagrostis canadensis, Canada bluejoint
Elymus canadensis, Canada wild rye
Muhlenbergia racemosa, upland wild-timothy
Panicum virgatum, switchgrass
Spartina pectinata, prairie cordgrass

Open Wetlands

These communities covered about another 5 percent of the state, and, along with mosses and members of the sedge family (Cyperaceae), grasses were and are an important component of wetlands.

Sedge meadows

Sedge meadows (Hipp 2008) covered about 3 percent of Wisconsin in presettlement times. However, this natural community grades into wet prairie and calcareous fen in the southern part of Wisconsin (where it was very common) and into open bog and poor fen in the northern part of the state. Changes in hydrology caused by draining and ditching and invasion by exotic plants have greatly reduced the extent and health of sedge meadows in Wisconsin, especially in the south. Reed canary grass (*Phalaris arundinacea*) may be dominant in grazed and/or ditched stands. The dominant plants in this open wetland community are tussock sedge (*Carex stricta*), other species of *Carex* such as *C. aquatilis*, *C. lacustris*, *C. oligosperma*, *C. pellita*, and *C. utriculata*, as well as bulrushes (*Scirpus atrovirens* and *S. cyperinus*), and Canada bluejoint. Cattails

(*Typha*) are also common, as may be sensitive fern (*Onoclea sensibilis*) and marsh fern (*Thelypteris palustris*), along with many wildflowers, such as water-horehounds (*Lycopus*), spotted joe-pye-weed (*Eutrochium maculatum*), boneset (*Eutrochium perfoliatum*), blue vervain (*Verbena hastata*), grass-leaved goldentop (*Euthamia graminifolia*), marsh bellflower (*Campanula aparinoides*), swamp milkweed (*Asclepias incarnata*), orange jewelweed (*Impatiens capensis*), and species of beggar-ticks (*Bidens*), smartweeds (*Persicaria*), and water docks (*Rumex*).

Bromus ciliatus, fringed brome
Calamagrostis canadensis, Canada bluejoint
Calamagrostis stricta, slim-stemmed bluejoint
Glyceria borealis, northern manna grass
Glyceria canadensis, rattlesnake manna grass
Glyceria grandis, reed manna grass
Glyceria striata, fowl manna grass
Muhlenbergia glomerata, marsh wild-timothy
Phalaris arundinacea,* reed canary grass
Phragmites australis, reed (there are also exotic populations)
Poa palustris, fowl meadow grass
Spartina pectinata, prairie cordgrass
Sphenopholis intermedia, satin wedge grass

Starlight peatlands sedge meadow, Jackson County, July 21, 2000.

Open bog, Norrie Lake, Marathon County, 2000.

Bogs

These cold, poorly drained peatlands developed in areas where the only source of nutrients was rainwater. Bog understories are composed mostly of sphagnum (*Sphagnum*) mosses, sedges (*Carex*), cotton grasses (*Eriophorum*; not a true grass), and "ericaceous" shrubs such as leatherleaf (*Chamaedaphne calyculata*), Labrador-tea (*Rhododendron groenlandicum*), and blueberries and cranberries (*Vaccinium* species). Orchids are characteristic of this habitat, but these bogs are poor in grasses.

Calamagrostis canadensis, Canada bluejoint
Glyceria canadensis, rattlesnake manna grass
Muhlenbergia glomerata, marsh wild-timothy
Muhlenbergia uniflora, bog muhly (on wet, sandy peat)
Phragmites australis, common reed (there are also exotic populations)

Calcareous fens and springs

A subset of open wetlands are calcareous fens, forested seeps, and spring runs in which nutrient-rich groundwater percolates through the substrate, allowing a distinctive and rich assemblage of calciphilic (calcium-loving) plants to flourish. Fens cover a minuscule proportion of Wisconsin's area, perhaps only 0.01 percent even in presettlement times, and certainly much less now due to changes in hydrology and invasive plants. This community is related to wet prairies and sedge meadows and shares many of the same species and threats—invasives such as reed canary grass, purple loosestrife (*Lythrum salicaria*), and glossy buckthorn (*Frangula alnus*). Some rare grasses occur in calcareous fens and springy areas.

Bromus ciliatus, fringed brome
Catabrosa aquatica, brook grass (springs)
Deschampsia cespitosa, tufted hair grass (springs)
Glyceria species, manna grasses
Muhlenbergia richardsonis, mat muhly
Panicum flexile, fen panic grass
Phalaris arundinacea,* reed canary grass
Poa paludigena, bog bluegrass
Poa palustris, fowl meadow grass
Sphenopholis intermedia, satin wedge grass
Torreyochloa fernaldii, Fernald's false manna grass
Torreyochloa pallida, pale false manna grass

Calcareous fen, Bass Lake, Waushara County, September 9, 2006.

Dave Seils in a fenny meadow along Lawrence Creek, with abundant large mature plants of tufted hair grass (*Deschampsia cespitosa*), Waushara County, July 21, 2012.

Truly Aquatic Grasses

Under the proper conditions, some grasses can not only survive temporary inundation but also thrive and produce "rafts" of floating leaves in cold, flowing, well-aerated water (Skawinski 2011) and even submersed leaves.

Alopecurus aequalis, short-awned foxtail
Glyceria borealis, northern manna grass
Glyceria grandis, reed manna grass
Glyceria septentrionalis, eastern manna grass
Leersia oryzoides, rice cut-grass
Spartina pectinata, prairie cordgrass

Torreyochloa fernaldii, Fernald's false manna grass
Torreyochloa pallida, pale false manna grass
Zizania aquatica, southern wild rice
Zizania palustris, northern wild rice

Prairie cordgrass (*Spartina pectinata*).

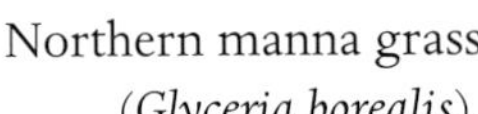

Northern manna grass (*Glyceria borealis*).

Forests

Although grasses are not dominant in shaded forest understories, there are a few shade specialists, generally with broad leaf blades.

Southern mesic (moist) forests

Southern moist forests are upland deciduous tree communities growing on rich, well-drained soils on sites not susceptible to fire. In presettlement times this type covered approximately 10 percent of Wisconsin. It is especially common in areas of dolostone bedrock such as in the Lake Michigan counties, in parts of the Driftless Area of the southwest, and in far western Wisconsin near the confluence of the Mississippi and St. Croix Rivers. The dominant tree species are sugar maple (*Acer saccharum*), with codominants basswood (*Tilia americana*), red oak (*Quercus rubra*), hackberry (*Celtis occidentalis*), bitternut hickory (*Carya cordiformis*), black walnut (*Juglans nigra*), white ash

(*Fraxinus americana*), and, near Lake Michigan, American beech (*Fagus grandifolia*). These forests typically have fine displays of spring ephemeral wildflowers such as spring-beauty (*Claytonia virginica*), trout-lilies (*Erythronium*), trilliums (*Trillium*), violets (*Viola*), bloodroot (*Sanguinaria canadensis*), mayapple (*Podophyllum peltatum*), and Dutchman's-breeches (*Dicentra cucullaria*); garlic mustard (*Alliaria petiolata*) is a serious invasive threat.

Agrostis perennans, autumn bent grass
Bromus latiglumis, ear-leaved brome
Bromus nottowayanus, Nottoway brome
Diarrhena obovata, obovate beak grain
Dichanthelium latifolium, broad-leaved panic grass
Elymus hystrix, bottlebrush grass
Festuca subverticillata, nodding fescue
Leersia virginica, white grass
Milium effusum, wood millet
Patis racemosa, black-seeded rice grass
Poa sylvestris, forest bluegrass

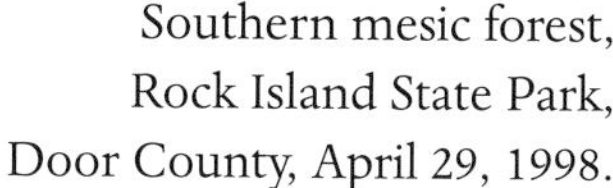
Southern mesic forest, Rock Island State Park, Door County, April 29, 1998.

Northern mesic forest with wood millet (*Milium effusum*) in sun fleck, Devils Island, Apostle Islands National Lakeshore, 1991.

Southern dry forests, oak openings, woodlands, and oak barrens

Southern dry forests are deciduous communities growing on dry, loamy to sandy upland soils; these sites are subject to occasional fires on a time scale of years or decades. In presettlement times, this community type covered around 4 percent of southern Wisconsin and intergraded with oak openings and savannas. Oaks are dominant and include white oak (*Quercus alba*), black oak (*Q. velutina*), red oak (*Q. rubra*), and bur oak (*Q. macrocarpa*); other common trees are black cherry (*Prunus serotina*), shagbark hickory (*Carya ovata*), black walnut (*Juglans cinerea*), and red maple (*Acer rubrum*). Common herbaceous understory species are Penn sedge (*Carex pensylvanica*), wild geranium (*Geranium maculatum*), white snakeroot (*Ageratina altissima*), hog-peanut (*Amphicarpaea bracteata*), and tick-trefoils (*Desmodium*). Spring ephemeral displays are absent from this community, and most herbs are "summer greens."

Presettlement oak savannas, openings, and barrens were semiforested deciduous communities growing on loamy to sandy upland soils and subject to occasional to frequent fires. The understory was open and grassy under a natural fire regime but with fire suppression became choked with smooth sumac (*Rhus glabra*), gray dogwood (*Cornus racemosa*), American hazelnut (*Corylus americana*), various cherries and plums (*Prunus* species), and brambles (*Rubus* species), as well as exotics such as honeysuckles (*Lonicera* species) and common buckthorn (*Rhamnus cathartica*). These oak communities grade into dry and moist prairies. In addition to the following species, many grass species characteristic of mesic to dry prairies may be common in sunny openings in these communities.

Brachyelytrum erectum, southern shorthusk
Bromus kalmii, Kalm's brome
Bromus latiglumis, ear-leaved brome
Danthonia spicata, poverty oat grass
Dichanthelium depauperatum, poverty panic grass
Dichanthelium latifolium, broad-leaved panic grass
Dichanthelium linearifolium, linear-leaved panic grass
Dichanthelium villosissimum var. *praecocius*, prairie panic grass
Dichanthelium xanthophysum, pale panic grass
Elymus villosus, downy wild rye
Patis racemosa, black-seeded rice grass
Sphenopholis obtusata, prairie wedge grass

Oak woodland, grading into opening, Nine Mile Island State Natural Area, Dunn County, June 28, 1989.

Oak opening near Eagle, Waukesha County.

Floodplain forests

These are hardwood forests that occur along rivers that flood periodically; they are best developed along the larger rivers in southern Wisconsin, but this community is also found essentially throughout the state. This community probably occupied about 1 percent of Wisconsin's area. Tree canopy dominants are silver maple (*Acer saccharinum*), river birch (*Betula nigra*), green ash (*Fraxinus pennsylvanica*), swamp white oak (*Quercus bicolor*), cottonwood (*Populus deltoides*), and box elder (*Acer negundo*). Common understory plants are wood-nettle (*Laportea canadensis*) and ostrich fern (*Matteuccia struthiopteris*). Climbing woody vines (lianas) such as Virginia creepers (*Parthenocissus* species), river grape (*Vitis riparia*), and eastern poison ivy (*Toxicodendron radicans*) are often common.

Bromus latiglumis, ear-leaved brome
Bromus pubescens, Canadian brome
Cinna arundinacea, common wood-reed
Diarrhena obovata, obovate beak grain
Echinochloa muricata, American barnyard grass

Echinochloa walteri, bottlebrush barnyard grass
Elymus riparius, riverbank wild rye
Elymus virginicus, Virginia wild rye
Elymus wiegandii, Wiegand's wild rye
Eragrostis frankii, sandbar love grass (on sandbars and mud flats)
Eragrostis hypnoides, creeping love grass (on sandbars and mud flats)
Festuca paradoxa, cluster fescue
Glyceria striata, fowl manna grass
Leersia lenticularis, catchfly grass
Leersia oryzoides, rice cut-grass
Leersia virginica, white grass
Muhlenbergia frondosa, wire-stemmed muhly

Floodplain forest, Colic Bayou, Wolf River, Waupaca County, August 23, 2000.

Floodplain forest understory dominated by *Leersia lenticularis*, Bertom Lake, Mississippi River, Grant County, August 19, 2012.

Northern mesic (moist) forests

Northern mesic forests dominated most of the northern half of Wisconsin. They were mixed forests, with both deciduous and coniferous trees codominant. Sugar maple (*Acer saccharum*) and hemlock (*Tsuga canadensis*) were the dominant trees in presettlement times, covering fully one-third of the state. Yellow birch (*Betula alleghaniensis*), white pine (*Pinus strobus*), basswood (*Tilia americana*), and white ash (*Fraxinus americana*) were also important forest trees. In the cooler Lake Michigan counties extending inland to Forest, Langlade, Shawano, and Waupaca Counties, beech (*Fagus grandifolia*) was an important tree species. Common herbs—with species diversity ranging from poor in pure hemlock stands to rich in pure maple stands—include club mosses (*Lycopodium* species), wood ferns (*Dryopteris* species), lady fern (*Athyrium filix-femina*), large-leaved aster (*Eurybia macrophylla*), wild sarsaparilla (*Aralia nudicaulis*), Canada mayflower (*Maianthemum canadense*), and bluebead lily (*Clintonia borealis*). Many of the spring ephemerals listed under southern mesic forests were common. While not rich in grasses, northern mesic forests do harbor several rare, elusive, native bluegrasses, as well as the very invasive wood bluegrass (*Poa nemoralis*).

Brachyelytrum aristosum, northern shorthusk
Bromus latiglumis, ear-leaved brome
Bromus pubescens, Canadian brome
Cinna latifolia, drooping wood-reed
Melica smithii, Smith's melic grass
Milium effusum, wood millet
Oryzopsis asperifolia, rough-leaved rice grass
Poa alsodes, grove bluegrass
Poa languida, languid bluegrass
Poa nemoralis,* wood bluegrass (exotic, invasive)
Poa saltuensis, woodland bluegrass
Poa wolfii, Wolf's bluegrass
Schizachne purpurascens, false melic grass

Northern mesic forest, hemlock-dominated stand, Chambers Island, Door County, May 9, 1998.

Beech-dominated stand, Chambers Island, Door County, May 9, 1998.

Northern dry forests

These are the pine forests that we often think of as covering most of northern Wisconsin; but in presettlement times this community type actually covered only 6–7 percent of the state, mostly on nutrient-poor sites with excessively drained sandy or rocky soils. Relict stands of this forest type extend south to the Iowa and Illinois borders and beyond, often on moist, north-facing hillsides or in river gorges. Dominant trees include white pine (*Pinus strobus*), red pine (*P. resinosa*), and jack pine (*P. banksiana*) and Hill's oak (*Quercus ellipsoidalis*); on the dry end this community type intergraded with pine barrens. Large acreages of northern dry forests were cut and burned during the catastrophic logging of the late nineteenth and early twentieth centuries. Much of this land was then colonized by white birch (*Betula papyrifera*), red oak (*Quercus borealis*), red maple (*Acer rubrum*), quaking aspen (*Populus tremuloides*), and bigtooth aspen (*P. grandidentata*). Vast acreages of open "barrens" were also planted to pine or naturally succeeded to densely stocked "dry" forests. Common herbs include bracken fern (*Pteridium aquilinium*), wood anemone (*Anemone quinquefolia*), partridgeberry (*Mitchella repens*), large-leaved aster (*Eurybia macrophylla*), wintergreen (*Gaultheria procumbens*), and shinleaf (*Pyrola elliptica*).

Brachyelytrum aristosum, northern shorthusk
Danthonia compressa, flattened oat grass
Danthonia spicata, poverty oat grass
Dichanthelium depauperatum, poverty panic grass
Oryzopsis asperifolia, rough-leaved rice grass
Piptatheropsis pungens, mountain rice grass
Schizachne purpurascens, false melic grass

Pine relict, Mirror Lake State Park, Sauk County, August 1997.

North Bay white cedar swamp, Door County, June 22, 2000.

Northern conifer swamps

Forested wetlands (swamps) are dominated by these conifers: northern white cedar (*Thuja occidentalis*), balsam fir (*Abies balsamea*), black spruce (*Picea mariana*), and tamarack (*Larix laricina*); tag alder (*Alnus incana* subsp. *rugosa*) is often a common understory shrub. These communities covered about 6–7 percent of Wisconsin's presettlement landscape, mostly in the northern part of the state but extending south to the bed of Glacial Lake Wisconsin in Jackson County and as small relicts elsewhere throughout the south. They are poor in grasses.

Agrostis perennans, autumn bent grass
Calamagrostis canadensis, Canada bluejoint
Cinna latifolia, drooping wood-reed
Glyceria striata, fowl manna grass
Poa palustris, fowl meadow grass

Pine Barrens

This savanna community is adapted to frequent fires and is characterized by an open to dense forest of scrubby jack pine (*Pinus banksiana*), often with admixtures of northern pin oak (*Quercus ellipsoidalis*) or red pine (*P. resinosa*). It covered 6–7 percent of Wisconsin's presettlement landscape, mostly on sandy soils in the central (Jackson, Juneau, and Wood Counties), far northwestern, far north-central (Vilas County), and far northeastern (Marinette County) parts of the state. Pine barrens may also be found on sand terraces along the lower Wisconsin River such as in the Lone Rock area. Many dry prairie grasses and wildflowers are present, especially in central and far northwestern Wisconsin barrens.

Agrostis scabra, northern tickle grass
Bromus kalmii, Kalm's brome
Danthonia spicata, poverty oat grass
Dichanthelium linearifolium, linear-leaved panic grass
Dichanthelium xanthophysum, pale panic grass
Digitaria cognata, fall witch grass
Elymus trachycaulus, slender wheatgrass
Festuca saximontana, Rocky Mountain fescue
Koeleria macrantha, June grass
Muhlenbergia racemosa, upland wild-timothy
Piptatheropsis canadensis, Canada mountain rice grass
Piptatheropsis pungens, mountain rice grass
Poa compressa,* Canada bluegrass
Poa pratensis,* Kentucky bluegrass
Schizachyrium scoparium, little bluestem

Bracken grassland, a variant of pine barrens: Aurora Barrens, Florence County, August 10, 1987.

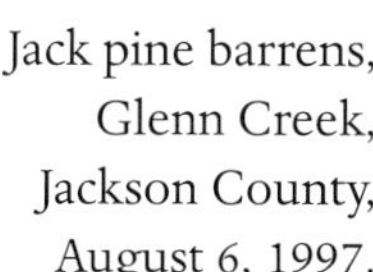

Jack pine barrens, Glenn Creek, Jackson County, August 6, 1997.

Great Lakes Shorelines

These areas total much less than 1 percent of the state's area but are rich in distinctive grasses, mostly due to the presence of open to stabilized sand dunes.

Great Lakes dunes and beaches

Wisconsin's Great Lakes shorelines feature a distinctive set of communities. Sand beaches are dynamic features that respond quickly to water level changes and severe storms. On Lake Superior, they are best developed in bays on the Bayfield Peninsula and in the Apostle Islands as sand spits. Along Lake Michigan, beaches are well developed at several sites in Door County, at Point Beach State Forest in Manitowoc County, and at Kohler-Andrae State Park in Sheboygan County. The dominant dune plant is the sand-binding American beach grass (*Ammophila breviligulata*). Frequent associates are common juniper (*Juniperus communis*), creeping juniper (*J. horizontalis*), beach pea (*Lathyrus japonicus*), and sand cherry (*Prunus pumila*).

Agrostis stolonifera,* bent grass
Ammophila breviligulata, American beach grass
Avenella flexuosa, wavy hair grass
Calamovilfa longifolia, sand reed
Elymus canadensis, Canada wild rye
Elymus lanceolatus, thick-spiked wheatgrass
Festuca occidentalis, western fescue
Festuca saximontana, Rocky Mountain fescue
Leymus arenarius,* lyme grass
Poa compressa,* Canada bluegrass

Invasive lyme grass (*Leymus arenarius*) on Lake Michigan beach.

Dunes, Point Beach State Park, Manitowoc County.

Rocky shores of Lake Superior

Wave-splashed sandstone shorelines and clay bluffs of the Apostle Islands and the Bayfield Peninsula (Lake Superior) may feature rugged to nearly horizontal ledges that are often moistened by wave-splash; several characteristic grasses are found here (Judziewicz and Koch 1993).

Agrostis scabra, northern tickle grass
Avenella flexuosa, tufted hair grass
Calamagrostis canadensis, Canada bluejoint
Calamagrostis stricta, slim-stemmed bluejoint
Deschampsia cespitosa, tufted hair grass
Poa compressa,* Canada bluegrass
Poa glauca, glaucous bluegrass
Sphenopholis intermedia, satin wedge grass
Trisctum spicatum, spike trisetum

Spike trisetum (*Trisetum spicatum*) on the sandstone shoreline of Lake Superior, Manitou Island, Apostle Islands, 1991.

Habitat of glaucous bluegrass (*Poa glauca*), North Twin Island, Apostle Islands, June 26, 2012.

Anthropogenic Open Areas

Grasses are abundant in open disturbed areas that are caused and maintained by human activity. Pastures, meadows, old fields, active agricultural fields, lawns, railroad rights-of-way, and roadsides frequently harbor good grass diversity in a successional setting. Although these are generally regarded as "weedy" habitats, about one-third of the grasses listed below are actually native.

Agrostis gigantea,* red top
Agrostis scabra, northern tickle grass
Alopecurus pratensis,* meadow foxtail
Anthoxanthum odoratum,* sweet vernal grass
Bromus inermis,* smooth brome
Dactylis glomerata,* orchard grass
Danthonia spicata, poverty oat grass
Dichanthelium acuminatum var. *fasciculatum*, twice-flowering panic grass

Dichanthelium columbianum, puberulent panic grass
Dichanthelium meridionale, slender panic grass
Digitaria species, crabgrasses and fall witch grass
Echinochloa crus-galli,* Eurasian barnyard grass
Elymus repens,* quack grass
Eragrostis spectabilis, purple love grass
Festuca rubra,* red fescue
Festuca trachyphylla,* hard fescue
Hordeum jubatum,* squirrel-tail barley
Lolium arundinaceum,* tall meadow fescue
Lolium perenne,* ryegrass
Lolium pratense,* clustered meadow fescue
Muhlenbergia mexicana, Mexican muhly
Panicum capillare, witch grass
Panicum dichotomiflorum,* fall panic grass (agricultural weed)
Phalaris arundinacea,* reed canary grass
Phleum pratense,* timothy
Poa bulbosa,* bulblet bluegrass
Poa compressa,* Canada bluegrass
Poa pratensis,* Kentucky bluegrass
*Setaria** species, foxtails (agricultural weeds)

Fall witch grass (*Digitaria cognata*) forming a purple carpet in a ditch, town of Breed, Oconto County.

Old field, Portage County.

Salt-tolerant exotic grasses

Areas where salt used on winter roads, sidewalks, and parking lots accumulates often have interesting weedy grasses from drier climates such as the dry saline interior American West or grasses more at home in oceanic coastal salt marshes. The most saline habitats include the margins of interstate highways, railroad switching yards, fairgrounds, shipyards (as in the city of Superior), and even unlikely places such as pickle factories where waste brine is dumped (as in the city of Oconto).

Agropyron cristatum,* crested wheatgrass
Aristida species, three-awned grasses (some native, others exotic)
Crypsis schoenoides,* prickle grass
*Digitaria** species, crabgrasses
Diplachne fusca var. *fascicularis*,* bearded sprangle-top
Distichlis spicata,* salt grass
Eragrostis capillaris, lace grass (but also native on rock outcrops)
Eragrostis cilianensis,* stink grass
Eragrostis pectinacea,* small love grass
Muhlenbergia asperifolia,* scratch grass
Poa arida,* plains bluegrass
Poa compressa,* Canada bluegrass
Puccinellia distans,* European alkali grass
Puccinellia nuttalliana,* Nuttall's alkali grass
Sclerochloa dura,* fairgrounds grass
Setaria pumila,* yellow foxtail
Sporobolus cryptandrus, sand dropseed
Sporobolus neglectus, small dropseed
Sporobolus vaginiflorus, poverty dropseed

Meter-wide strip of European alkali grass (*Puccinellia distans*) on wet sidewalk margin on UW–Stevens Point campus, August 6, 2012.

"Waste ground" along the railroad in Stevens Point, Portage County, with bearded sprangle-top (*Diplachne fusca* subsp. *fascicularis*; small, light green grass just below center) locally common, August 17, 2012.

Ornamental grasses

Grasses have become popular as ornamental plants. A detailed account of all the grasses grown ornamentally in Wisconsin is beyond the scope of this book. The reader is referred to works that will aid in the selection of ornamental grasses suitable for our climate and that will not escape and become invasive (Darke 1999; Greenlee 1992; King and Oudolf 1998; Oaks 1990). The following grasses are currently available in local nurseries and catalogs. They are rated as hardy (zones 2–5), meaning that they are known to survive over winter in Wisconsin. We have denoted with an asterisk (*) those species that are *not* hardy as perennials but are grown as annuals or are overwintered indoors.

Exotic grasses sold as ornamentals

Achnatherum calamagrostis (L.) Beauv., silver spear grass
Achnatherum extremorientale (H. Hara) Keng, silver spear grass
Alopecurus pratensis L., meadow foxtail; cultivars 'Aureo-variegatus', 'Glauca'
Andropogon saccharoides Sw., silver bluestem
Arrhenatherum elatius (L.) P. Beauv. ex J. Presl and C. Presl, variegated oat grass
Bouteloua dactyloides (Nutt.) J. T. Columbus, buffalo grass
Bouteloua gracilis (Kunth) Lag. ex Griffiths, blue grama
Briza media L., perennial quaking grass

Calamagrostis ×acutiflora (Schrad.) DC., feather reed grass; cultivars 'Avalanche', 'El Dorado', 'Karl Foerster', 'Stricta'
Calamagrostis brachytricha Steud. [*Deyeuxia brachytricha*], Korean feather reed grass
Calamagrostis epigeios (L.) Roth, Chee reed grass, feathertop; cultivar 'Hortorum'
Chasmanthium latifolium (Michx.) H. O. Yates, woodland wild-oats, northern sea-oats, spangle grass, shingle grass; cultivar 'River Mist'
Cenchrus purpurascens Thunb. [*Pennisetum alopecuroides*], fountain grass
Cenchrus spicatus (L.) Cav. [*Pennisetum glaucum*], pearl millet
Coix lacryma-jobi L.,* Job's-tears
Cortaderia selloana (Schult. and Schult. f.) Asch. and Graebn., pampas grass; cultivar 'Pumila'
Cymbopogon citratus (DC.) Stapf,* lemon grass
Dichanthelium clandestinum (L.) Gould, deer-tongue grass
Elymus magellanicus (E. Desv.) A. Love [*Agropyron magellanicum* (E. Desv.) Hack.], blue Magellan grass
Eragrostis trichodes (Nutt.) A. W. Wood, sand love grass; cultivar 'Bend'
Fargesia rufa T. P. Yi, red bamboo
Festuca amethystina L., sheep's fescue, large blue fescue, tufted fescue
Festuca glauca Lam., blue sheep's fescue, blue fescue, gray fescue
Glyceria maxima (Hartm.) Holmb., reed manna grass
Hakonechloa macra (Munro) Honda, hakone grass; cultivars 'Albo-striata', 'All Gold', 'Aureola', 'Namomi', 'Nicholas', 'Stripe It Rich'
Helictotrichon sempervirens (Vill.) Pilg., blue oat grass; cultivar 'Blue Sapphire'
Holcus mollis L., cultivar 'Alba-variegata'
Hordeum jubatum L., squirrel-tail barley
Lagurus ovatus L., hare's-tail
Leymus arenarius (L.) Hochst., blue lyme grass; sand ryegrass; blue beach grass; cultivar 'Glaucus'
Melica altissima L., purple Siberian melic; cultivar 'Atropurpurea'
Melica ciliata L., hairy melic grass, silky melic grass
Miscanthus floridulus (Labill.) Warb. ex K. Schum. and Lauterb., giant Chinese silver grass
Miscanthus sacchariflorus (Maxim.) Benth., Amur silver grass, silver banner grass
Miscanthus sinensis Andersson, Chinese silvergrass or eulalia; cultivars 'Adagio', 'Arabesque', 'Autumn Lights', 'Blondo', 'Bluetenwunder',

'Border Bandit', 'Burgunder', 'Cabaret', 'Cosmopolitan', 'Dixieland', 'Flamingo', 'Gold Feather', 'Goliath', 'Gracillimus', 'Graziella', 'Grosse Fontaine', 'Huron Blush', 'Huron Sentinel', 'Huron Sunrise', 'Kaskade', 'Little Fountain', 'Little Kitten', 'Little Zebra', 'Malepartus', 'Morning Light', 'Nippon', 'Puenktchen', 'Purpurascens', 'Rigoletto', 'Sarabande', 'Silver Feather', 'Silver Spider', 'Siren', 'Strictus', 'Undine', 'Variegatus', 'Yaku Jima', 'Zebrinus'

Muhlenbergia reverchonii Vasey and Scribn., autumn embers muhly

Molinia caerulea (L.) Moench, purple moor grass; cultivars 'Skyracer', 'Moorflame', 'Variegata'

Molinia littoralis Host, tall moor grass

Phalaris arundinacea L., cultivar 'Picta' ribbon grass

Phyllostachys aurea Carrière ex Rivière and C. Rivière, golden bamboo

Pleioblastus pygmaeus (Miq.) Nakai, dwarf bamboo

Pleioblastus viridistriatus (Regel) Makino, striped dwarf bamboo

Sesleria autumnalis (Scop.) F. W. Schultz, autumn moor grass

Sesleria caerulea (L.) Ard., blue moor grass

Sesleria heufleriana Schur, green moor grass

Spodiopogon sibiricus Trin., frost grass

Sporobolus airoides (Torr.) Torr., alkali dropseed

Sporobolus wrightii Munro ex Scribn., giant sacaton grass

Stipa tirsa Steven, horsetail feather grass

Stipa ucrainica P. A. Smirn., Ukrainian feather grass

Tridens flavus (L.) Hitchc., purple-top

Tripidium ravennae (L.) H. Scholz [*Erianthus ravennae* (L.) P. Beauv., *Saccharum ravennae*], Ravenna grass, plume grass, hardy pampas grass

Tripsacum dactyloides L., gama grass

Zoysia japonica Steud., Korean lawngrass (reported from Dunn County, Wisconsin [Manual of North American Grasses 2014], but specimen not located)

Native grasses sold as ornamentals

Ammophila breviligulata, American beach grass

Andropogon gerardii, big bluestem, cultivar 'Silver Sunrise'

Avenella flexuosa (L.) Drejer [as *Deschampsia flexuosa*], wavy hair grass

Bouteloua curtipendula (Michx.) Torr., side-oats grama

Bouteloua hirsuta Lag., hairy grama

Deschampsia caespitosa (L.) P. Beauv., tufted hair grass; cultivars 'Northern Lights', 'Goldgehange', 'Goldstaub'

Elymus canadensis L., Canada wild rye
Elymus hystrix L. [as *Hystrix patula*], bottlebrush grass
Elymus villosus Muhl. ex Willd., silky or downy wild rye
Eragrostis spectabilis (Pursh) Steud., purple love grass
Koeleria macrantha (Ledeb.) Schult., June grass, crested hair grass
Milium effusum L., golden wood millet, Bowles' golden grass; cultivar 'Aureum'
Panicum virgatum L., switchgrass; cultivars 'Badlands', 'Cheyenne', 'Cloud Nine', 'Haense Herme', 'Heavy Metal', 'North Winds', 'Prairie Fire', 'Rotstrahlbusch', 'Shenandoah', 'Squaw', 'Warrior'
Schizachyrium scoparium (Michx.) Nash, little bluestem; cultivars 'Prairie Blues', 'Blaze'
Sorghastrum nutans (L.) Nash, Indian grass; cultivars 'Indian Steel', 'Sioux Blue'
Spartina pectinata Link, prairie cordgrass
Sporobolus heterolepis (A. Gray) A. Gray, prairie dropseed

Below are some of the common grasses cultivated ornamentally in Wisconsin:

Feather reed grass (*Calamagrostis* ×*acutiflora*).

Pearl millet (*Cenchrus spicatus*).

Fountain grass (*Cenchrus purpurascens*).

Chinese silver grass or eulalia (*Miscanthus sinensis*).

Golden bamboo (*Phyllostachys aurea*); hardy in Stevens Point, Portage County, August 18, 2012.

Blue oat grass (*Helictrotrichon sempervirens*).

Keys to the Grass Genera of Wisconsin

What Features Must I Look at to Identify an Unknown Grass?

Learning to identify grasses is challenging and requires a significant investment of time. Yet, you will feel a real sense of accomplishment as you learn to recognize more and more genera both by spikelet structure and, with experience, by the overall look or "gestalt" of the plants, even when you are still meters away from the plant and the details of the spikelets are essentially invisible to you.

In our experience it is best to examine the spikelets of a new grass carefully with a hand lens or, better yet, a dissecting microscope. The most important characters to note when trying to identify an unknown grass are:

The leaves: Are the margins of the sheaths (the basal portions that embrace the culms) open and overlapping, or are they fused? What are the type (membranous, hairy) and length of the ligules? How wide are the blades? Are they flat or inrolled?

The inflorescence: Is it a panicle (with branching and rebranching), a contracted spike-like panicle, a true spike (narrow, and with the spikelets placed directly on the central axis), or a collection of spikes with a finger-like or tree-like branching form?

The spikelets: How many florets are in each one? Are the glumes short or long, similar to each other, or unlike in size and shape? Do they have awns? How many nerves do they have?

The florets: Do the florets (which are dominated by the large enfolding lemma) fall separately from each other, or does the entire spikelet fall from its stalk (pedicel) as a unit? Are the florets compressed laterally (like the body of a fish, such as a bass), dorsally (like the human body), or terete (like a snake)?

The lemmas: Are they are alike or radically different in shape and size? Are some sterile, that is, not producing fruits? Are they awned (provided with a needle-like projection)? If they are awned, are the awns produced from the base of the lemma, from the middle of its back, from between two apical teeth, or at the very tip of lemma, with no associated teeth?

The nerves of the lemmas: How many are there? (Do not forget to count the midrib.) Are they conspicuous and raised, or are they flush with the surface and hard to see? Do they converge at the apex (as in most grass genera), or do they run parallel to each other, such as in love grass (*Eragrostis*) and manna grass (*Glyceria*)?

Abbreviations Used in the Photographs

AN anther
AW awn
BL blade
CA caryopsis
FL floret (also FL1, FL2, etc.)
LE lemma
LG lower glume
LI ligule
NO node
PA palea
RA rachilla
SH sheath
SP spikelet
ST stigma
UG upper glume

The Ten Commonest Grasses in Wisconsin

Wisconsin has 232 species of grasses, but for the Wisconsin neophyte agrostologist just beginning to learn grasses, a very few abundant species will be the first ones to "jump out," demanding notice and identification. So it may be useful to list what we consider the ten most obvious common grasses that average Wisconsinites would notice around their house or during a ten-minute walk around their neighborhood, ranked roughly in order of abundance and conspicuousness. With the partial exception of common reed, none of these species is native to Wisconsin!

Poa pratensis (Kentucky bluegrass): Look at your lawn! This cool-season species blooms in the spring and fall.
Bromus inermis (smooth brome): A dominant species that lines upland roadside and ditches and abandoned agricultural fields. It is most evident in summer and fall.
Phalaris arundinacea (reed canary grass): A dominant, midsize grass of wetlands, especially in central and southern Wisconsin.
Elymus repens (quack grass): A dominant weed of ditches and gardens.
Poa compressa (Canada bluegrass): Dominates dry open fields and roadsides. It has a smaller, "wimpier" inflorescence than Kentucky bluegrass.
Agrostis stolonifera (redtop): A very common roadside and old field grass with steeple-shaped reddish panicles with whorled branches.
Phleum pratense (timothy): Never dominant, but this pasture species with distinctive pencil-like inflorescences is found in many habitats.
Setaria pumila (yellow foxtail): Abundant on roadsides in the late summer and fall. It looks like a "bristly timothy." There are several foxtail species in Wisconsin, but this is the commonest species.
Digitaria ischaemum (smooth crabgrass): The smallest grass on this list, this sprawling species with finger-like inflorescence branches is impossible to miss in your garden or in sidewalk or pavement cracks. There are several crabgrass species in Wisconsin, but this is the commonest species.
Phragmites australis (common reed): Not as widespread as the other species on this list, but this huge, colonial wetland grass with big, fluffy fall panicles is dominant where found.

Grasses with Unusual Characters

The following grasses with distinctive characters are keyed out later in the text, but we believe that it is useful to list them here separately. Those marked by an asterisk (*) are exotic; that is, they are not native to Wisconsin.

Strongly clump-forming bunchgrasses

Avenella flexuosa, wavy hair grass
Dactylis glomerata,* orchard grass
Deschampsia cespitosa, tufted hair grass

*Lolium arundinaceum** and *L. pratense*,* meadow fescues
Panicum virgatum, switchgrass
Sporobolus heterolepis, prairie dropseed

Glaucous (waxy bluish-whitened) foliage

Alopecurus pratensis,* meadow foxtail (slightly)
Andropogon gerardii, big bluestem (when young)
Elymus, several species of wild rye, especially *E. canadensis*, *E. lanceolatus*, *E. repens*,* *E. riparius*, and *E. trachycaulus* (strongly)
Leymus arenarius,* lyme grass (strongly)
Lolium arundinaceum,* tall meadow fescue (slightly)
Milium effusum, wood millet (slightly)
Pascopyrum smithii,* western wheatgrass (strongly)
Poa arida,* plains bluegrass (strongly)
Poa glauca, glaucous bluegrass (strongly)
Schizachyrium scoparium, little bluestem (when young)

Sheaths swollen with hidden inflorescences

Dichanthelium clandestinum,* deer-tongue grass
Leersia oryzoides, rice cut-grass
Sporobolus, dropseed (all species except *S. heterolepis*)
Triplasis purpurea, purple sand grass

Flowering in early spring, sometimes before May 1

Bromus tectorum,* cheat grass
Festuca octoflora, six-weeks fescue
Oryzopsis asperifolia, rough-leaved rice grass
Poa alsodes, grove bluegrass
Poa annua,* annual bluegrass
Poa pratensis,* Kentucky bluegrass
Schizachne purpurascens, false melic grass

Grasses of deep forest shade, often with broad leaves

Agrostis perennans, autumn bent grass (narrow leaves)
Brachyelytrum aristosum and *B. erectum*, shorthusks
Bromus latiglumis, ear-leaved brome
Bromus nottowayanus, satin brome
Bromus pubescens, Canadian brome
Cinna arundinacea and *C. latifolia*, wood-reeds

Diarrhena obovata, obovate beak grain
Dichanthelium latifolium, broad-leaved panic grass
Elymus hystrix, bottlebrush grass
Festuca subverticillata, nodding fescue
Leersia lenticularis, catchfly grass
Leersia virginica, white grass
Melica smithii, Smith's melic grass
Milium effusum, wood millet
Muhlenbergia, muhly grasses (several species)
Oryzopsis asperifolia, rough-leaved rice grass
Patis racemosa, black-seeded rice grass
Piptatheropsis pungens, mountain rice grass
Poa alsodes, grove bluegrass
Poa languida, bluegrass
Poa nemoralis,* wood bluegrass
Poa paludigena, bog bluegrass
Poa saltuensis, woodland bluegrass
Poa sylvestris, forest bluegrass
Poa wolfii, Wolf's bluegrass
Schizachne purpurascens, false melic grass

Tumbleweeds, the mature inflorescences breaking off and blown away as a unit

Agrostis hyemalis, southern tickle grass
Agrostis scabra, northern tickle grass
Chloris verticillata,* windmill grass
Digitaria cognata, fall witch grass
Eragrostis capillaris, lace grass
Eragrostis spectabilis, purple love grass
Muhlenbergia asperifolia,* scratch grass
Panicum capillare, witch grass

Shiny foliage

Anthoxanthum hirtum, sweet grass
Bromus latiglumis, ear-leaved brome
Bromus nottowayanus, satin brome
Diarrhena obovata, obovate beak grain
Festuca subverticillata, nodding fescue
Glyceria grandis, reed manna grass

Oryzopsis asperifolia, rough-leaved rice grass
Patis racemosa, black-seeded rice grass
Spartina pectinata, prairie cordgrass
Sphenopholis intermedia, slender wedge grass

Sheath or blade auricles ("ears") present

Bromus latiglumis, ear-leaved brome
Elymus virginicus, Virginia wild rye
Hordeum vulgare,* barley
*Lolium arundinaceum** and *L. pratense*,* meadow fescues

Wintergreen leaves

Dactylis glomerata,* orchard grass
Dichanthelium, twice-flowering panic grasses (many species)
*Lolium arundinaceum** and *L. pratense*,* meadow fescues
Oryzopsis asperifolia, rough-leaved rice grass

Native annuals

Aristida, three-awned grass (all species except *A. purpurascens*)
Cenchrus longispinus, sandbur
Digitaria cognata, fall witch grass
Echinochloa muricata, American barnyard grass
Echinochloa walteri, bottlebrush barnyard grass
Eragrostis frankii, sandbar love grass
Eragrostis hypnoides, creeping love grass
Festuca octoflora, six-weeks fescue
Panicum capillare, witch grass
Panicum flexile, fen panic grass
Panicum philadelphicum, Philadelphia panic grass
Sporobolus neglectus, small dropseed
Sporobolus vaginiflorus, poverty dropseed
Triplasis purpurea, purple sand grass
Zizania aquatica, southern wild rice
Zizania palustris, northern wild rice

Exotic annuals

Apera interrupta, silky bent grass
Avena fatua, wild oats
Avena sativa, oats

Beckmannia syzigachne, American slough grass
Bromus tectorum, cheat grass (and all other six exotic species, except *B. inermis*)
Crypsis schoenoides, prickle grass
Cynodon dactylon, Bermuda grass
Digitaria ischaemum, smooth crabgrass
Digitaria sanguinalis, northern crabgrass
Diplachne fusca subsp. *fascicularis*, bearded sprangle-top
Echinochloa crus-galli, European barnyard grass
Echinochloa esculenta, Japanese millet
Eleusine indica, goose grass
Eragrostis species, love grasses, including *E. capillaris*, *E. cilianensis*, *E. mexicana*, *E. minor*, and *E. pectinacea*
Eriochloa villosa, woolly cup grass
Festuca myuros, rat-tail fescue
Festuca octoflora, six-weeks fescue
Hordeum species, barleys (all three species)
Panicum dichotomiflorum, elbow grass
Panicum miliaceum, broomcorn millet
Poa annua, annual bluegrass
Polypogon interruptus, ditch polypogon
Polypogon monspeliensis, rabbit's-foot grass
Sclerochloa dura, fairgrounds grass
Secale cereale, rye
Setaria species, foxtails (all six species)
Sorghum bicolor, broomcorn or sorghum
Triticum aestivum, wheat
Ventenata dubia, ventenata grass
Zea mays, corn

Keys to the Grass Genera of Wisconsin

This set of keys to genera of grasses is based on inflorescences, spikelets, and some vegetative characters. Some grasses can be identified before spikelets are fully developed or in late fruiting stages as spikelets break apart or fall from the inflorescence branches, but the identification of most species is based on features of glumes and florets, requiring well-developed spikelets. The numbers accompanying the illustrations in the following keys refer to the genus numbers in the field guide section.

Key 1. Key to certain genera and to other keys for the remaining genera

1. Spikelets encased in a spiny or bony involucre, or subtended by bristles, or on a cob surrounded by a leafy husk

2. Spikelets in two or more rows on a thickened cob, surrounded by leaves; exotic; rare escape from cultivation .... *Zea mays* (the only species) 86

2. Spikelets encased in a spiny or bony involucre, or subtended by bristles

3. Spikelets encased in a spiny or bony involucre

4. Involucre forming a hard bur armed with sharp spines; plants annual; weedy in dry sand .... *Cenchrus longispinus* (the only species) 21

4. Involucre bony but not spiny; low stoloniferous perennials; exotic, rare .... *Bouteloua dactyloides* (one of four species in Wisconsin) 14

3. Spikelets subtended (surrounded at the base) by bristles; exotic

5. Bristles less than 2 cm long, persistent after spikelets drop; small to large annuals .... *Setaria* (six species) 74

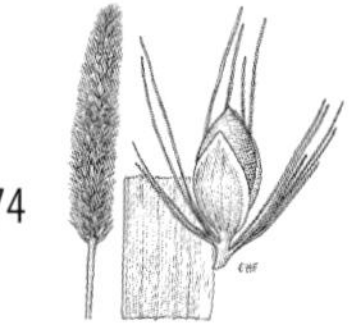

5. Bristles 2–4 cm long, falling attached to the spikelets; robust cultivated perennials .... *Cenchrus americanus* and *C. compressus* (cultivated species)

1. Spikelets not encased in a spiny or bony involucre or on a cob in a leafy husk, although part of an inflorescence may be enclosed in a leaf sheath in some species

6. Culms (stems) more than 2 m tall .... Key 2

6. Culms less than 2 m tall (or specimen incomplete, height unknown)

7. Leaf sheaths closed (fused at least 1 cm above base) .... Key 3 Key 3

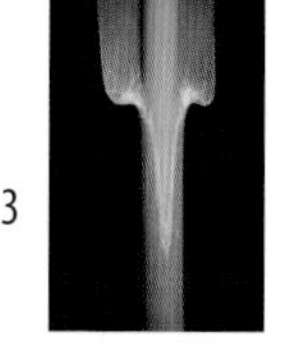

7. Leaf sheaths open (overlapping their entire length) or fused only at the base (or specimen lacking leaf sheaths)

8. Inflorescence of one or more spikes with spikelets attached directly to the main axis or to major branches, or of racemes with spikelets on short pedicels, or of a dense spike-like panicle

9. Inflorescence of two or more spikes, racemes, or spike-like branches arising directly from the main axis of the inflorescence and not rebranching .... Key 4 Key 4

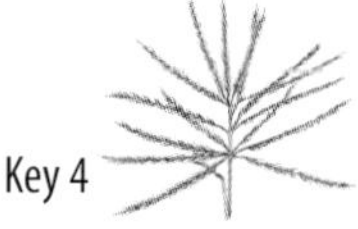

9. Inflorescence a single unbranched spike, with all spikelets sessile along the inflorescence axis (or in two genera with a sessile spikelet at each node flanked by one or two stalked spikelets), or a very dense cylindrical spike-like panicle or raceme with branches and pedicels shorter than the spikelets
   10. Inflorescence a single unbranched spike with all spikelets sessile along the inflorescence axis (or in two genera with a sessile spikelet at each node flanked by one or two stalked spikelets) .... Key 5 — Key 5
   10. Inflorescence a very dense cylindrical spike-like panicle, the branches and pedicels shorter than the spikelets .... Key 6 — Key 6

8. Inflorescence a panicle (typically with major branches spreading at time of pollination), the main branches rebranching and longer than the spikelets
   11. Lemmas or glumes with conspicuous awns ...... Key 7 — Key 7
   11. Lemmas and glumes awnless, or with short awns concealed within the spikelets
      12. Spikelets with just a single fertile floret
         13. Mature floret rigid and usually shiny .... Key 8 — Key 8
         13. Mature floret membranous and dull, with same texture as glumes .... Key 9 — Key 9
      12. Spikelets with two or more fertile florets
         14. Florets two to five (usually three or four) per spikelet (including rudimentary florets) .... Key 10 — Key 10
         14. Florets consistently four or more per spikelet (including rudimentary florets) .... Key 11 — Key 11

Key 2. Grasses that are at least 2 m tall

This key includes grasses that often are more than 2 m tall. Shorter specimens of these grasses also key out in other keys. Under unusual conditions some other grasses could reach 2 m in height, but they are not included in this key.

1. Inflorescence of several spikes or of narrow spike-like racemes
   2. Inflorescence of several pinnately arranged spikes; spikelets crowded into one-sided spikes; wet prairies and wetlands *Spartina pectinata* (the only species) 77
   2. Inflorescence of spike-like racemes radiating from nearly the same point; spikelets not crowded into one-sided spikes

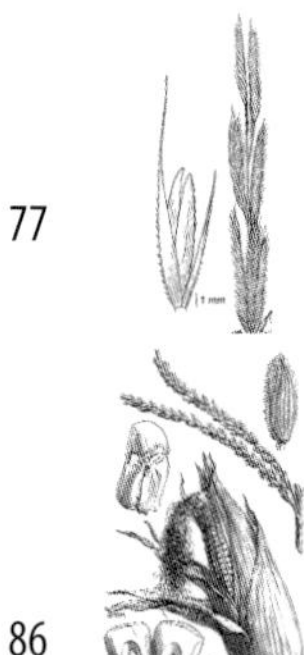

      3. Large "tassel" of spike-like racemes bearing paired male spikelets borne at top of main culm (enclosed by sheaths of upper leaves earlier in season); exotic, extensively cultivated, but rarely self-seeded .... *Zea mays* (the only species) 86
      3. Spike-like racemes of bisexual, female, or sterile spikelets borne in clusters of two to twenty at the end of branches

         4. Racemes two to six in a cluster, eventually disarticulating, with scattered short hairs; sessile spikelets bisexual, pedicelled spikelets male; native, abundant in mesic prairies .... *Andropogon gerardii* (the only species) 5

         4. Racemes ten to twenty in a cluster, not disarticulating, bearing long, silky, white hairs; all spikelets pedicelled and bisexual; exotic, nearly sterile, but spreading by rhizomes .... *Miscanthus sacchariflorus* (the only species) 54

1. Inflorescence a panicle

   5. Spikelets (when mature) with long, silky, hairs; native and exotic, in wetlands *Phragmites australis* (the only species) 65
   5. Spikelets glabrous or with small, inconspicuous hairs

      6. Upper panicle branches erect or ascending, bearing elongated, terete, awned female spikelets; lower panicle branches spreading, bearing male spikelets; pulpy aquatic to semiaquatic annuals; native, in wetlands .... *Zizania* (two species) 87
      6. Panicle branches essentially alike; upland plants or rhizomatous wetland plants

7. Spikelets with several florets; sheaths fused; wetland perennials with vigorous rhizomes; exotic, rare in wetlands ....*Glyceria maxima* (plus five other species) 43
7. Spikelets with one or two fertile florets; sheath overlapping; upland annuals or perennials
   8. Spikelets with two fertile florets, the lower floret male, bearing a bent awn near the base, the upper floret perfect, bearing a short, straight awn; exotic, rare .... *Arrhenatherum elatius* (the only species) 9
   8. Spikelets with one fertile floret; bent awn, if present, arising at apex of floret
      9. Spikelets all alike, on pedicels; awns absent; glumes thin, the lower about one-third as long as the spikelet; exotic, in farm fields .... *Panicum miliaceum* (plus five other species) 59
      9. Spikelets of two kinds; sessile spikelets bisexual; pedicelled spikelets awned, male, sterile, or represented by a ciliate pedicel; glumes thick and hard
         10. Pedicelled spikelet male or sterile; ligule of soft hairs; exotic ..... *Sorghum* (two species and one hybrid) 76
         10. Pedicelled spikelet usually absent, the pedicel ciliate; ligule stiff and cartilaginous; native, abundant in prairies .... *Sorghastrum nutans* (the only species) 75

Key 3. Grasses with closed leaf sheaths (fused at least 1 cm above the base)

For convenience, these genera also appear in the "open sheath" keys.

1. Florets two per spikelet, the lemmas three-nerved; rare in springy places .... *Catabrosa aquatica* (the only species) 20
1. Florets three to six per spikelet, the lemmas five-nerved
   2. Spikelets in small, dense clusters near the ends of few, elongate inflorescence branches; exotic; common ...... *Dactylis glomerata* (the only species) 29
   2. Spikelets evenly dispersed throughout a small to large panicle
      3. Lemmas awnless

4. Lemmas with nerves converging toward the apex; leaf blades with M-shaped crease about one-third of the way from the base toward the apex; abundant weed .... *Bromus inermis*
4. Lemmas with nerves parallel, not converging toward the apex; leaf blades lacking an M-shaped crease; native species
   5. Lemmas all similar in shape; glumes one-nerved; common wetland genus .... *Glyceria* 43 
   5. Uppermost lemma convolute (inrolled), club-shaped; glumes three- to five-nerved; rare in dry-to-mesic sites, far southern Wisconsin .... *Melica nitens* 52

3. Lemmas awned
   6. Lemmas awned from near middle of back; rachilla densely hairy; rare exotic .... *Avenula pubescens* (the only species) 12 
   6. Lemmas awned from at or near the apex, or awnless; rachilla glabrous or puberulent
      7. Callus of floret bearded; glumes purple, the lemmas green .... *Schizachne purpurascens* (the only species) 70 
      7. Callus of floret glabrous; glumes and lemmas uniformly green or purple
         8. Leaf sheaths glabrous to pubescent, but not retrorsely scabrous; ligules usually up to 2.5 mm (never more than 4 mm) long; lemmas five- to seven-nerved; common .... *Bromus* (thirteen species) 17 
         8. Leaf sheaths retrorsely scabrous; ligules 2–5 mm long; lemmas seven-nerved; rare, hardwood forests in far northern Wisconsin .... *Melica smithii* (there is one other species) 52

Key 4. Inflorescence with two or more spikes or spike-like, densely flowered branches arising directly from axis of inflorescence

1. All or most spikes, racemes, or spike-like branches radiating palmately—or nearly so—from a more or less common point at the tip of the inflorescence axis

  2. Plants erect perennials 1–3 m tall; spikelets 7–10 mm long, awned

    3. Racemes three to five, not obscured by white hairs; native, common in wet to dry prairies .... *Andropogon gerardii* (the only species) 5

    3. Racemes five to fifteen, obscured by abundant long, silky, white hairs; exotic, occasional in ditches and wetlands ... *Miscanthus sacchariflorus* (the only species) 54

  2. Plants sprawling or creeping weedy annuals less than 0.5 m tall; spikelets less than 7 mm long, awnless

    4. Spikelets one-flowered; rare exotic .... *Cynodon dactylon* (the only species) 27

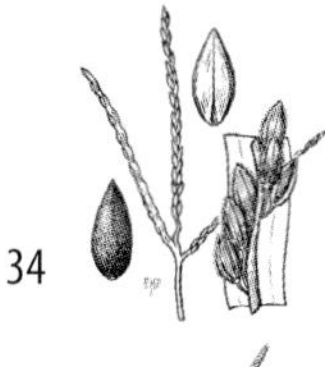

    4. Spikelets two- to six-flowered

      5. Spikelets 1.5–3.5 mm long, two-flowered, the lower floret sterile; exotic; common weeds .... *Digitaria* (three species) 34

      5. Spikelets 4–6 mm long, three- to six-flowered, all florets fertile; exotic, uncommon, mostly in southern Wisconsin .... *Eleusine indica* (the only species) 38

1. Spikes, racemes, or spike-like branches arranged singly along the inflorescence axis

  6. Spikelets strongly laterally flattened

    7. Floret one per spikelet (no rudimentary florets present); plants coarse, 1–2 m tall; native, wetlands .... *Spartina pectinata* (the only species) 77

    7. Florets two to many per spikelet (rudimentary florets may be present); plants less than 0.6 m tall

8. Terminal inflorescence branches digitate (finger-like), numerous, with an additional whorl of branches a short distance beneath; rare exotic, southern Wisconsin .... *Chloris verticillata* (the only species) 23

8. Terminal inflorescence branches not digitate, solitary to numerous, if more than one, then spaced out along the rachis

   9. Spikelets with one awnless fertile lower floret and one three-awned rudimentary upper floret; plants erect; mostly native, dry-to-mesic prairies .... *Bouteloua* (four species) 14

   9. Spikelets with five to numerous florets, these all alike; plants sprawling; exotic, occasional in saline areas .... *Diplachne fusca* (the only species) 35

6. Spikelets terete to dorsally flattened

   10. Spikelets awned

      11. Inflorescence a pyramidal panicle of racemes; spikelets ovoid, hispid; native and exotic; wetlands .... *Echinochloa* (four species) 37

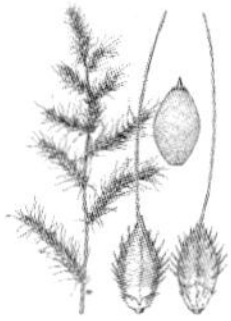

      11. Inflorescence an elongate aggregation of alternate racemes; spikelets narrowly linear, lanceolate, scabrous; common, dry prairies .... *Schizachyrium scoparium* (the only species) 71

   10. Spikelets awnless

      12. Spikelets 4.5–5 mm long, the fertile floret with a tiny but distinct basal knob; exotic, uncommon in disturbed areas and fields, mostly in southern Wisconsin .... *Eriochloa villosa* (the only species) 41

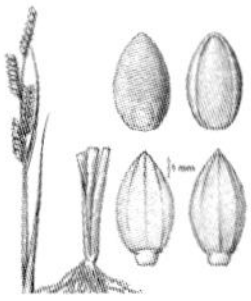

      12. Spikelets 1.5–3.5 mm long, the fertile floret lacking a basal knob

13. Spikelets 2.5–3.5 mm long, rotund, very strongly dorsally flattened; exotic, uncommon in ditches and wetlands .... *Beckmannia syzigachne* (the only species) 13

13. Spikelets 1.5–2.5 mm long, ovate, weakly dorsally flattened; native, dry open areas, occasional in southern and central Wisconsin .... *Paspalum setaceum* (the only species) 61

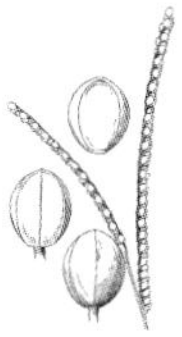

Key 5. Inflorescence a true spike, with all spikelets placed directly on the main axis

1. Lemmas with a minute spiny keel and margins, otherwise glabrous, gradually prolonged into a long awn; exotic, escaped from cultivation or erosion control plantings .... *Secale cereale* (the only species) 73 
1. Lemmas glabrous to pubescent, but without a spiny keel and margins, awned or awnless
   2. Plants 1–2 m tall, the foliage strongly bluish; exotic, mostly on sandy Lake Michigan beaches .... *Leymus arenarius* (the only species) 50 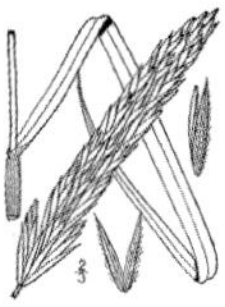
   2. Plants at most 1 m tall, the foliage not strongly bluish; native and exotic in various habitats
      3. Glumes 3–6 mm wide, less than three times as long as wide .... *Triticum aestivum* (the only species) 84 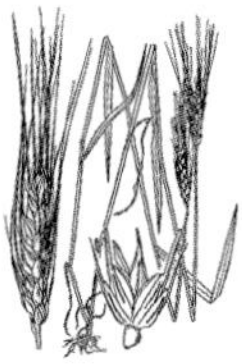
      3. Glumes less than 3 mm wide, more than three times as long as wide
         4. Spikelets two to three at each node, the glumes four to six at each node
            5. Spikelets two at each node, total number of glumes not more than four; native and exotic .... *Elymus* (in part; there are twelve species) 39 
            5. Spikelets three at each node, total number of glumes six
               6. Larger lemmas with bodies 4–6 mm long; spike easily disarticulating at maturity; exotic .... *Hordeum* (in part; there are two other species) 47

6. Larger lemmas with bodies 8–12 mm long; spike not disarticulating at maturity
   7. Lemma awn recurved to divergent, as robust as the glume awns; native and exotic .... *Elymus* (in part; there are twelve species)
   7. Lemma awn straight, much more robust than glume awns; exotic, sparingly escaped from cultivation .... *Hordeum vulgare* (in part; there are two other species) 47

4. Spikelets one at each node, the glumes not more than two at each node
   8. Glume solitary, the spikelets many-flowered, placed edgewise against the rachis; exotic, common weed .... *Lolium perenne* (there are two other species) 51 
   8. Glumes two, the spikelets few- to many-flowered, placed sidewise against the rachis
      9. Glumes with bodies 3–4 mm long, their midrib asymmetric and strongly keeled; spike congested, strongly two-ranked; exotic, rare .... *Agropyron cristatum* (the only species) 1 
      9. Glumes with bodies at least 4 mm long, their midrib symmetric and not strongly keeled; spike congested or not, but not strongly two-ranked
         10. Lemma awns strongly recurved or spreading at maturity .... *Elymus trachycaulus* (there are eleven other species in the genus) 39 
         10. Lemma awns straight or absent
            11. Lemmas densely hairy; rare native of Lake Michigan dunes .... *Elymus lanceolatus* (there are eleven other species in the genus) 39 
            11. Lemmas glabrous
               12. Leaf blades more or less broad and flat, with obscure nerves .... *Elymus* (in part; there are twelve species)
               12. Leaf blades inrolled when dry, the nerves prominent and raised; exotic, occasional .... *Pascopyrum smithii* (the only species) 60

Key 6. Inflorescence a dense, spike-like panicle

1. Spikelets awnless
   2. Inflorescence ovoid, 1–4 cm long; rare exotic, last collected in 1972 .... *Crypsis schoenoides* (the only species) 26
   2. Inflorescence narrowly cylindrical, 5–40 cm long; natives
      3. Floret one per spikelet; plant 0.5–1 m tall; spikelets 10–15 mm long; native, locally common on Great Lakes dunes .... *Ammophila breviligulata* (the only species) 4
      3. Florets two per spikelet; plant less than 0.5 m tall; spikelets less than 5 mm long
         4. Peduncle finely pubescent right below the inflorescence; spikelets dull, disarticulating above the glumes, which are 3–4.5 mm long; dry areas, often in barrens .... *Koeleria macrantha* (the only species) 48
         4. Peduncle glabrous right below the inflorescence; spikelets shiny, disarticulating below the glumes, which are 2.5–3 mm long; wetlands or less commonly uplands .... *Sphenopholis* (two species) 78
1. Spikelets awned
   5. Florets two to four per spikelet
      6. Florets consistently two per spikelet, both florets with awns 3–5 mm long; plants hairy; native, rare on rocky shores of Lake Superior .... *Trisetum spicatum* (the only species) 83
      6. Florets two to four per spikelet, florets with awns less than 3 mm long; plants more or less smooth; exotic weeds
         7. Spikelets 7–9 mm long, the florets unlike, the lowest two small and male, the uppermost floret larger and female; locally common in northern Wisconsin .... *Anthoxanthum odoratum* (there is one other species) 6
         7. Spikelets 3–5 mm long, the florets all alike; rare .... *Cynosurus cristatus* (the only species) 28
   5. Floret one per spikelet
      8. Glumes 6–9 mm long, the body of the lemma 5–7 mm long; native, forests in central and northern Wisconsin .... *Oryzopsis asperifolia* (the only species) 58

8. Glumes less than 5 mm long, the body of the lemma less than 4 mm long; native or exotic
   9. Glumes awned; lemmas awnless or awn-tipped; exotic
      10. Glumes with awns 1–2 mm long; lemmas awnless; abundant . . . . *Phleum pratense* (the only species) 64

      10. Glumes with awns 2–10 mm long; lemmas with awns ca. 0.5 mm long; rare . . . . *Polypogon* (two species) 68

   9. Glumes awnless, lemmas awned; mostly wetlands

      11. Spikelets short-hairy all over, disarticulating below the glumes; native and exotic . . . . *Alopecurus* (four species) 3

      11. Spikelets long-hairy only at the base of the floret, disarticulating above the glumes; native . . . . *Calamagrostis stricta* (there are two other species) 18

Key 7. Grasses with panicles, the spikelets with conspicuous awns

This key includes grasses with awns on the glumes and/or lemmas. Awns are usually a continuation of the midrib of the lemmas or glumes without a border of soft tissue. A tapered narrow apex of a glume or lemma does not constitute an awn if soft tissue continues to the apex. In some species the small awns arise near the base of lemmas, partially concealed by hairs, or awns are jointed at the base and eventually drop; these grasses are also included in Keys 8–11.

1. Glumes 18–25 mm long . . . . *Avena* (two species) 10
1. Glumes less than 15 mm long

   2. Awns three-forked; native and exotic, dry, often sandy open areas . . . . *Aristida* (six species) 8
   2. Awns unbranched, one per floret
      3. Glumes hard; lemmas thin and transparent; bisexual spikelets sessile in short racemes, paired with one or two male or sterile spikelets or a ciliate pedicel

4. Pedicelled spikelet male or sterile; ligule of soft hairs; exotic .... *Sorghum* (two species and one hybrid) 76

4. Pedicelled spikelet usually absent, the pedicel ciliate; ligule stiff and cartilaginous; native, prairies ..... *Sorghastrum nutans* (the only species) 75

3. Glumes thin, soft; lemmas rigid or thin but not transparent; spikelets pedicelled
   5. Spikelets with a single fertile floret
      6. Floret thin, more or less membranous, not harder than the glumes, dull, usually laterally compressed; awn not jointed at the base
         7. Spikelets unisexual; female spikelets awned, elongate, terete, on erect to ascending branches in the upper portion of the panicle; male flowers pendant, awnless, on spreading lower branches; tall, pulpy annuals; native; wetlands .... *Zizania* (two species) 87

         7. Spikelets bisexual and alike; panicle not differentiated into erect and spreading branches
            8. Lemma surrounded at base by hairs 2–4 mm long, partially concealing a delicate dorsal awn; mostly wetlands .... *Calamagrostis* (three species) 18
            8. Lemma glabrous or hairs less than 1 mm long; awn conspicuous, arising near the tip of the lemma or glumes
               9. Spikelets 8–13 mm long, excluding awns; leaf blades finely hairy; upland forests .... *Brachyelytrum* (two species) 15
               9. Spikelets 2–4 mm long, excluding awns; blades not strongly tapered at base

                  10. Glumes with two tiny terminal teeth and awns 2–10 mm long; spikelets falling with glumes attached; exotic, rare .... *Polypogon* (two species) 68
                  10. Glumes without terminal teeth, tapering to an awn 1–3 mm long, or glumes awnless, the awn attached to the lemma; florets falling separately, leaving the glumes attached to the panicle branches

11. Awn 4–10 mm long, attached below tip of lemma; glumes awnless; rachilla prolonged as a bristle beyond floret base; exotic, rare ...... *Apera interrupta* (the only species) 7 
11. Awn 1–10 mm long, attached to the tip of lemma or glumes; rachilla not prolonged as a bristle; native .... *Muhlenbergia* (twelve species) 56

6. Floret rigid, harder than the glumes, usually shiny, terete or dorsally compressed; awn jointed at base in several species
    12. Spikelets dorsally compressed, hispid, with one or two tapering awns; lower sterile floret present, resembling the second glume (spikelets with three apparent glumes); native and exotic, common in wetlands .... *Echinochloa* (four species) 37 
    12. Spikelets terete to laterally compressed, glabrous, with one abruptly distinct awn, jointed at tip of hard, ovoid floret; sterile floret and apparent third glume absent
        13. Awn persistent, twisted near base, stout
            14. Awn 10–20 cm long; body of lemma 10–25 mm long; native and exotic .... *Hesperostipa* (two species) 45 
            14. Awn 2–4 cm long; body of lemma 4–7 mm long; exotic, rare, open dry areas .... *Nassella viridula* (the only species) 57 
        13. Awn deciduous, straight, rather delicate
            15. Panicle raceme-like; leaves nearly all basal, green throughout winter; common in upland woods, central and northern Wisconsin ...... *Oryzopsis asperifolia* (the only species) 58 
            15. Panicle open; leaves produced along stem, deciduous
                16. Leaves 1–2 mm wide, dull; lemmas 2.5–4 mm long; dry, often coniferous forests .... *Piptatheropsis* (two species) 66

16. Leaves 8–15 mm wide, shiny; lemmas 5–7 mm long; hardwood forests ...... *Patis racemosa* (the only species) 62

5. Spikelets with two or more fertile florets

17. Spikelets with just two florets; uncommon to rare species

18. Spikelets disarticulating below the glumes, hairy throughout, the upper floret with a tiny fishhook-shaped awn; exotic, rare, not collected since 1975 .... *Holcus lanatus* (the only species) 46

18. Spikelets disarticulating above the glumes, glabrous except for the bearded callus; if the upper floret is awned, then the awn is not fishhook-shaped

19. Lemma awns 1.5–2 mm long; rare in wetlands, northern and central Wisconsin ..... *Deschampsia cespitosa* (the only species) 31

19. Lemma awns 3–20 mm long

20. Lemmas 8–15 mm long, the awns 15–20 mm long; exotic, rare in northern Wisconsin ..... *Avenula pubescens* (the only species) 12

20. Lemmas 2–7 mm long, the awns 3–10 mm long

21. Lower lemma with prominent bent awn; upper lemma with short, straight awn; robust (1–2 m tall); exotic, rare ..... *Arrhenatherum elatius* (the only species) 9

21. Both lemmas with fairly prominent awns; plants less than 1 m tall

22. Lower lemma with awn produced from the back near the base; upper lemma with awn less than 8 mm long; native, rare in dry open ground, northern Wisconsin ..... *Avenella flexuosa* (the only species) 11

22. Lower lemma with awn produced from the tip; upper lemma with awn 10–16 mm long; exotic, rare .... *Ventenata dubia* (the only species) 85

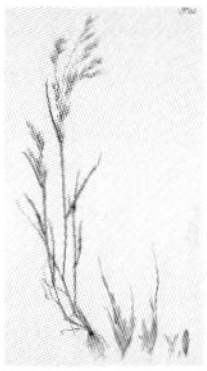

17. Spikelets with three to ten florets; many species common

23. Glumes as long as or slightly longer than the florets; basal leaves curling in common species; beard of hairs present at summit of leaf sheath; common in mostly dry open areas . . . . *Danthonia* (two species) 30

23. Glumes much shorter than the spikelet; leaves straight, not curling; beard of hairs absent at summit of leaf sheath

24. Lemma awn arising from the apex; leaf sheath margins free

25. Lemma with three minute awns arising from the apex; exotic, rare in southern Wisconsin . . . . *Tridens flavus* (the only species) 81

25. Lemma with just one tapering, terminal awn; native and exotic . . . . *Festuca* (eight species) 42

24. Lemma awn arising from between two minute to large apical teeth; leaf sheath margins fused

26. Lemma awn arising from between two large teeth about one-quarter of the length of the lemma; glumes purple, florets green . . . . *Schizachne purpurascens* (the only species) 70

26. Lemma awn arising from between two small teeth much less than one-quarter of the length of the lemma; spikelets either all purple or green, not bicolored

27. Leaf sheaths glabrous to pubescent but not retrorsely scabrous; ligules usually up to 2.5 mm (never more than 4 mm) long; lemmas five- to seven-nerved; common . . . . *Bromus* (thirteen species) 31

27. Leaf sheaths retrorsely scabrous; ligules 2–5 mm long; lemmas seven-nerved; rare, hardwood forests in far northern Wisconsin . . . . . *Melica smithii* (there is one other species) 52

Key 8. Grasses with panicles and just one rigid fertile floret per spikelet, without conspicuous awns

1. Fertile floret not subtended by sterile floret(s) (two minute sterile florets less than 1 mm long present in *Phalaris*)
   2. Floret terete or dorsally compressed, not subtended by minute sterile florets; native, rich upland forests .... *Milium effusum* (the only species) 53

   2. Floret laterally compressed, subtended by two minute hairy sterile florets less than 1 mm long; abundant exotic of open wetlands ... *Phalaris* (two species) 63

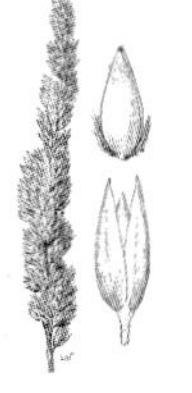

1. Fertile floret subtended by a sterile floret whose lemma resembles a "third glume"
   3. Spikelets hispid; native and exotic, common in wetlands .... *Echinochloa* (four species) 37

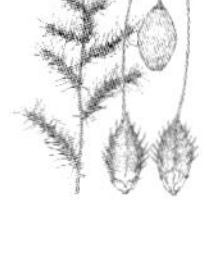

   3. Spikelets smooth or hairy but not hispid
      4. Ligules membranous
         5. Fertile floret rigid, the apex of the lemma tipped with tiny prickles; first glume 1–2 mm long; rare, sandbars and sloughs along the lower Wisconsin River ...... *Coleataenia rigidula* (the only species) 25

         5. Fertile floret leathery, glabrous; first glume less than 0.5 mm long; common in open sandy ground .... *Digitaria cognata* (there are two other species in *Digitaria*) 34

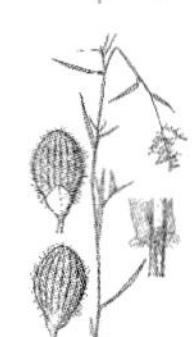

      4. Ligules hairy, or base membranous, hairy above
         6. Spikelets at least slightly pubescent (use hand lens) .... *Dichanthelium* (in part; there are sixteen species) 33
         6. Spikelets glabrous
            7. Terminal panicle small, typically less than 10 cm tall, the main branches forked at the base; usually perennials with remnant old, overwintering leaves present at base of this year's growth; typically flowering and fruiting from May to July; native .... *Dichanthelium* (in part; there are sixteen species)

7. Terminal panicle large, 10–40 cm tall, the main branches simple at the base; annuals or perennials, but with no remnants of old, overwintering leaves present at the base of this year's growth; typically flowering and fruiting from July to September; native and exotic .... *Panicum* (six species) 59

Key 9. Grasses with panicles and just one membranous fertile floret per spikelet, without conspicuous awns

1. Florets three, the lower two prominent and male, the uppermost one somewhat smaller and fertile; moist open areas .... *Anthoxanthum hirtum* (there is one other species) 6

1. Florets one or two

2. Florets two, the lower one sterile and glume-like; native, open sandy areas .... *Digitaria cognata* (there are two other species) 34

2. Floret solitary

3. Spikelet lacking glumes; palea three-nerved; often in wetlands .... *Leersia* (three species) 49

3. Spikelet with glumes; palea two-nerved

4. Lemma with a delicate, easily overlooked awn from its prominently bearded base ... *Calamagrostis* (three species) 18

4. Lemma awnless, or awned from the apex; base of lemma bearded or not

5. Ligule hairy

6. Base of lemma strongly bearded .... *Calamovilfa longifolia* (the only species) 19

6. Base of lemma glabrous ..... *Sporobolus* (six species) 79

5. Ligule membranous
   7. Inflorescence a contracted panicle, and/or glumes gently curving (sigmoid), not straight .... *Muhlenbergia* (twelve species) 56

   7. Inflorescence an open panicle; glumes straight

      8. Spikelets deciduous; rachilla minutely prolonged beyond the floret; lemma minutely awned from just below its apex .... *Cinna* (two species) 24
      8. Spikelets persistent; rachilla not prolonged beyond the floret; lemmas awnless or gradually awned from the apex

         9. Lemma shorter than the glumes, nearly or completely glabrous at base, awnless; lower panicle branches whorled .... *Agrostis* (five species) 2
         9. Lemma body distinctly longer than the glumes, glabrous to bearded at the base, awnless or tapering into awns up to 10 mm long; panicle branches borne singly at the nodes .... *Muhlenbergia* (twelve species)

Key 10. Grasses with panicles and usually three to four florets per spikelet, without conspicuous awns

1. Glumes 18–25 mm long; exotic, cultivated in part .... *Avena* (two species) 10
1. Glumes less than 18 mm long
   2. Florets two per spikelet

      3. Glumes more or less similar in size and shape; spikelets disarticulating above the glumes; rare, moist calcareous woods and bluffs, eastern Wisconsin ...... *Graphephorum melicoides* (the only species) 44

      3. Glumes unequal in size

         4. Glumes equal in shape (although the upper is larger than the lower); spikelets disarticulating above the glumes and between the florets; lemmas blunt; leaf sheath margins fused; rare, springs .... *Catabrosa aquatica* (the only species) 20

4. Glumes unequal in shape, the lower narrowly lanceolate, the upper somewhat obovate; spikelets disarticulating below the glumes; leaf sheath margins free; lemmas acute; fairly common, mostly in wetlands .... *Sphenopholis* (two species) 78

2. Florets three to five per spikelet
   5. Uppermost floret a club-shaped rudiment; leaf sheath margins fused; lemmas seven- to nine-nerved; rare, southwestern Wisconsin .... *Melica nitens* (there is one other species) 52

   5. Uppermost floret reduced but shaped just like the lower florets; leaf sheath margins free; lemmas three- to nine-nerved
      6. Lemmas three-nerved
         7. Lemmas usually with a tuft of cobwebby hairs at the base; leaf blade tips boat-shaped ...... *Poa* (in part; there are fifteen species) 67

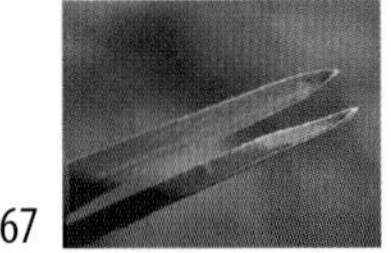

         7. Lemmas lacking a tuft of cobwebby hairs at the base; leaf blade tips acute
            8. Lemmas pubescent; bases of inflorescences mostly hidden in leaf sheaths; native, rare in dry open sandy ground, mostly southern Wisconsin .... *Triplasis purpurea* (the only species) 82

            8. Lemmas glabrous; inflorescences exserted, not hidden in leaf sheaths
               9. Mature florets bottle-like, protruding from spikelet; native, rare in mesic floodplain forests, southern Wisconsin .... *Diarrhena obovata* (the only species) 32

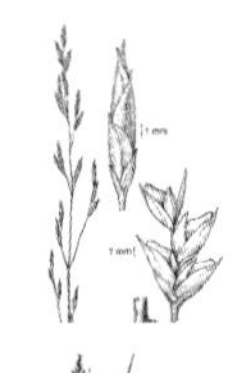

               9. Mature florets not bottle-like and protruding from spikelet; exotic, rare in northern Wisconsin .... *Molinia caerulea* (the only species) 55

      6. Lemmas five- to nine-nerved
         10. Lemmas seven- to nine-nerved; panicle small, compact, and one-sided; rare exotic of disturbed ground, eastern Wisconsin .... *Sclerochloa dura* (the only species) 72

         10. Lemmas five-nerved; panicle various, typically large and open

11. Florets three, the lower two prominent and male, the uppermost one somewhat smaller, more rigid and female; native, moist open areas .... *Anthoxanthum hirtum* 6

11. Florets two to five, all alike in shape and texture (although uppermost floret may be reduced in size)

12. Lemmas obtuse, with nerves parallel; exotic, often in wet saline areas .... *Puccinellia* (two species) 69

12. Lemmas acute, with nerves converging toward the apex

13. Lemmas with a cobwebby callus and/or hairs along the nerves; leaf blade apices obtuse and boat-shaped; native and exotic .... *Poa* (fifteen species) 67

13. Lemmas lacking a cobwebby callus or hairs along the nerves; leaf blade apices acute

14. Leaf blades inrolled, less than 3 mm wide; native and exotic .... *Festuca* (eight species) 42

14. Leaf blades flat, 3–8 mm wide; exotic ...... *Lolium* (in part; there are three species) 51

Key 11. Grasses with panicles and consistently four or more fertile florets per spikelet, without conspicuous awns

1. Plants 1–3 m tall, the leaves at least 2 cm wide; mature spikelets nearly concealed by abundant hairs on the rachilla; native and exotic, wetlands .... *Phragmites australis* (the only species) 65
1. Plants less than 1 m tall, the leaves less than 2 cm wide; spikelets glabrous or nearly so
2. Spikelets 20–40 mm long and nearly as wide; lemmas many-nerved; rare escape from cultivation .... *Chasmanthium latifolium* (the only species) 22

2. Spikelets less than 20 mm long and not more than 5 mm wide; lemmas three- to seven-nerved
   3. Lemmas round, spreading perpendicular to the rachis; exotic, rare in northern Wisconsin .... *Briza media* (the only species) 16

   3. Lemmas lanceolate to ovate, strongly erect and appressed to the rachis (as in most grasses)
      4. Ligules hairy
         5. Paleas falling together with the lemmas (as in most grasses); spikelets unisexual, the male larger than the female; exotic, rare in wet saline areas .... *Distichlis spicata* (the only species) 36

         5. Paleas persistent on the rachilla after the lemmas have fallen; spikelets perfect (in nearly all species), all alike; native and exotic, common .... *Eragrostis* (ten species) 40

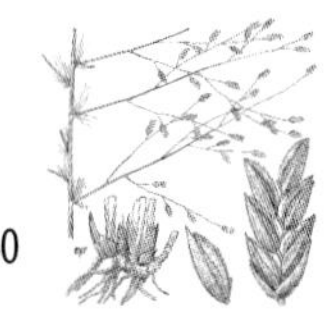

      4. Ligules membranous
         6. Spikelets in small, dense clusters near the ends of a few, elongate inflorescence branches; leaf sheaths fused; exotic; common .... *Dactylis glomerata* (the only species) 29

         6. Spikelets evenly dispersed throughout a small to large panicle; leaf sheaths fused or free
            7. Lemma nerves converging at the minutely two-toothed summit, which is often awned; leaf sheaths fused; wetlands or uplands .... *Bromus* (twelve species) 17

            7. Lemma nerves parallel or convergent, but awnless summit not minutely toothed; leaf sheaths partly to wholly fused
               8. Upper glume with one distinct nerve; leaf sheaths totally fused; common, wetlands, mostly native .... *Glyceria* (six species) 43

               8. Upper glume with three to five distinct nerves; leaf sheaths fused only near the base; uncommon, wetlands, exotic or native

9. Lemma nerves five, obscure and converging at the apex; exotic in saline areas statewide . . . . *Puccinellia* (two species) 69

9. Lemma nerves seven to nine, prominent and parallel, not converging at the apex; native in non-saline wetlands in northern Wisconsin . . . . *Torreyochloa* (two species) 80

Field Guide to Wisconsin Grasses

1. *AGROPYRON*, CRESTED WHEATGRASS

Greek, *agrios*, "wild," and *pyros*, "wheat"

The golden, densely flowered, narrow bottlebrush- or comb-like spikes are distinctive. Several Eurasian species; the following is well established in the western United States and spreading via roads and railroads to the Midwest. Differentiated from related genera (*Elymus*, *Hordeum*, *Leymus*, *Pascopyrum*, *Secale*, and *Triticum*) by the following combination of characters: tufted perennials; one spikelet per node, but spikelets crowded; glumes eccentrically keeled; several florets per spikelet, disarticulating between florets.

Agropyron cristatum (L.) Gaertn., crested wheatgrass

cristatum, Latin for "crested" or "comb-like"

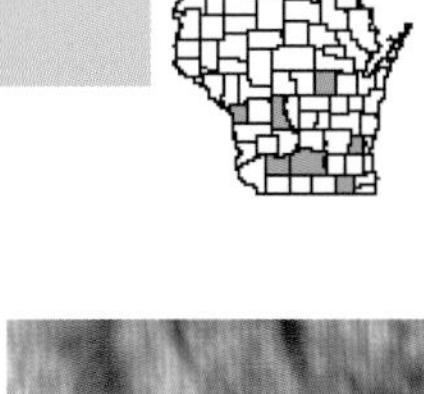

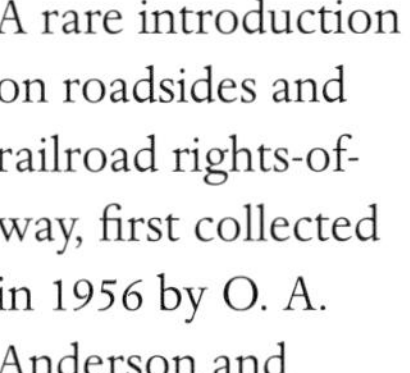

A rare introduction on roadsides and railroad rights-of-way, first collected in 1956 by O. A. Anderson and Thomas G. Hartley.

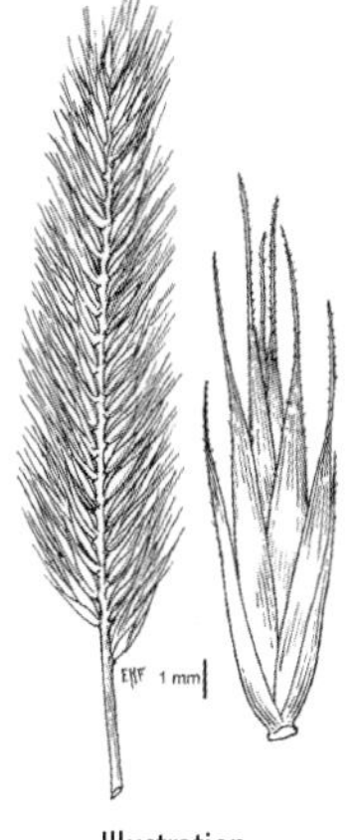

Illustration

Inflorescence showing side and edgewise views

Inflorescences

2. *AGROSTIS*, BENT OR TICKLE GRASS

Greek for "field"

Common grasses with panicles of small, single-flowered spikelets, the glumes scabrous on their keels, the lemmas awnless and hairless; palea delicate or absent (Shinners 1943). Rhode Island bent (*Agrostis capillaris* L.) is a European turf grass that has escaped in nearby Illinois and the Upper Peninsula of Michigan and may eventually be found in Wisconsin. It would key to *A. gigantea* or *A. stolonifera* but differs in its shorter ligules (less than 2 mm long) and leaf blades (less than 4 mm wide). 150 species; cosmopolitan.

1. Palea present, at least half as long as the lemma; anthers 1–1.5 mm long
   2. Plants erect (although rhizomatous); leaf blades 4–10 mm wide; inflorescence branches divergent; spikelets usually purplish; common weed .... *A. gigantea*
   2. Plants decumbent and creeping; leaf blades 1.5–4 mm wide; inflorescence branches strongly ascending; spikelets greenish; occasional weed .... *A. stolonifera*
1. Palea absent or tiny, less than one-quarter as long as the lemma; anthers about 0.5 mm long
   3. Panicle greenish, the branches 3–6 cm (rarely up to 12 cm) long, often branching below the middle, smooth to sparingly scabrous; leaf blades 1.5–4 mm wide, flat; edges and shade of moist woods .... *A. perennans*
   3. Panicle reddish, the branches 6–15 cm long, usually branching beyond the middle, scabrous; leaf blades 0.5–2 mm wide, inrolled; dry open areas, shores
      4. Lemmas 1–1.3 mm long; occasional .... *A. hyemalis*
      4. Lemmas 1.3–1.7 mm long; common .... *A. scabra*

Agrostis gigantea Roth, redtop

gigantea, "very large," from Latin *giganteus*, "of or belonging to the giants"

A Eurasian exotic; an early introduction for hay (Lapham 1853). One of our most abundant weeds, redtop is found not only in all types of disturbed habitats such as fields, pastures, and roadsides but also in prairies, rock outcrops, and woodlands, where it may appear native. The pyramidal panicles with numerous horizontally radiating branches at the lower nodes are distinctive.

Illustration

Ligular area of leaf

Habit

Spikelet

Agrostis hyemalis **(Walter) Britton, Sterns and Poggenb., southern tickle grass**

hyemalis, "of the winter," flowering in winter

Occasional in wet to more commonly dry areas in central and northern Wisconsin; much less common than *Agrostis scabra*.

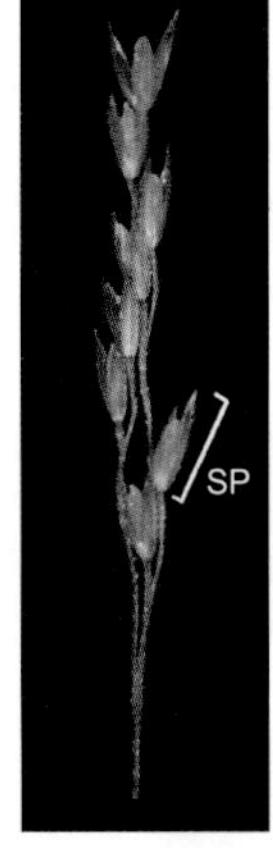

Section of inflorescence

Illustration

Spikelet

Agrostis perennans **(Walter) Tuck., autumn bent grass**

perennans, "perennial"

Common in a variety of habitats but most characteristic of bare ground in shady forest understories, as well as moist forest trails and margins. The regularly forking main panicle branching pattern is distinctive.

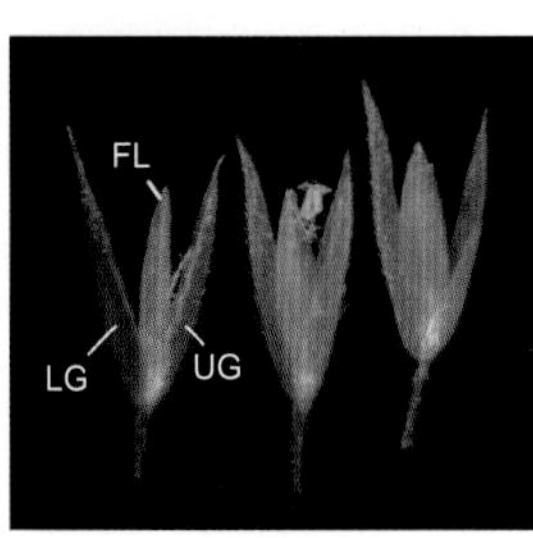

Spikelet

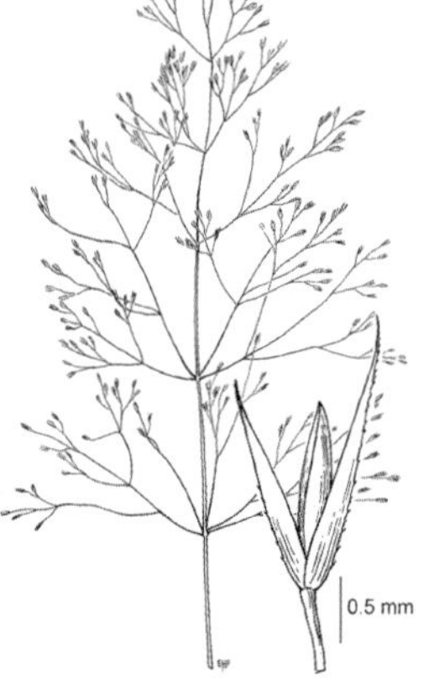

Illustration

Inflorescence, showing forking branches

Agrostis scabra Willd., northern tickle grass

scabra, "rough"

Widespread in a wide variety of habitats; locally abundant in dry open areas and characteristic of moist, open, sandy, or peaty shorelines.

Illustration

Agrostis stolonifera L., creeping bent grass

stolonifera, "bearing stolons or runners"

A Eurasian exotic, first collected in 1897. Locally common on shores, beaches, streamsides, and roadsides. Bent grass is well named. Dense colonies of this species have all of the leaves "bent" or swept in the same direction, like well-combed hair.

Habit, with expanded inflorescence

Habit, with unexpanded inflorescence

Illustration

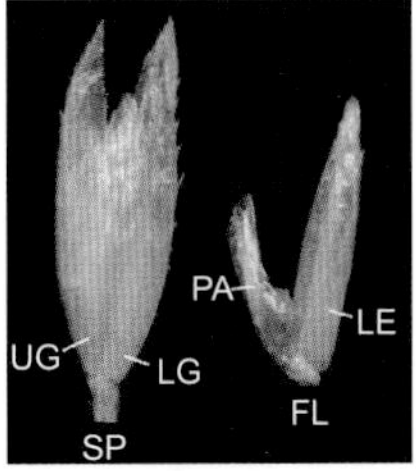

Spikelet and floret

3. *ALOPECURUS*, MEADOW FOXTAIL

Greek *alopex*, "fox," and *oura*, "tail"

Occasional wetland or pasture grasses with dense, pencil-like panicles resembling the much commoner timothy (*Phleum*) but differing in their awnless, folded glumes, delicately awned florets, yellowish anthers, and spikelets that fall from the inflorescence axis. Thirty-six species of the temperate Northern Hemisphere and South America.

1. Spikelets 4–6 mm long (excluding awns), the awns exserted 4–6 mm beyond the glume tips ... *A. pratensis*
1. Spikelets 2–3.5 mm long (excluding awns), the awns exserted 0–3 mm beyond the glume tips
    2. Awns usually included in florets or exserted no more than ca. 1 mm beyond the glume tips, the awns inserted near the middle of the back of the lemma; fairly common .... *A. aequalis*
    2. Awns exserted 2–3 mm beyond the glume tips, the awns inserted near the base of the lemma; uncommon to rare
        3. Spikelets 2–2.5 mm long; anthers ca. 0.5 mm long .... *A. carolinianus*
        3. Spikelets 2.5–3 mm long; anthers 1.5–2 mm long .... *A. geniculatus*

Alopecurus aequalis Sobol., short-awned foxtail

aequalis, "equal"

Fairly common in open wetlands, including on muddy shores and in shallow water, sometimes with floating leaves; our only native species in the genus.

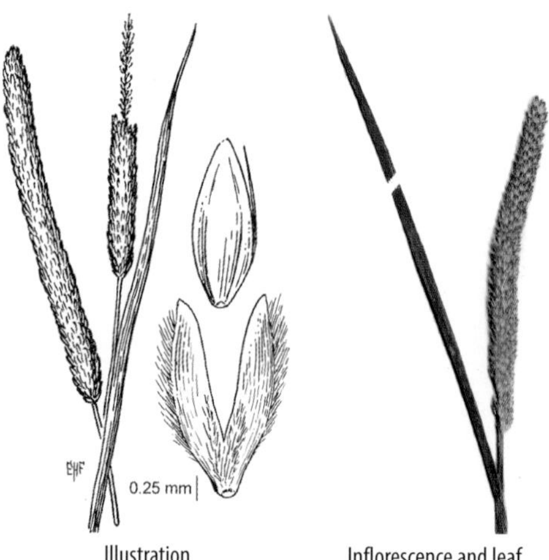

Illustration Inflorescence and leaf

Alopecurus carolinianus Walter, Carolina foxtail

carolinianus, "from the Carolinas"

Roadsides and fallow fields; a native of the Great Plains and Mississippi Valley, exotic here, first collected in 1912.

Illustration

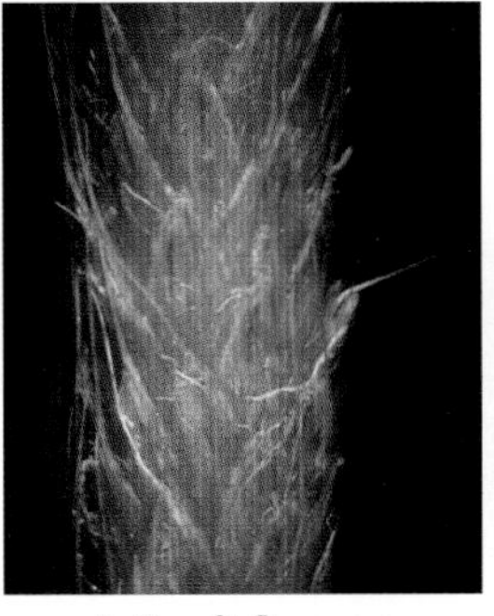

Section of inflorescence

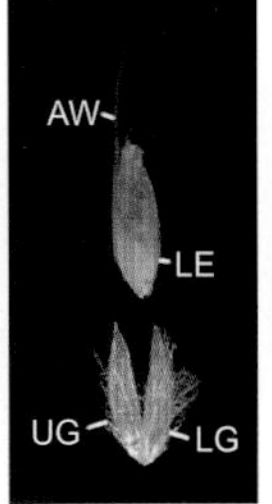

Spikelet

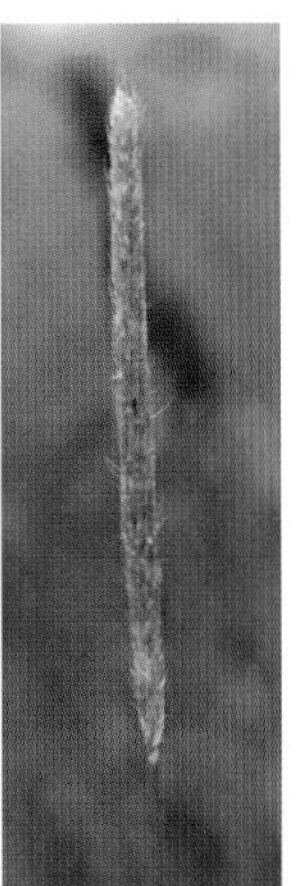

Inflorescence

Alopecurus geniculatus L., marsh foxtail

geniculatus, "bent sharply like a knee"

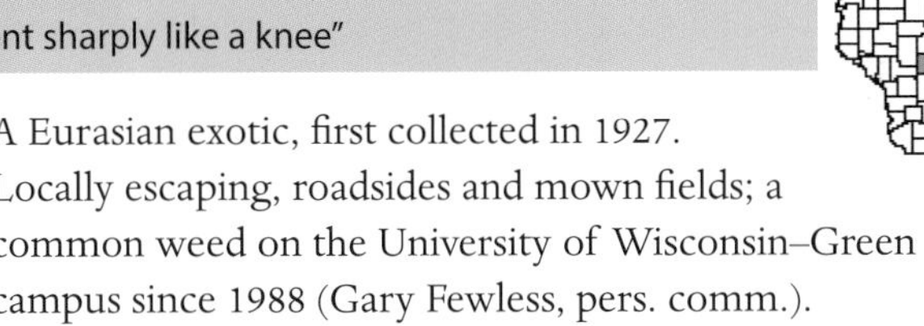

A Eurasian exotic, first collected in 1927. Locally escaping, roadsides and mown fields; a common weed on the University of Wisconsin–Green Bay campus since 1988 (Gary Fewless, pers. comm.).

Illustration

Inflorescence

Alopecurus pratensis **L., meadow foxtail**

pratensis, "of meadows"

A Eurasian exotic, first collected in 1915. Occasional in wetlands and fields; less common than *Alopecurus aequalis* and often found in drier habitats than that species, but spreading into wet areas near Green Bay and perhaps a threat (Gary Fewless, pers. comm.).

Illustration

Inflorescence, showing yellowish-white anthers

Inflorescences

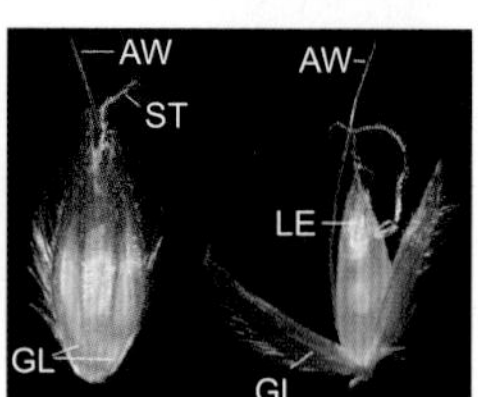

Spikelet

4. *AMMOPHILA*, BEACH GRASS

Greek *ammos*, "sand," and *fileiu*, "love"

Coarse Great Lakes beach and dune perennials from vigorous rhizomes with spike-like panicles of large spikelets, the single floret with a short tuft of hairs. Two species of Europe, Africa, and eastern North America.

Ammophila breviligulata Fernald, American beach grass

breviligulata, *brevis*, "short," and *ligula*, "tongue"

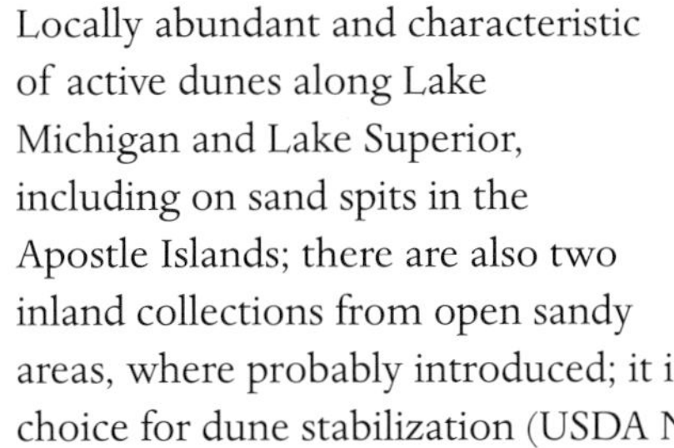

Illustration

Locally abundant and characteristic of active dunes along Lake Michigan and Lake Superior, including on sand spits in the Apostle Islands; there are also two inland collections from open sandy areas, where probably introduced; it is an excellent choice for dune stabilization (USDA NRCS 2014).

Portion of inflorescence

Habit

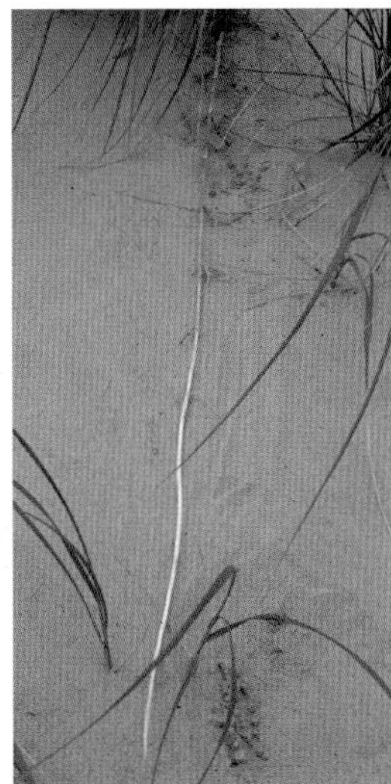

Rhizome

5. *ANDROPOGON*, BIG BLUESTEM

Greek, *andros*, "man's," and *pogon*, "beard," probably because of the hairy appearance produced by the seed heads by the sterile glumes

Robust, summer-flowering prairie grasses with several densely flowered racemes arranged in an approximately digitate (finger-like) fashion at the end of tall culms; 120 species; cosmopolitan. In addition to the following familiar species, two other bluestems come close to Wisconsin and may eventually be found here. Sand bluestem (*Andropogon hallii* Hack.) is a western species that occurs as an introduction as far east as the banks of the Mississippi River in Minnesota near Winona. It is quite similar to big bluestem but is a strongly rhizomatous plant with grayish-waxy foliage and shorter spikelet awns less than 10 mm long. Broomsedge bluestem (*A. virginicus* L.) is a common weedy native of the southern United States that has been collected many times in the Chicago area (Swink and Wilhelm 1994). It has sessile (main) spikelets only 3–4 mm long with straight, not twisted, awns. It resembles little bluestem (*Schizachyrium scoparium*) but has inflorescences whose bases are included in the swollen, spathe-like upper leaf sheaths.

Andropogon gerardii Vitman, big bluestem

gerardii, for Louis Gérard (1733–1819), French botanist from the late 1700s to early 1800s who was the first to describe this species technically

The tallest and most dominant prairie, savanna, and woodland grass, preferring mesic sites but also frequent in dry or wet ones too; now found statewide because of plantings for erosion control and prairie restoration, and appearing to spread northward along highways.

Illustration

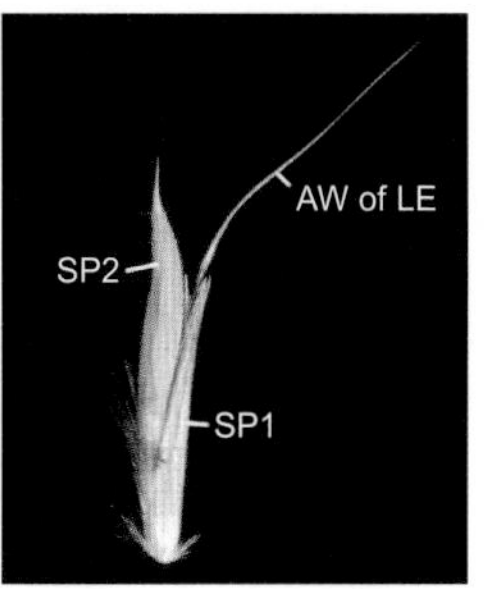

Spikelet pair; upper left spikelet is pedicelled (stalked), while lower right spikelet is sessile (no stalk)

Leaf ligular area

Section of inflorescence

Single inflorescence

Inflorescences

6. *ANTHOXANTHUM*, SWEET VERNAL GRASS

Greek, "flower" and "yellow"

Fragrant perennials with open to pencil-like contracted panicles of awnless spikelets, each with two large male or sterile florets below the single perfect floret. Fifty to seventy species; circumboreal.

1. Panicle open; glumes about equal in length; lower two florets male; native, fairly common ...... *A. hirtum*
1. Panicle contracted, pencil-like; glumes unequal in length, the lower shorter than the upper; lower two florets sterile; exotic, most common in the Lake Superior region .... *A. odoratum*

Anthoxanthum hirtum **(Schrank) Y. Schouten and Veldkamp, northern sweet grass**

hirtum, "hairy"

Illustration

Occasional to fairly common in meadows, moist prairies, and sedge meadows and on moist, sandy roadsides. It is one of the earliest native grasses to flower, usually in late April or early May. The tough, shiny leaves are used ceremonially by various tribes and used to weave fragrant "sweet grass braids." Our plants are subsp. *arctica* (J. Presl) G. Weim.; Fassett (1951) treats this as *Hierochloe odorata*.

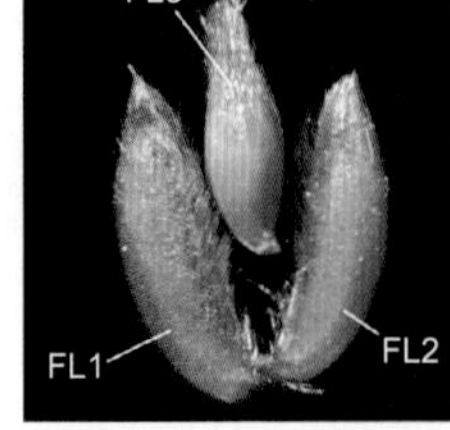

Spikelet

Inflorescence

Sweet grass braid by Joan Elias

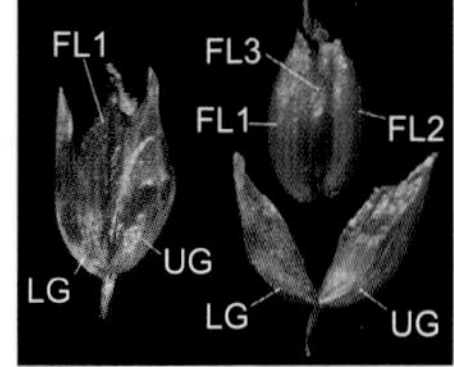

Florets, showing two large lower male florets and one smaller bisexual upper floret

Northern sweet grass, typical roadside colony

***Anthoxanthum odoratum* L., sweet vernal grass**
odoratum, "fragrant"

A European exotic, listed by Lapham (1853), but first collected in 1956 by Henry C. Greene and John T. Curtis.

Occasional on roadsides and in meadows and old fields, most common in the Lake Superior counties; the leaves are fragrant.

Illustration

Young inflorescence

Mature inflorescences

7. *APERA*, SILKY BENT GRASS

Greek for "unmaimed," application obscure

Rare annuals resembling bent or tickle grass (*Agrostis*) but with delicately awned lemmas and well-developed paleas. Three Eurasian species.

Apera interrupta (L.) P. Beauv., silky bent grass

interrupta, "interrupted in some fashion"

Illustration

A rare roadside European weed, first collected in waste ground in 1981 by Judziewicz in Oconto County, along with *Puccinellia distans* and *Ventenata dubia* (Solheim and Judziewicz 1984); since 2008 common on the University of Wisconsin–Green Bay campus in pea gravel at the edge of trails (Gary Fewless, pers. comm.).

Portion of inflorescence

Portion of inflorescence

8. *ARISTIDA*, THREE-AWNED GRASS

Latin, *arista*, "awn" or "beard"

Tufted plants with fine leaves, the spikelets very slender and single-flowered, the lemma with a sharp-pointed base (callus) and a three-awned summit; 250–350 species, cosmopolitan.

1. Awns twisted into a 5–12 mm long united "column" at the lemma tip (before diverging into three equal awns); sand barrens, central and southwestern Wisconsin .... *A. tuberculosa*
1. Awns separate from each other at the apex of the lemma, no column present
    2. Awns easily breaking (disarticulating) from the summit of the lemma at maturity; rare species .... *A. desmantha*
    2. Awns not disarticulating from the summit of the lemma at maturity
        3. Lower glume 15–25 mm long (excluding awn if present), three- to five-nerved; lemma body 10–20 mm long; awns 4–7 cm long .... *A. oligantha*
        3. Lower glume 3–15 mm long (excluding awn if present), one-nerved; lemma body 4–10 mm long; awns up to 3 cm long

4. Central awn spiraled one or two times near its base
   5. Lemma body 8–12 mm long, with central awn 10–20 mm long, the lateral awns 5–13 mm long, spreading; glumes unequal; common, often abundant on gravel or sand of road shoulders .... *A. basiramea*
   5. Lemma body 6–8 mm long, with central awn 5–10 mm long, the lateral awns 1–3 mm long, erect; glumes nearly equal; rare, southern Wisconsin .... *A. dichotoma*
4. Central awn not spiraled near base (may be slightly curving)
   6. Annual; lower glume 4–9 mm long, slightly shorter than upper glume; lower leaf sheaths glabrous; uncommon, mostly central Wisconsin .... *A. longespica*
   6. Perennial; lower glume 10–15 mm long, slightly longer than upper glume; lower leaf sheaths hairy; uncommon, mostly southwestern Wisconsin .... *A. purpurascens*

Aristida basiramea **Engelm. ex Vasey, fork-tipped three-awned grass**

basiramea, "branching at the base"

Common late summer– and early fall–blooming species of dry open areas generally: prairies, barrens, and rock outcrops; locally abundant in dry sand or gravel of roadsides and in quarries and pits. Our most widespread and common species of *Aristida*.

Illustration

Habit

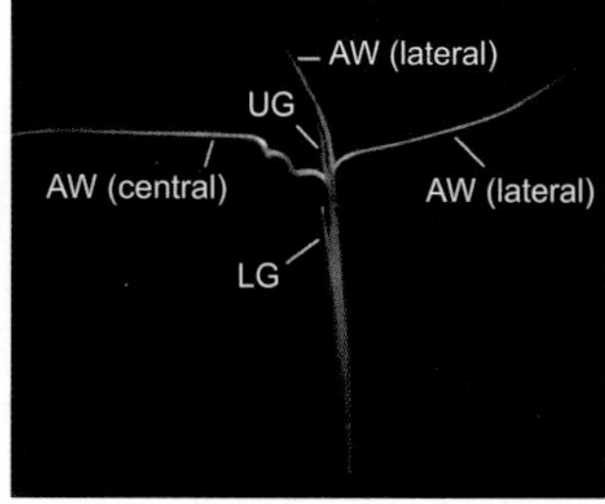

Spikelet

Aristida desmantha Trin. and Rupr., curly three-awned grass

desmantha, Greek, *desme*, "bundle," and *anthos*, "flower"

Illustration

This species of the southern Great Plains is only known from a single collection made at La Crosse in July 1887 by L. H. Pammel and housed at the Utah State University herbarium; it is not certain whether the collection was made from a native or an introduced population. Mary Barkworth kindly checked its identification (27 July 2012), made in 1995 by *Aristida* specialist Kelly Allred.

Aristida dichotoma Michx., Curtiss's three-awned grass

dichotoma, "forked in pairs"

SPECIAL CONCERN

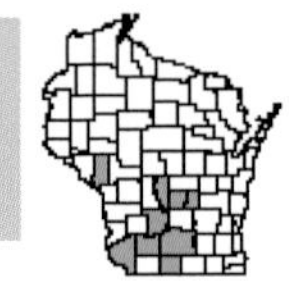

A rare species of dry prairies on open sandstone and granite outcrops; our plants are var. *curtissii* A. Gray. Fassett (1951) treats this as *Aristida basiramea* var. *curtissii*.

Illustration

Section of inflorescence

Spikelet

Aristida longespica Poir., Kearney's three-awned grass

longespica, longus, "long, extended," *spica*, "spike"

Occasional, sandy prairies, oak openings, dry roadsides, and railroad rights-of-way. Our taxon is var. *geniculata* (Raf.) Fernald; if recognized at the species level, its correct name is *Aristida necopina* Shinners. Fassett (1951) treats this as *A. intermedia*.

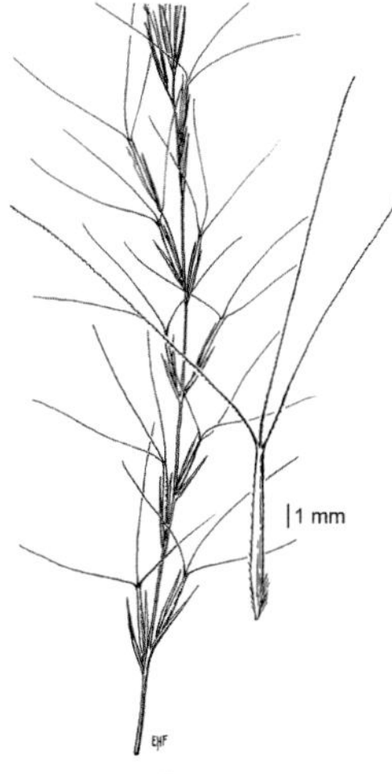

Illustration

UG

LG

Habit

Spikelet

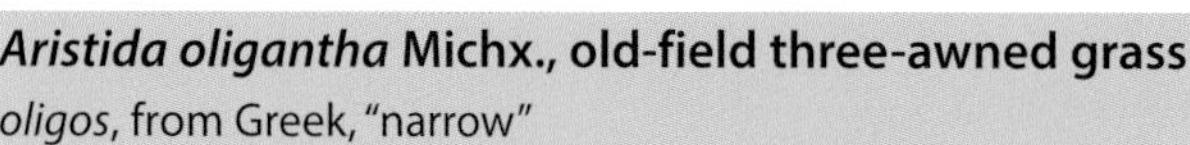

Aristida oligantha Michx., old-field three-awned grass

oligos, from Greek, "narrow"

Uncommon in dry sandy or gravelly sites, often along railroads; regarded as native, although nearly all of our collections are from disturbed areas.

Habit

Habit

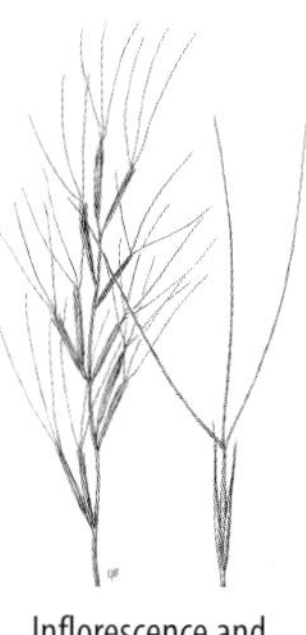

Inflorescence and spikelet illustration

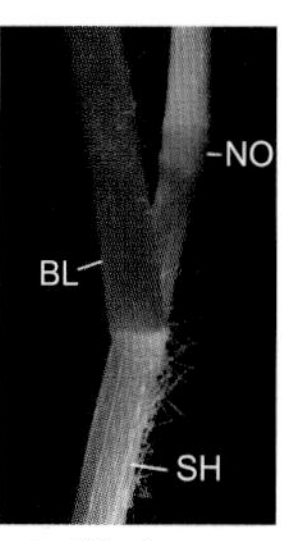

Leaf ligular area

Aristida purpurascens Poir., arrow-feathered three-awned grass

purpurascens, "becoming purple or purplish"

Uncommon in dry prairies, on sandstone cliffs, and along railroads.

Illustration

Spikelet showing long awns

Spikelet detail

Aristida tuberculosa Nutt., dune three-awned grass

tuberculosa, "with tubercles or bumps"

The extremely long awns make this our most striking species of the three-awned grasses; occasional in sand barrens, dry prairies, and railroad rights-of-way. A good place to view this species is at Spring Green Reserve in Sauk County.

Illustration

Habit

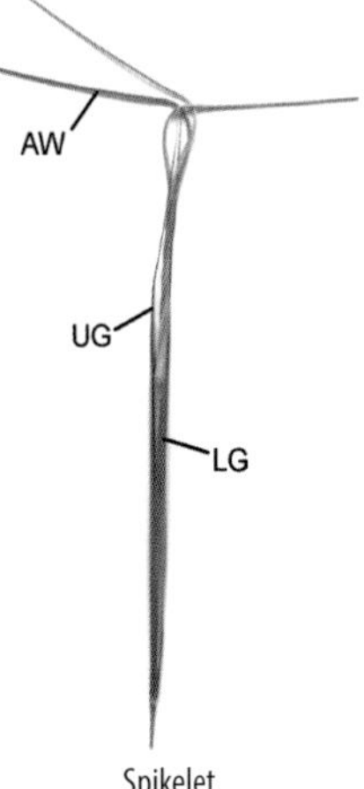

Spikelet

9. *ARRHENATHERUM*, TALL OAT GRASS

Greek for "man" and "awn" in reference to awned staminate floret

A robust clump-forming perennial with oat-like spikelets (although not as large as those of *Avena*); two florets, the lower male and with a bent awn, the upper perfect with a short straight awn; six Eurasian species.

Arrhenatherum elatius (L.) P. Beauv. ex J. Presl and C. Presl, tall oat grass

elatius, "exalted"

A Eurasian exotic, first collected in 1889; occasional along railroads and roadsides before 1950, but more infrequently after that, and last collected in 1997.

Inflorescence

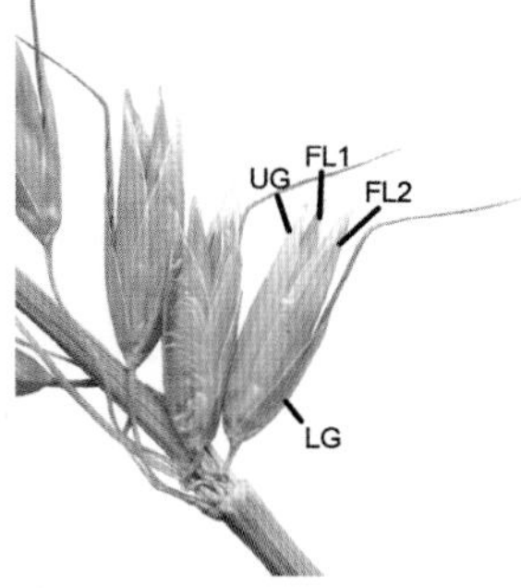

Spikelets

Illustration

Habit

10. *AVENA*, OATS

Latin for "oats"

Familiar cultivated cereals with panicles of large, dangling spikelets; glumes very large and with numerous prominent green nerves; lemmas several, sometimes with a twisted awn from the middle; twenty-nine Old World species.

1. Lemmas with strongly twisted, stout awns .... *A. fatua*
1. Lemmas with straight, weak awns, or awnless .... *A. sativa*

Avena fatua L., wild oats

fatua, "foolish, insipid, worthless"

A Eurasian exotic, first collected in 1908. An uncommon weed almost exclusively restricted to railroad yards and rights-of-way.

Illustration

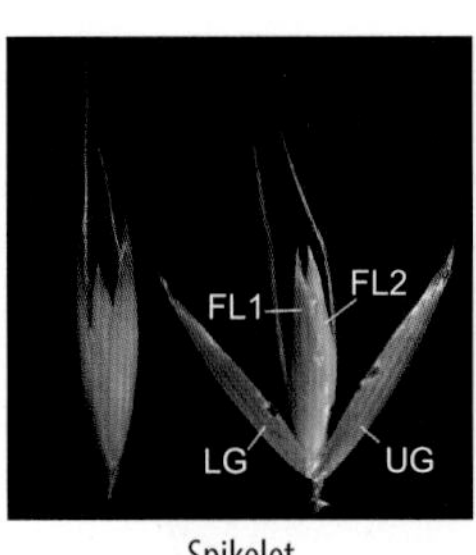

Spikelet

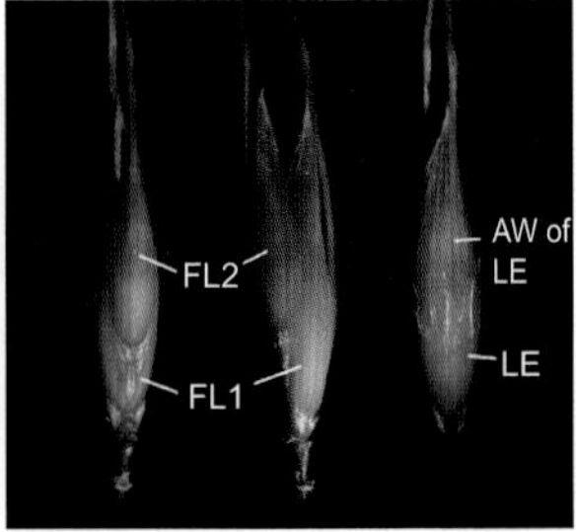

Florets

Avena sativa L., oats

sativa, "sown"

A Eurasian exotic, first collected in 1894. Commonly cultivated crop that occasionally escapes to roadsides.

Illustration

Spikelets

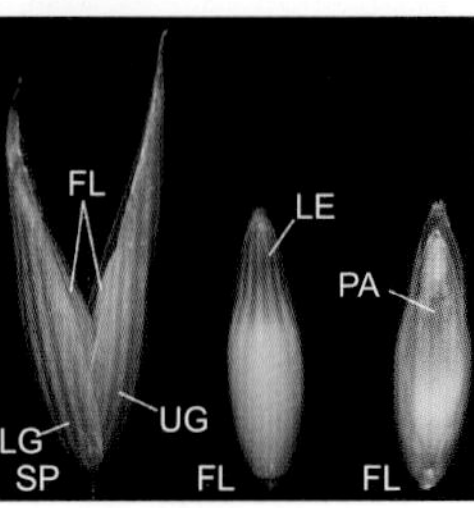

Spikelets

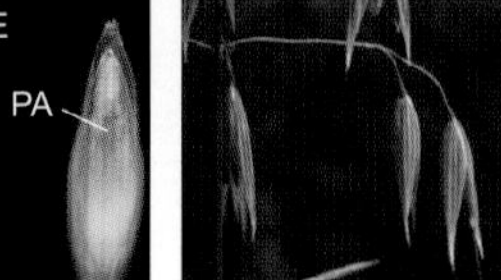

Inflorescence showing large, dangling spikelets

11. *AVENELLA*, WAVY HAIR GRASS

like *Avena*, Latin for "oats"

Clump-forming perennials with abundant fine leaves and panicles of small spikelets with two florets, the lemmas thin with a bent and twisted awn arising below the middle (Valdes and Scholz 2006); six species, five Eurasian, ours a North American native.

Avenella flexuosa (L.) Drejer, wavy hair grass

flexuosa, "flexible"

SPECIAL CONCERN

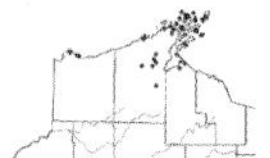

Illustration

Dry open dunes and sand barrens, locally common on beaches and dunes of Lake Superior (including Apostle Islands), Lake Michigan (Door County and west side of Green Bay), and somewhat inland on sand in the Bayfield Peninsula (Moquah Barrens), and near Green Bay in Marinette and Oconto Counties. Fassett (1951) treats this as *Deschampsia flexuosa*.

Portion of inflorescence

Clump-forming habit in field

12. *AVENULA*, ALPINE OAT GRASS

like *Avena*, Latin for "oats"

Clump-forming perennials with closed leaf sheaths and narrow panicles of multiflowered spikelets with the larger glume longer than the two to seven florets; rachilla hairy; lemmas from the middle of the back; thirty European species plus one in western North America.

Avenula pubescens (Huds.) Dumort., downy alpine oat grass

pubescens, "with soft, downy hair"

A European exotic, first collected in 2008 by Steve C. Garske in Ashland County; so far, a rare weed of roadsides and pine plantations. This species superficially resembles a gigantic version of poverty oat grass (*Danthonia spicata*) but differs in its larger spikelets with the lemmas bearded at the base but glabrous on their backs.

Illustration

Leaf ligular area

Young inflorescences

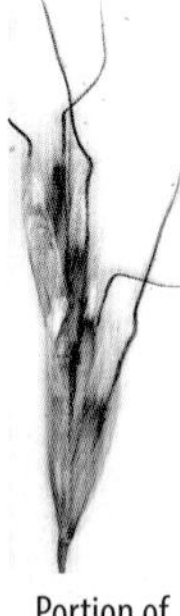

Portion of inflorescence

Spikelets

Downy alpine oat grass, habit

13. *BECKMANNIA*, SLOUGH GRASS

for Johann Beckmann (1739–1811), professor of botany at Göttingen

A distinctive grass with densely crowded rows of round, strongly dorsally flattened spikelets like a roll of tiny coins. At first glance it could be mistaken for a species of the much commoner barnyard grass (*Echinochloa*), but it differs in that the spikelets are smooth, not rough-hairy, and flattened, not plump. Two species of North America and Eurasia.

Beckmannia syzigachne (Steud.) Fernald, American slough grass

syzigachne, Greek, *syzygos*, "joined," and *achene*, "achene"

Illustration

A Great Plains species that is native east to western Minnesota; perhaps exotic in Wisconsin, where it was first collected in 1896. It is uncommon in disturbed wet areas such as ditches and muddy shores, most frequently near Superior and on the swampy western shore of Green Bay; seemingly less common now than in the early twentieth century; in other parts of the United States it is sometimes grown for hay (USDA NRCS 2014).

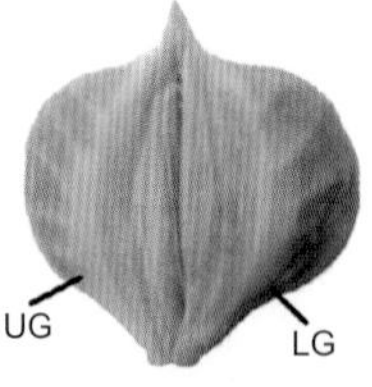

Spikelet

Section of inflorescence, showing strongly overlapping spikelets like coins in a roll

14. *BOUTELOUA*, GRAMA GRASS

for Claudio Bouteloua (1774–1842), a Spanish horticulturist

Grasses with the crowded spikelets distinctively arranged in one to several, one-sided, comb-like, reflexed or dangling racemes; the spikelet structure is quite complex, with a lower fertile awnless floret and several smaller, sterile, awned florets. The rare buffalo grass (*Bouteloua dactyloides*) is exceptional in being dioecious (separate male and female plants), the female spikelets enclosed in the husk of a nonspiny bur (Columbus 1999). Forty species, North and South America.

1. Inflorescence with twenty-five to fifty dangling spikes that detach from the rachis at maturity; plants 30–50 cm tall; common, prairies and barrens, southern Wisconsin .... *B. curtipendula*
1. Inflorescence with one to five divergent spikes that remain attached to the rachis at maturity; plants 5–30 cm tall; uncommon to rare
    2. Plants creeping, dioecious (either male or female), the female spikelets enclosed in a small (nonspiny) bur, the male spikelets 5–15 mm long; rare adventive .... *B. dactyloides*
    2. Plants erect, the spikelets all alike and 20–40 mm long
        3. Rachis of spike not prolonged as a naked bristle beyond the uppermost spikelets; rare adventive .... *B. gracilis*
        3. Rachis of spike prolonged as a naked bristle projecting 3–10 mm beyond the uppermost spikelets; occasional, prairies and barrens, southern Wisconsin .... *B. hirsuta*

Bouteloua curtipendula (Michx.) Torr., side-oats grama

curtipendula, Latin, "short-hanging"

Common in dry-to-mesic prairies, savannas, and bluffs; sometimes escaping from prairie plantings, but normally a fairly "conservative" species that does not easily spread much from cultivation.

Illustration

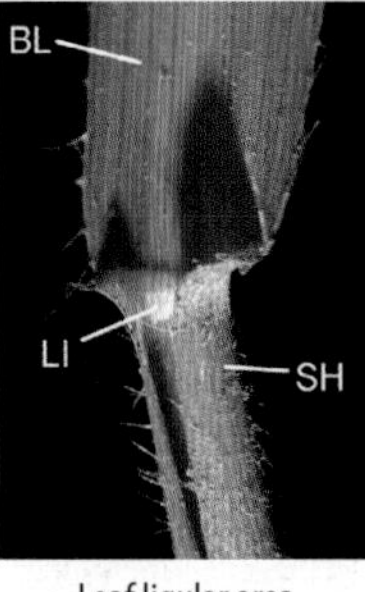

Leaf ligular area

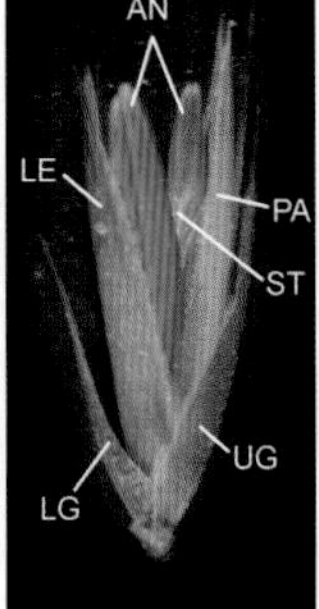

Spikelet

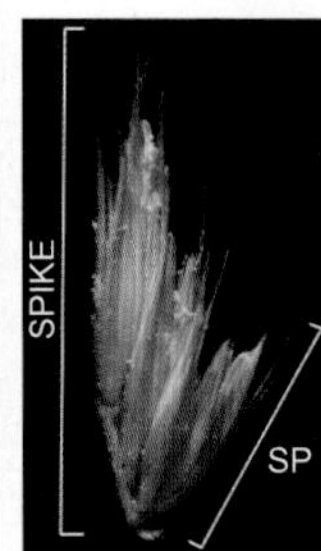

Spike

Inflorescence showing two rows of short, dangling spikes

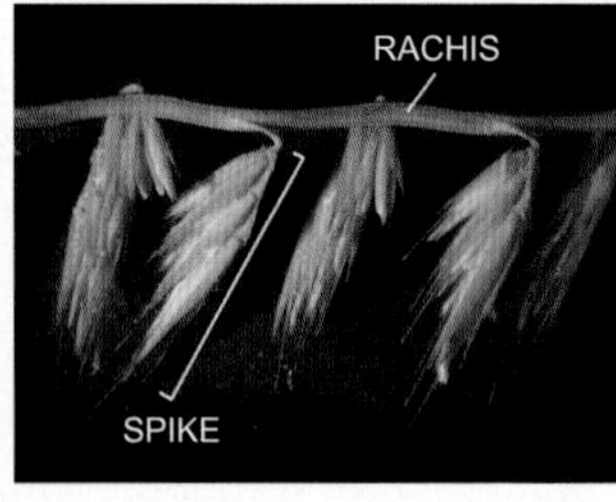

Detail of inflorescence showing four spikes

Side-oats grama, habit

Bouteloua dactyloides **(Nutt.) J. T. Columbus, buffalo grass**

dactyloides, from Greek *dactylos*, "finger," and *oid*, "like," maybe because of its leaf shape or for its resemblance to *Dactylis*, genus of orchard grass

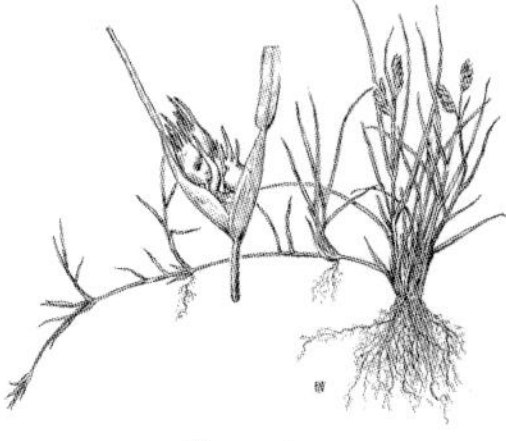

Illustration

First collected in 1956 in railroad yards in La Crosse by Thomas G. Hartley; a rare adventive in dry ground in southwestern Wisconsin; abundant and dominant in short-grass prairies of the western Great Plains. Long known as *Buchloe dactyloides* (Nutt.) Engelm. (Columbus 1999).

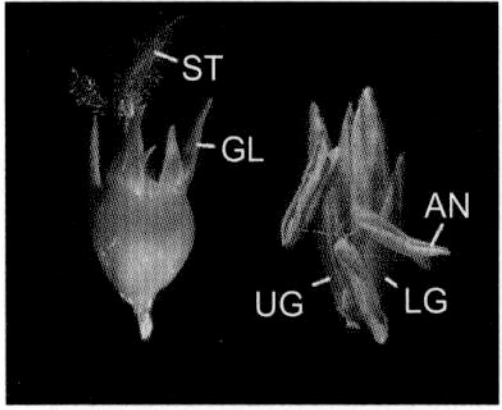

Female bur and an enclosed spikelet

Male spikes

Bur-like female spike

Male inflorescence

Bouteloua gracilis **(Kunth) Lag. ex Griffiths, blue grama**

gracilis, "thin, slender"

Illustration

An uncommon introduction from farther west, where it is native on the Great Plains as far east as central Minnesota; in Wisconsin, mostly along railroads, where first collected in 1936 and last collected in 1977 in the "Whiting Triangle" in Portage County (Freckmann 1978); this population was not located again in 2012.

Single spike showing short bristle tip (*right*)

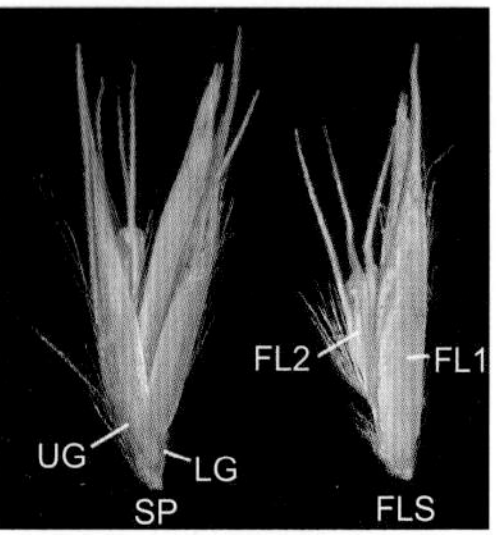

Spikelet and floret

Blue grama, habit showing two spikes

Bouteloua hirsuta **Lag., hairy grama**

hirsuta, "covered with hair"

Occasional in dry prairies and barrens, including rock outcrops, sandy outwash plains, and sand terraces along major rivers in southern and central Wisconsin. The "hangnail"-like rachilla extension is distinctive.

Illustration

Single spike showing long bristle tip (*top*)

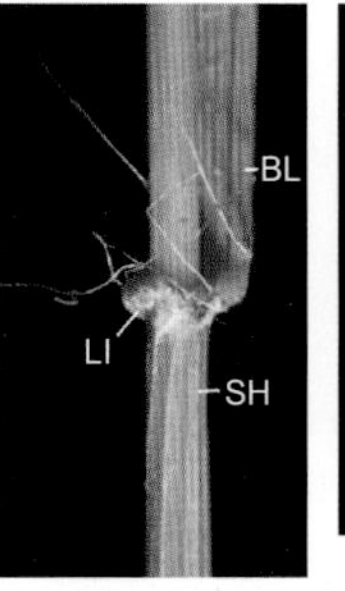

Leaf ligular area

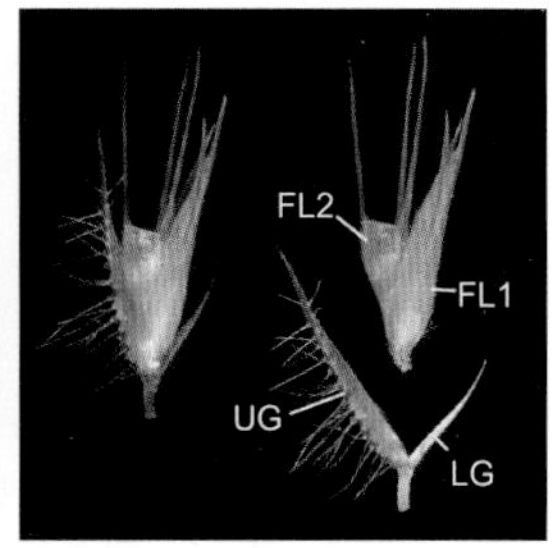

Spikelet and floret

15. *BRACHYELYTRUM*, SHORT-HUSKED GRASS

from Greek words for "short" and "husk," referring to minute glumes

Common perennial, spring-flowering, woodland grasses with relatively broad, finely hairy leaves; inflorescence a sparse, contracted panicle of slender, single-flowered spikelets with tiny glumes and long-awned lemmas. Three species, one from eastern Asia, the other two North American.

1. Lemmas scabrous but not hispid, with hairs less than 0.5 mm long; mostly northern Wisconsin . . . . *B. aristosum*
1. Lemmas hispid with harsh-feeling hairs 0.5–1 mm long, at least in the central portions of the lemma; mostly southern Wisconsin . . . . *B. erectum*

Brachyelytrum aristosum (Michx.) Trel., northern shorthusk

aristosum, "bearded, furnished with awns"; "bristly"

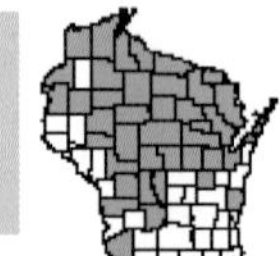

Common, moist, usually mesic forests; in all but the southern counties.

Brachyelytrum erectum (Schreb. ex Spreng.) P. Beauv., southern shorthusk

erectum, "erect"

Common, moist, usually mesic forests in central and northern counties.

Illustration

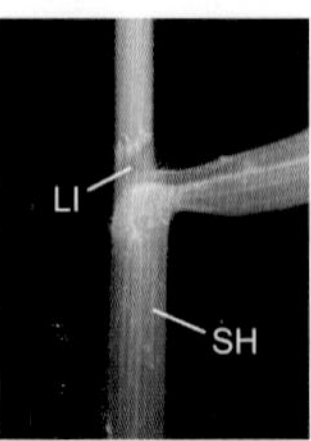

Leaf ligular area

Inflorescence

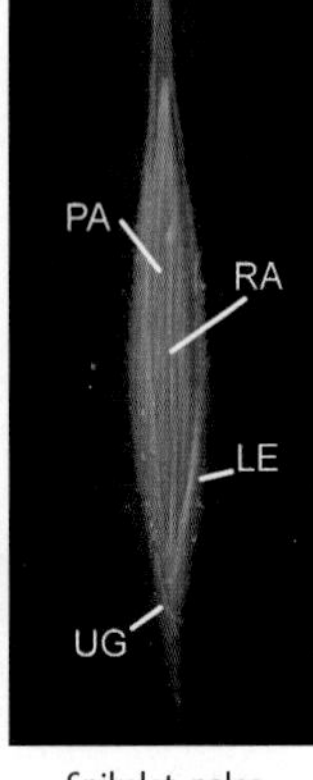

Spikelet, palea side view

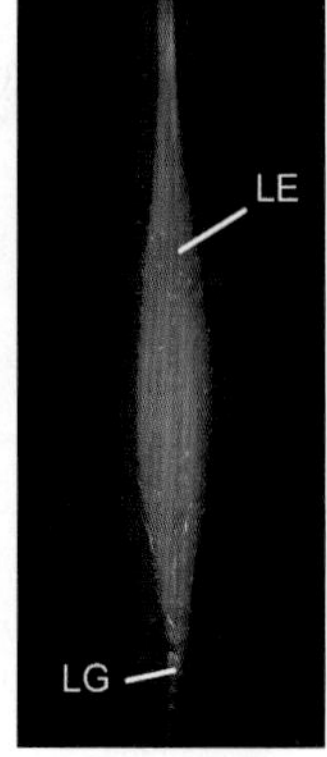

Spikelet, lemma side view

Southern shorthusk, habit in dense shade

16. *BRIZA*, QUAKING GRASS

from the Greek word for a type of grain

Small grasses with panicles of broad, dangling, many-flowered spikelets. Nineteen species of Eurasia and one species in South America.

Briza media L., perennial quaking grass

media, "intermediate"

A Eurasian exotic, first collected in 2009 by Steven C. Garske in disturbed ground in Ashland County. Our species resembles rattlesnake manna grass (*Glyceria canadensis*) but has lemmas that spread almost perpendicularly from the rachilla and are obscurely rather than strongly nerved.

Habit

Inflorescence

17. *BROMUS*, BROME GRASS

Greek, *bromos*, an ancient name for the oat

Familiar grasses with closed leaf sheaths and panicles of several-flowered, awnless or usually awned spikelets; if awned, the awns arise between two small teeth at the tip of the lemma. By far the commonest species is the upland weed *Bromus inermis*; the commonest native species are *B. ciliatus*, *B. kalmii*, and *B. pubescens*. One hundred to four hundred species; cosmopolitan.

1. First glume distinctly one-nerved; second glume three-nerved, rarely five-nerved
  2. Annual weed with awns 15–30 mm long .... *B. tectorum*
  2. Perennials (native and weedy) with awns absent or up to 10 mm long
    3. Leaf blades with M-shaped crease about one-third to half of the way from the base; plants strongly rhizomatous; lemmas often flushed with purple when young, usually awnless, or occasionally with awns up to 3 mm long; abundant weed .... *B. inermis*
    3. Leaf blades often lacking an M-shaped crease; plants in tufts or small clumps, not rhizomatous; lemmas greenish when young, with awns 3–10 mm long; native (except in the rare *B. erectus*)
      4. Panicle branches erect; leaf blades inrolled (involute); anthers 5–7 mm long; rare adventive .... *B. erectus*
      4. Panicle branches ascending to divergent, but not erect; leaf blades flat; anthers 1–5 mm long; widespread native species
        5. Plants with seven to thirteen overlapping leaves that cover all of the nodes; summit of sheath has prominent auricles and a dense band of hairs .... *B. latiglumis*
        5. Plants not as leafy, with only four to six leaves, the uppermost one or two nodes exposed and not covered by overlapping sheaths; summit of sheath lacking prominent auricles, with or without a dense band of hairs
          6. Backs of lemmas glabrous, the margins long-hairy; glumes glabrous; anthers 1–1.8 mm long; wetlands, often in fens .... *B. ciliatus*
          6. Lemmas uniformly hairy; glumes usually hairy at least on the keel; anthers 2–5 mm long; usually rich upland hardwood forests
            7. Upper leaf surfaces shiny .... *B. nottowayanus*
            7. Upper leaf surfaces dull .... *B. pubescens*
1. First glume distinctly three- to five-nerved; second glume five- to seven-nerved
    8. Backs of lemmas uniformly hairy; glumes usually hairy, at least on the midrib; awns usually straight (except in the rare *B. hordeaceus*)
      9. Awns 5–15 mm long; inflorescence branches much shorter than the spikelets; rare adventive annual .... *B. hordeaceus*
      9. Awns 2–4 mm long; inflorescence branches longer than the spikelets; common perennial of dry areas, often in barrens .... *B. kalmii*
    8. Backs of lemmas glabrous; glumes also glabrous; awns variously curved, not straight; uncommon to rare weeds

10. Lemmas equaling or slightly shorter than the paleas; leaf sheaths glabrous . . . . *B. secalinus*
10. Lemmas exceeding the paleas; leaf sheaths (except sometimes the uppermost ones) finely hairy
    11. Lemmas broad in profile, 2.5–4 mm wide, with a wide membranous border; awns prominent, strongly divergent . . . . *B. squarrosus*
    11. Lemmas narrow in profile, 1.5–2.5 mm wide, without a hyaline border; awns absent to small, never strongly divergent
        12. Longest awns longer than their respective lemmas, and at least twice as long as the lowest lemma awn . . . . *B. arvensis*
        12. Longest awns about as long as their lemmas, or shorter and less than twice as long as the lowest lemma awn . . . . *B. racemosus*

Bromus arvensis L., field brome

arvensis, "of planted fields"

A Eurasian exotic, first collected in 1936 by S. C. Wadmond. An occasional weed of mostly dry gravelly or sandy sites; mostly in southern Wisconsin. Fassett (1951) and Barkworth et al. (2007) treat this as *Bromus japonicus*; we include that species within *B. arvensis*.

Illustration

Leaf ligular area

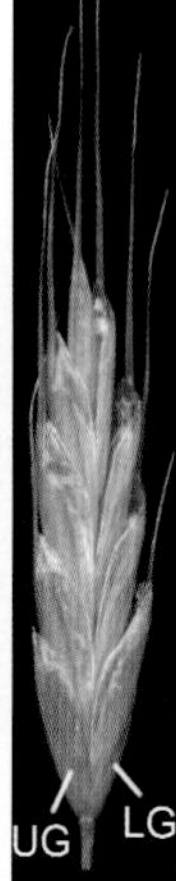

Spikelet

Young and mature spikelets

Bromus ciliatus L., fringed brome

ciliatus, Latin, *cilium*, "small hairs"

Common statewide in marshes, bogs, fens, swamps, sedge meadows, and moist to wet prairies; occasionally in drier open habitats in northern Wisconsin.

Illustration

Habit

Spikelet

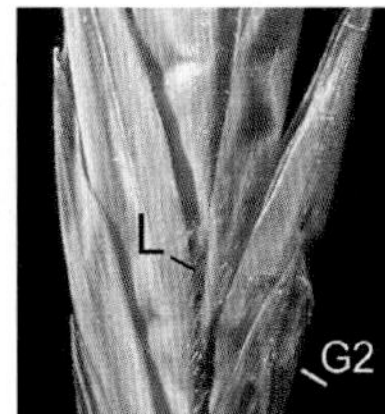

Detail of glumes and lemmas (showing marginal hairs)

Bromus erectus Huds., erect brome

erectus, "erect"

A European exotic, first collected in 1914 (Shinners 1940) and lastly in 1936, although noted by Freckmann along a lakeside railroad in Milwaukee County as late as 1960 or 1961; rare in dry open sites, perhaps not currently present in Wisconsin.

Bromus hordeaceus L., soft brome

hordeaceus, "of or relating to barley"

Illustration

A European exotic, first collected in 1883. A rare adventive of disturbed sites such as railroad ballast; last collected in 1970, so perhaps not currently present in Wisconsin. Fassett (1951) treats this as *Bromus mollis.*

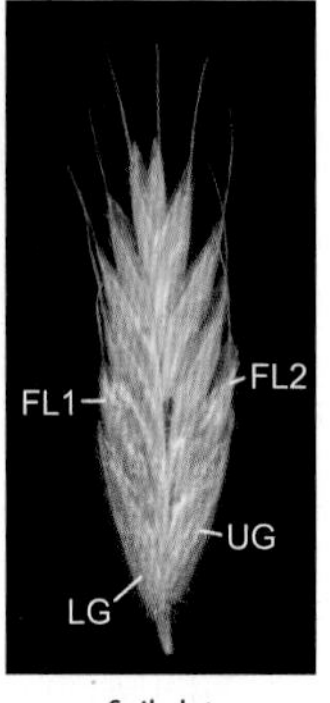

Spikelet

Habit

Bromus inermis Leyss., smooth brome

inermis, "not spiny"

Illustration

A European exotic, first collected in 1907 (Brues and Brues 1911). An abundant weed in upland fields and meadows and on roadsides. This is our most common brome and one of our commonest grasses. The leaf blades have a characteristic M-shaped crease about one-third to half of the way up from their bases; no other grass has this trait. Some specimens from Door County (Bailey's Harbor) have longer awns and pubescent lemmas and approach *Bromus pumpellianus* Scribn., a western US species that is disjunct along the northern shores of Lake Michigan in Michigan.

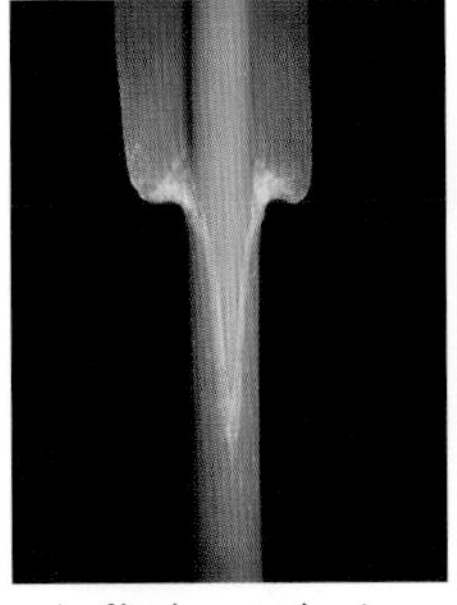
Leaf ligular area, showing closed sheath

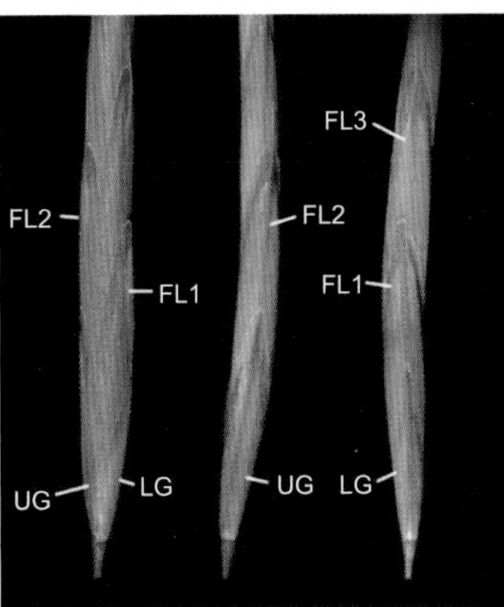

Spikelets

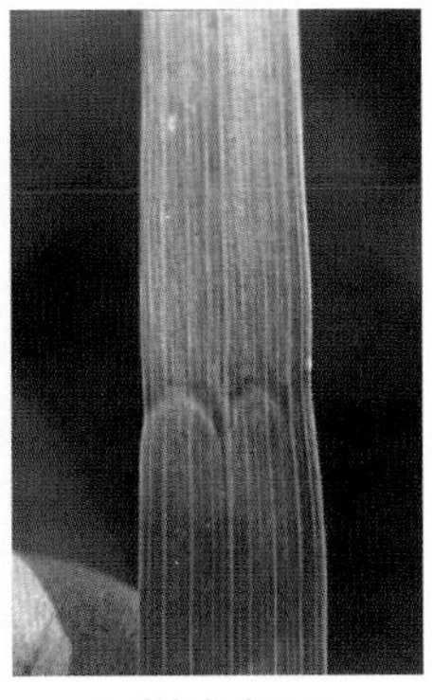
Leaf blade showing M-shaped crease

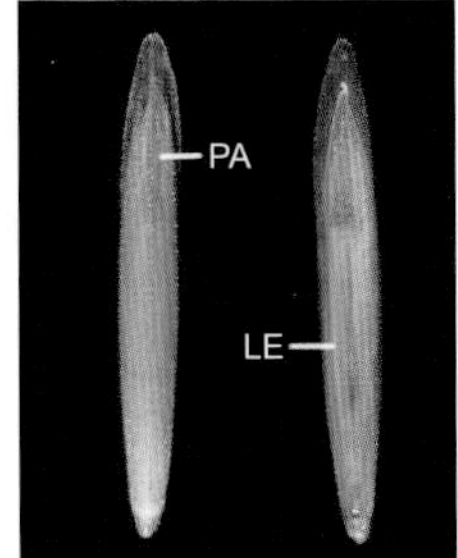

Florets

Young expanded inflorescence

Older inflorescence with spikelets showing one-sided "flagging"

Bromus kalmii A. Gray, prairie brome

kalmii, for Pehr Kalm (1716–1779), Swedish botanist

Common in dry prairies, barrens, savannas, and woodlands; less common in wet to mesic prairies and old fields. Our commonest native brome.

Illustration

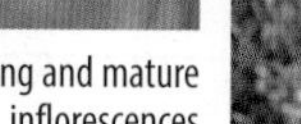

Young and mature inflorescences

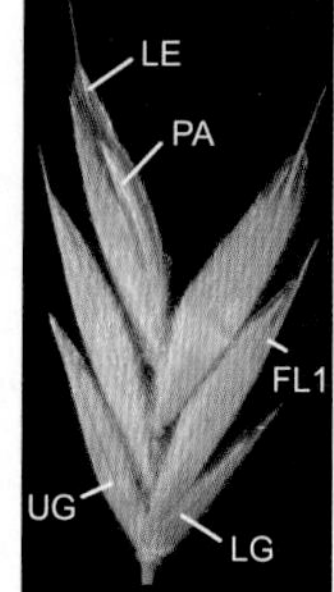

Spikelet

Bromus latiglumis (Shear) Hitchc., ear-leaved brome

latiglumis, *lati*, "wide," *glumis* referring to the "glume" or the small, sterile bracts

Fairly common in wet to dry deciduous forests. The sheath auricles ("ears") at the apex of the leaf sheath are characteristic. This species tends to be larger and to flower later (August) than the closely related *Bromus pubescens* (June). Fassett (1951) treats *B. latiglumis* as *B. purgans*, in part; also known as *B. altissimus* Pursh.

Illustration

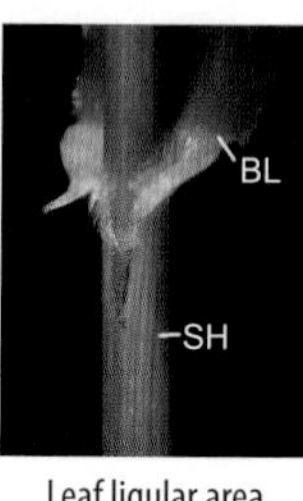

Leaf ligular area showing prominent auricles ("ears")

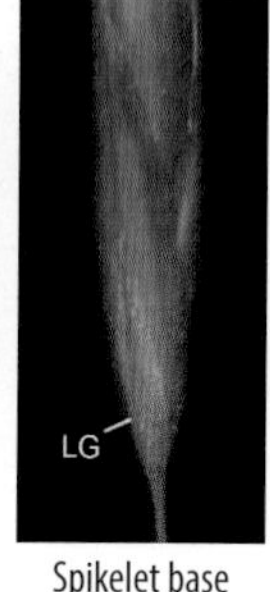

Spikelet base showing glumes

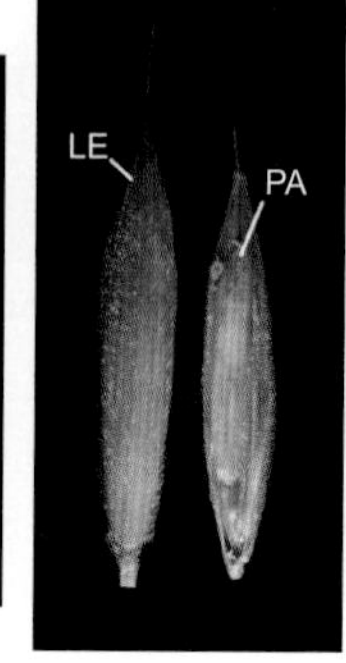

Florets

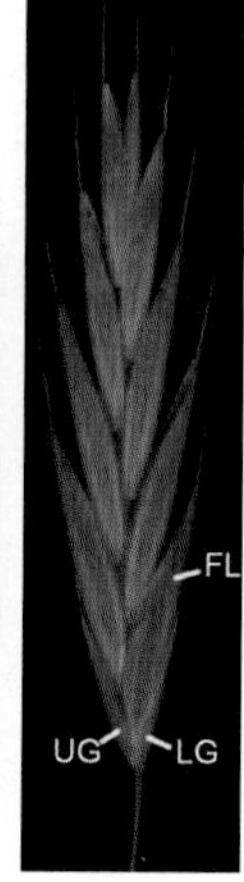

Spikelet

Ear-leaved brome, habit

Bromus nottowayanus Fern., satin brome
nottowayanus, for the Valley of the Nottoway River, Sussex County, Virginia

Occasional, mesic hardwood forests, absent from the far north. Closely related to Canadian brome (*Bromus pubescens*), differing in the glossy "satiny sheen" (Voss and Reznicek 2012) of the functional upper leaf surface. It also tends to have a small, discrete tuft of hairs at the summit of the sheath opposite the ligule; the sheath summit of *B. pubescens* lacks this tuft or is hairy throughout.

Bromus pubescens **Muhl. ex Willd., Canadian brome**

pubescens, "with soft, downy hair"

Like the preceding species (to which it is related; see discussion), occasional in wet to dry mesic, mostly deciduous forests.

Illustration

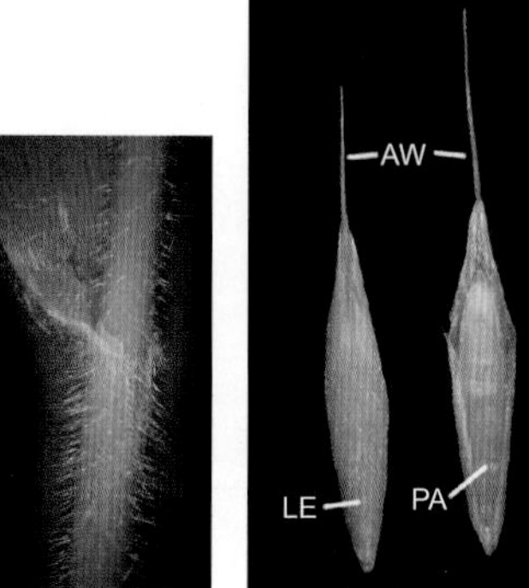

Leaf ligular area

Florets

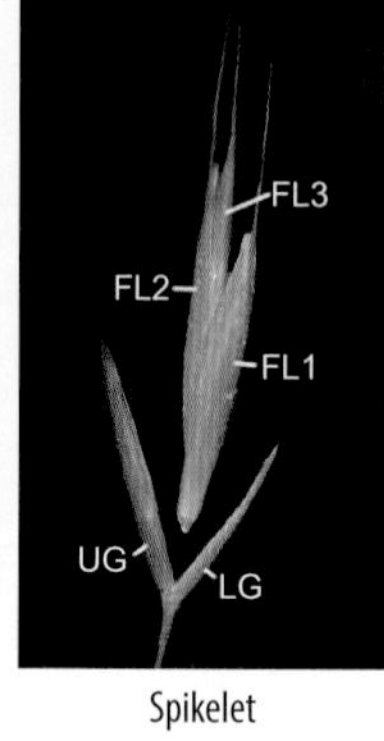

Spikelet

Bromus racemosus **L., European brome**

racemosus, Latin for "having a raceme" (a cluster of flowers each on its own stalk and arranged along a single central stem)

A European exotic, first collected in 1915; an uncommon weed of disturbed areas statewide. Fassett (1951) and Barkworth et al. (2007) treat this as *Bromus commutatus.*

Habit

Bromus secalinus **L., rye brome**

secalinus, "like rye"

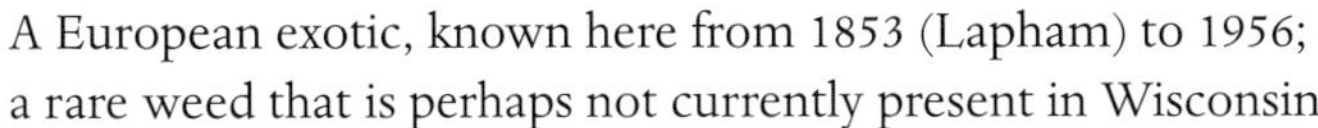

A European exotic, known here from 1853 (Lapham) to 1956; a rare weed that is perhaps not currently present in Wisconsin.

Illustration

Habit

Spikelet

Bromus squarrosus **L., corn brome**

squarrosus, squarrose, "spreading horizontally; curved at the ends"

A European exotic, first collected in 1973 by Raymond F. Schulenberg. A rare weed that has been found not only in disturbed areas but more recently (1996) in a dry prairie in Eau Claire County.

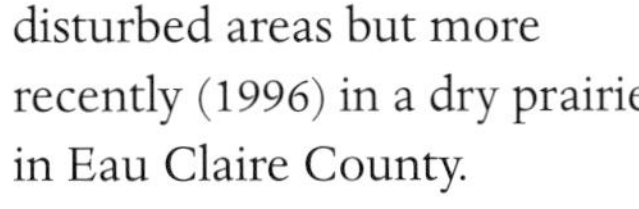

Illustration

Inflorescence

Bromus tectorum **L., cheat grass**

tectorum, "of roofs"

Illustration

A European exotic, first collected in 1910. A common annual weed, most characteristic of sandy or gravelly roadsides and railroad rights-of-way; spring flowering. One of the worst weeds in the western United States, where it colors the hills and mountainsides red when it dies.

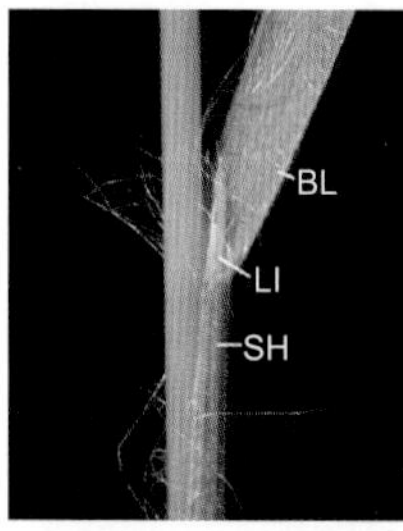

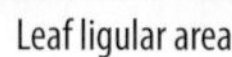

Leaf ligular area

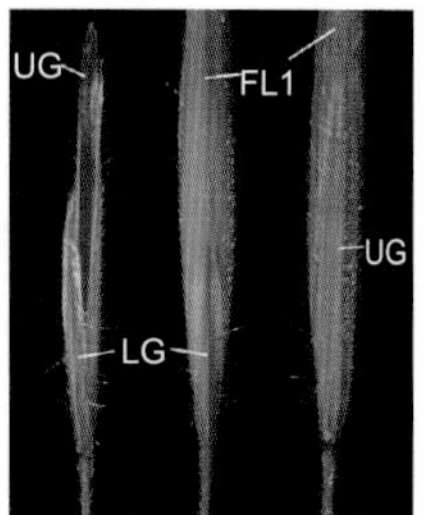

Spikelet

Habit

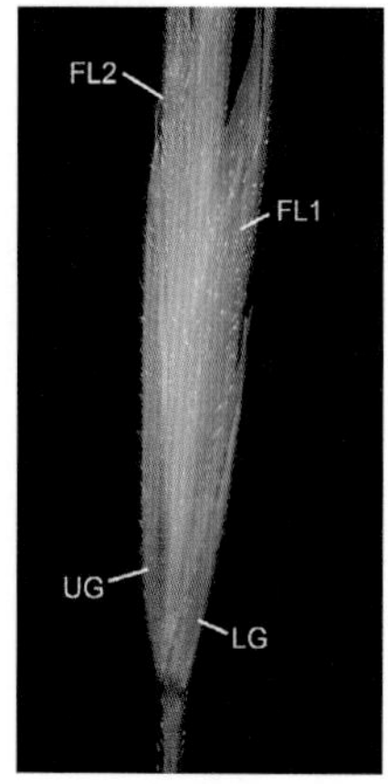

Glumes

18. *CALAMAGROSTIS*, BLUEJOINT

from the Greek mythological figure Kalamos, the son of Maiandros (aka Meander), god of the Meander River, and *agrostis*, "grass"

Tall, slender rhizomatous perennials with panicles that soon contract to become spike-like; spikelets single-flowered, the lemma with a dense basal beard and sporting a small awn from the middle of its back. *Calamagrostis canadensis* is our most abundant and widespread native wetland grass (Shinners 1943). *Calamagrostis ×acutiflora* (Schrad.) DC, feather reed or "Karl Foerster" grass, is a common, robust (to over 1 m tall), densely clump-forming cultivar; it has contracted panicles of spikelets 4–5 mm long, and the lemmas have exserted awns 3–4 mm long. One hundred species; cosmopolitan except in the lowland tropics.

1. Glumes and callus hairs significantly longer than the lemma; spikelets 4.5–5.5 mm long; rare adventive .... *C. epigeios*
1. Glumes and callus hairs about as long as the lemma; spikelets 2–4 (rarely 4.5) mm long; native
    2. Panicle open at time of flowering; lemmas membranous and smooth; callus hairs about as long as lemma; common, wetlands .... *C. canadensis*
    2. Panicle contracted at time of flowering, with erect branches; lemmas scabrous, somewhat rigid; callus hairs about half as long as lemma; common, occasional, wetlands (meadows and fens), as well as dry barrens .... *C. stricta*

Calamagrostis canadensis (Michx.) P. Beauv., Canada bluejoint

canadensis, of or referring to Canada

Illustration

Our most abundant native wetland grass, often codominant with sedges (especially the tussock sedge *Carex stricta* Lam.) in sedge meadow communities. Occasionally found in upland sites, for example, in crevices in otherwise dry rock outcrops. A variable species often confused with the invasive reed canary grass (*Phalaris arundincea*), Canada bluejoint differs in its narrow leaf blades (4–8 mm wide) and slightly smaller spikelets (3–5 mm long) with a small but dense beard of hairs at the base of the delicately awned floret. As with reed canary grass, the inflorescence is at first an expanded panicle that later contracts into a spike-like appearance.

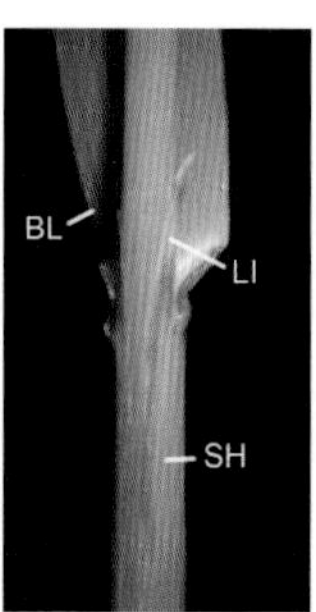

Leaf ligular area

Portion of young inflorescence

Portion of mature inflorescence showing callus hairs

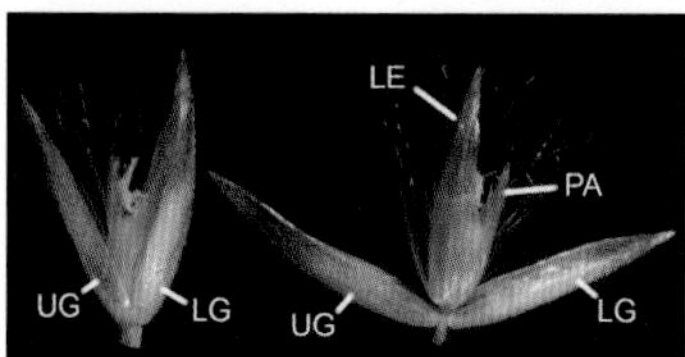

Spikelets

Canada bluejoint, young expanded inflorescence

Canada bluejoint, mature contracted inflorescence

Reed canary grass (*Phalaris arundinacea*, *left*) compared with Canada bluejoint (*Calamagrostis canadensis*, *right*)

Calamagrostis epigeios **(L.) Roth, Chee reed grass**

epigeios, *epi*, "over," *geios*, "earth," hence "over the earth"

Inflorescence

Rare European adventive, first collected in 1960; known from one collection made in moist sand in Pierce County by Miles Johnson.

Illustration

Calamagrostis stricta (Timm) Koeler, slim-stemmed reed grass

stricta, "erect, tight, rigid"

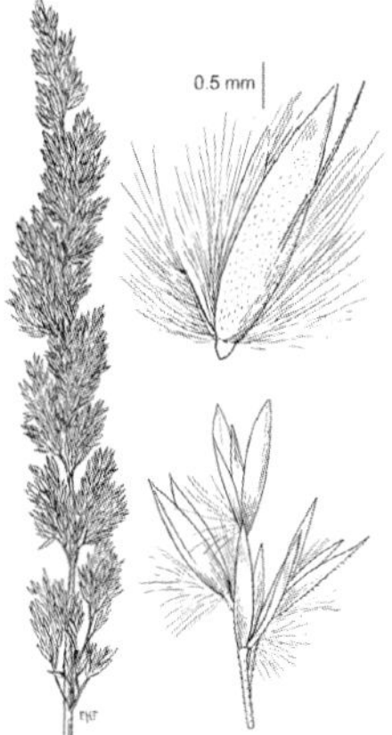

Illustration

An uncommon calciphile, most frequent near the dolomitic shores of Lake Michigan in Door County and in fens and sedge meadows elsewhere in the state, but also occasionally found in noncalcareous or drier upland sites. Both the typical subspecies and subsp. *inexpansa* (A. Gray) C. W. Greene occur in Wisconsin; we do not differentiate them here. Fassett (1951) treats this as two species: *Calamagrostis inexpansa* and *C. neglecta.*

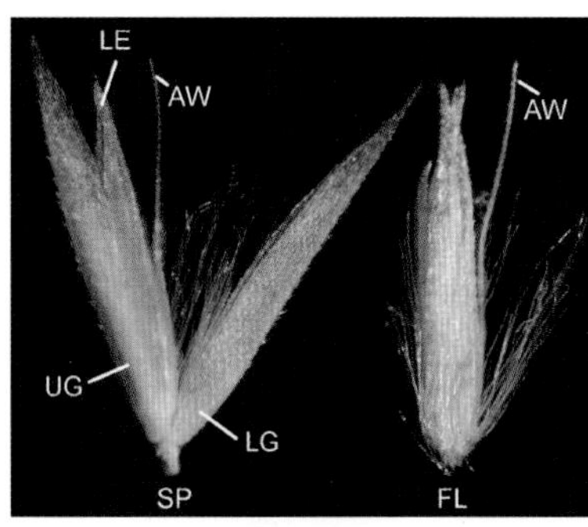

Spikelet and floret

Habit showing contracted inflorescences

19. *CALAMOVILFA*, SAND REED

from the Greek mythological figure Kalamos, the son of Maiandros (aka Meander), god of the Meander River, and *vilfa*, an early name applied to a genus of grass

Midsized rhizomatous perennials with long, involute leaf blades and hairy ligules; the spikelets are single-flowered, with papery glumes and awnless florets with a dense basal callus beard (Shinners 1943). *Calamovilfa* differs from *Calamagrostis* in its hairy rather than membranous ligules and awnless spikelets; from *Phalaris* it also differs in its hairy ligules, as well as its much narrower inrolled leaves. Five North American species.

Calamovilfa longifolia **(Hook.) Scribn., prairie sand reed**

longifolia, *longus*, "long, extended," *folius*, "leaves"

THREATENED (IN PART)

Illustration

Occasional, dry open sandy sites; var. *longifolia* is a western US taxon found mostly in sand barrens and terraces along the Mississippi, lower Wisconsin, and St. Croix Rivers. Variety *magna* Scribn. and Merr. is found mainly on sand dunes along Lake Michigan. There it often grows with the commoner American beach grass (*Ammophila breviligulata*), a shorter species with smaller spikelets and a contracted spike-like panicle.

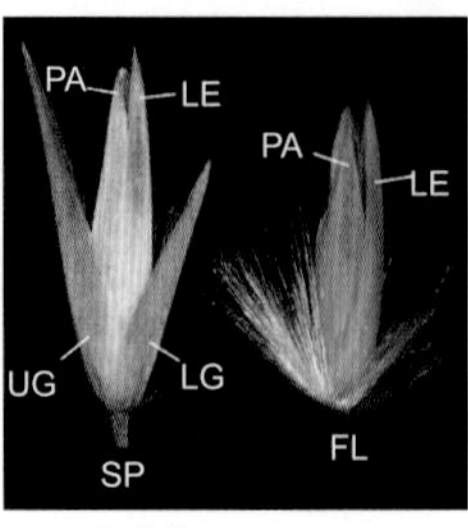

Spikelet, var. *magna*

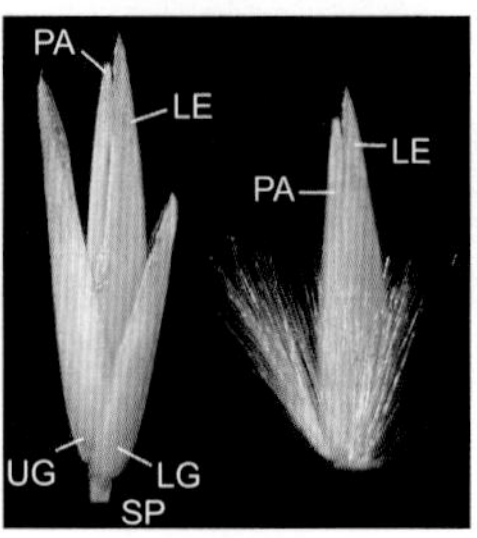

Spikelet, var. *longifolia*

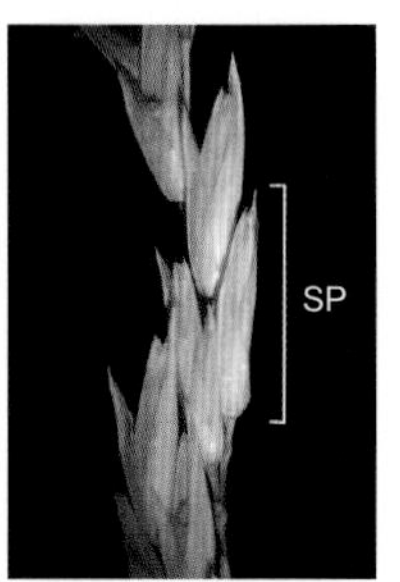

Section of inflorescence, var. *longifolia*

Prairie sand reed, Lake Michigan dune habitat of var. *magna*

20. *CATABROSA*, BROOK GRASS

Greek meaning "eating," referring to eroded glumes

Small, rare, somewhat succulent aquatic perennials; spikelets in panicles, with two distant florets, the glumes and lemmas conspicuously blunt. Two species of the Northern Hemisphere and South America.

Catabrosa aquatica (L.) P. Beauv., brook grass

aquatica, "growing in or near water"

ENDANGERED

Rare, cold springs and streamsides; a western species that "skips over" Minnesota and Michigan on its way to Wisconsin. First collected in 1934 by N. C. Fassett near Hudson, St. Croix County and

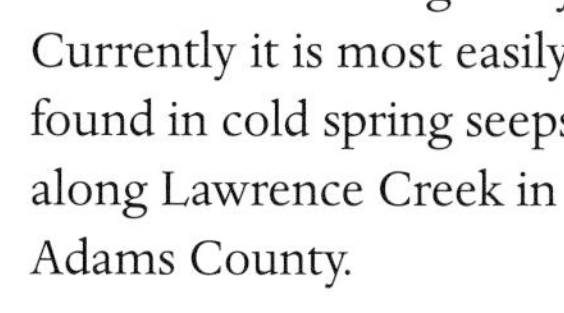

not located there again by Judziewicz in 1993. Currently it is most easily found in cold spring seeps along Lawrence Creek in Adams County.

Illustration

Portion of inflorescence

Portion of inflorescence

21. *CENCHRUS*, SANDBUR, FOUNTAIN GRASS, AND PEARL MILLET

ancient Greek name of *Setaria italica*

Small troublesome weedy annuals to robust tussock grasses cultivated for ornament and seed; inflorescence a spike-like panicle with dense clusters of reduced branches forming fascicles including three to 130 bristles and one to twelve spikelets, these falling as a unit, in some species very spiny. *Cenchrus* now includes *Pennisetum* (Chemisquy et al. 2010). One hundred to 150 species; cosmopolitan, tropical to warm temperate areas.

Cenchrus longispinus (Hack.) Fernald, field sandbur

longispinus, longus, "long, extended," and *spina,* "spine"

Illustration

Locally common annual in dry open sand. The bane of barefoot beachgoers! Along with rice cut-grass (*Leersia oryzoides*), our most painful grass to encounter.

Inflorescence

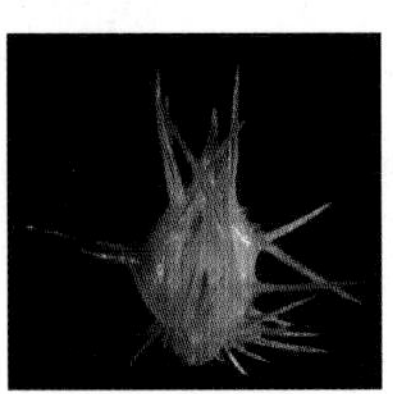
Spiny bur enclosing spikelet

22. *CHASMANTHIUM*, WILD OATS, SHINGLE GRASS

Greek *chasme*, "gaping," and *anthe*, "flower"

A broad-leaved woodland grass with large, attractive, strongly laterally flattened, multiflowered spikelets with several empty basal lemmas that resemble extra glumes. Five species of warm temperate southeastern North America.

Chasmanthium latifolium (Michx.) H. O. Yates, wild oats, shingle grass

latifolium, Latin, "wide to broad leaves"

Illustration

A rare American adventive, the single Wisconsin collection made under a bird feeder in Forest County in 1984 by Jesse Zirbel. Sometimes cultivated ornamentally for its large spikelets; more records are to be expected.

Spikelets

Mature inflorescence

Plant with young inflorescence

23. *CHLORIS*, WINDMILL OR FINGER GRASS

for the Greek goddess of flowers

Small perennials with several to many digitate (finger-like) terminal spikes (plus a subterminal whorl or two), the spikelets several-flowered, with both glumes and lemmas awned. It is a tumbleweed, the mature inflorescence breaking off and sent rolling by the wind. Fifty-five to sixty species; pantropical to warm temperate.

Chloris verticillata Nutt., windmill or finger grass

verticillata, "whorled"

Illustration

This weedy exotic from the Great Plains, superficially resembling a "prickly" crabgrass, was first collected by Judziewicz (in 2012) from a park on the Wisconsin River in Muscoda, Grant County.

Section of inflorescence

Spikelet

Habit

24. *CINNA*, WOOD-REED

Greek, a name used by Dioscorides for a kind of grass

Woodland perennials with somewhat drooping or one-sided panicles of single-flowered spikelets, these falling entire from their pedicels at maturity; rachilla slightly prolonged past the floret as a minute sterile rudiment; lemma with a minute short awn produced from just below its tip. This genus could be confused with *Agrostis* (but differing in its minutely awned lemmas and the presence of a sterile rachilla prolongation) and *Sphenopholis* (with two-flowered, awnless spikelets). Four species; circumboreal.

1. Spikelets 4.5–7 mm long; panicle branches ascending, at least at the base; statewide except in the far north, usually in deciduous floodplain forests . . . . *C. arundinacea*
1. Spikelets 2.5–4 mm long; panicle branches generally drooping; statewide except in the far south, in mesic to wet forests . . . . *C. latifolia*

Cinna arundinacea L., common wood-reed

arundinacea, "reed-like"

Fairly common in wet to mesic deciduous forests, mainly in southern Wisconsin.

Illustration

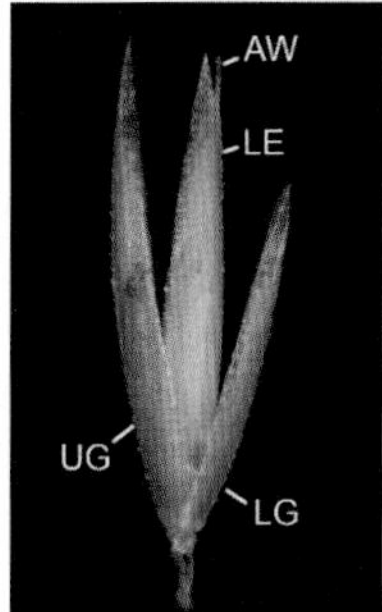

Spikelet

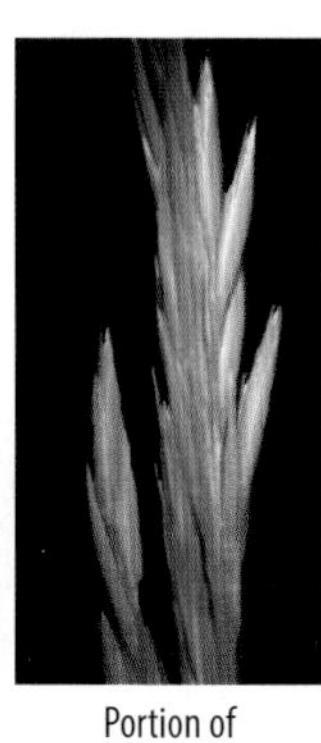

Portion of inflorescence

Inflorescence showing branches slightly ascending at base

Cinna latifolia (Trevir. ex Göpp.) Griseb., drooping wood-reed

latifolia, Latin, "wide to broad leaves"

Common in moist forests statewide, except in the southernmost counties.

Illustration

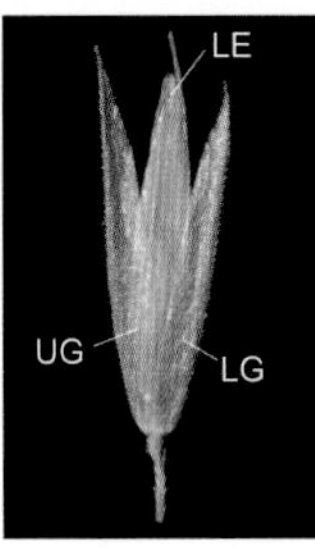

Spikelet

Inflorescence showing "flagging" drooping branches

25. *COLEATAENIA*, RED-TOPPED PANIC GRASS

Greek *koleos*, "sheath," and *tainia*, "ribbon" or "band"

Based upon molecular evidence, this genus is a recent segregate of *Panicum* and is difficult to distinguish from it morphologically—about the only difference is that the fertile florets of *Coleataenia* have minute, almost microscopic prickle hairs near their apices, while these hairs are absent in *Panicum*. For user convenience, this genus also keys out under *Panicum*. Eleven species of North and South America.

Coleataenia rigidula (Bosc ex Nees) LeBlond, red-topped panic grass or Munro grass

rigidula, Latin, "becoming rigid"

Portion of inflorescence

This species is known from sandbars, the edges of sloughs, and dry old fields along the margins of the lower Wisconsin River. The first collections were made in 1948 by Elizabeth Kirk Jones in both Crawford and Richland Counties, and the last by William Tans, Robert Read, and Kenneth Lange in Sauk County in 1974. It was not located by Judziewicz again during a search in 2012. A prominent, clump-forming grass up to 1.5 m tall whose nearest sites of abundance are along the Illinois River several hundred miles to the south, it is not clear whether this mysterious species is native to Wisconsin or even whether it stills occurs here. Barkworth et al. (2007) treat this as *Panicum rigidulum*, and it has only recently been recognized by its current name (Weakley et al. 2011).

Portion of inflorescence

26. *CRYPSIS*, PRICKLE GRASS

cryp, "hidden, secret, covered"

Low annuals with small, compact ovoid panicles of single-flowered, slightly prickly spikelets. Eight Eurasian species.

Crypsis schoenoides (L.) Lam., prickle grass

schoenoides, *schoenos*, from *schoinos*, "a rush or reed," and *oides*, "like"

Habit

A Eurasian exotic, collected from 1936 to 1972 in highly disturbed, often saline sites. Fassett (1951) treats this as *Heleochloa schoenoides*.

Habit

27. *CYNODON*, BERMUDA GRASS

Greek, *kyon*, "dog," and *odous*, "tooth," referring to the hard, sharp scales on the rhizomes and stolons

Low stoloniferous perennials with several digitate (finger-like) spikes; spikelets small, single-flowered, awnless, the lower glume semicircular, the upper glume straight. Nine Old World species of tropical and warm temperate regions.

Cynodon dactylon (L.) Pers., Bermuda grass

dactylon, Greek, *dactylos*, "finger"

Illustration

A rare introduction, first collected near Birge Hall in Madison, Dane County, in 2011 by Theodore Cochrane. It is a dominant warm-weather lawn grass in the central and southern United States, and it would not be surprising if it becomes more common in Wisconsin in the coming decades.

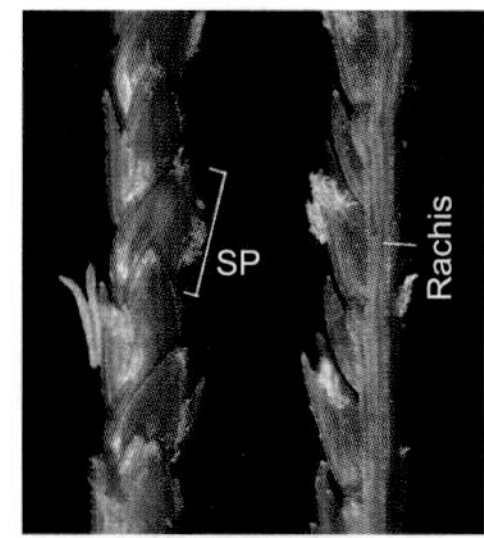

Section of inflorescence

Habit showing extensive stolons or runners

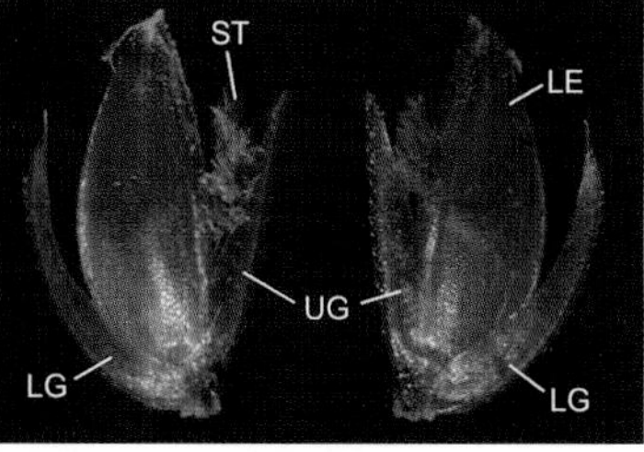

Spikelets

28. *CYNOSURUS*, DOG'S-TAIL

Greek for "dog's tail," referring to the flower spike

A small perennial with a spike-like panicle; spikelets of two types: sterile spikelets with several narrow florets surrounding fertile spikelets with two to four broader florets. The sterile spikelets are placed broadside to the axis of the inflorescence as in quack grass (*Elymus repens*), but in that common species the spikelets are all alike. Eight Eurasian species.

Cynosurus cristatus L., dog's-tail

cristatus, Latin, "crested or comb-like"

A rare adventive, known only from a single collection from Lake Park in Milwaukee made by C. T. Brues in 1908, "doubtlessly introduced in commercial grass seed mix" (Brues and Brues 1911).

Illustration

Inflorescence

29. *DACTYLIS*, ORCHARD GRASS

Greek, *dactylos*, "finger"

A tallish, clump-forming grass with keeled sheaths that have fused margins; panicle large and "leggy" looking, with several stiff spreading bare branches at the ends of which are dense clusters of several-flowered spikelets with very narrow florets. One Eurasian and northern African species.

Dactylis glomerata L., orchard grass

glomerata, "clustered"

An exotic, first collected in 1879. Roadsides, pastures, fields, and lightly wooded areas (hence the common name). This attractive species is one of the most widespread and common weedy grasses, but it seemingly never becomes dominant in native habitats. Its leaves are slightly succulent and persist over winter in a green state.

Illustration

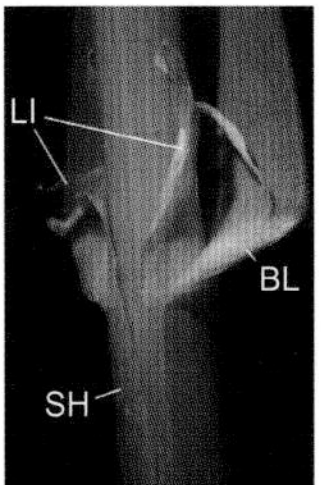

Leaf ligular area

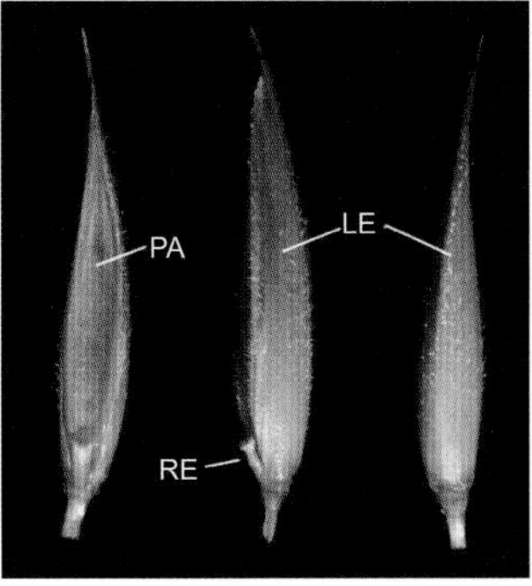

Florets

Spikelets in dense clusters at ends of branches

Inflorescences

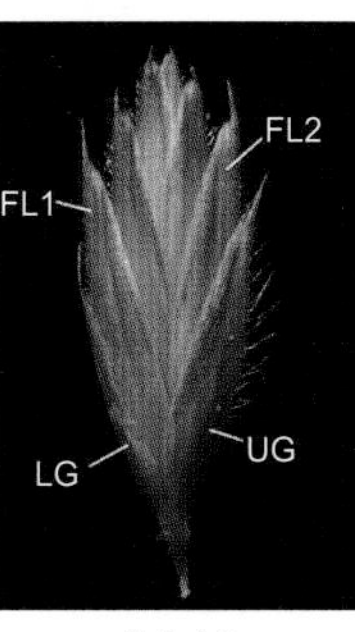

Spikelet

Orchard grass inflorescences

30. *DANTHONIA*, POVERTY OAT GRASS

for Étienne Danthoine (1739–1794), botanist of Marseilles

Small tufted perennials with hairy ligules and contracted panicles of several-flowered spikelets with relatively large glumes and hairy lemmas sporting prominent "kinked" awns from between apical teeth. Twenty species; cosmopolitan except in Australia.

1. Longest pedicels one or two times as long as the spikelets; lower panicle branches spreading or drooping when mature; dead basal leaves straight; lemma lobes 2–4 mm long; rare, upland forests in far northeastern Wisconsin .... *D. compressa*
1. Longest pedicels less than the length of the spikelets; all panicle branches erect when mature; dead basal leaves curling; lemma lobes 0.5–2 mm long; abundant, dry open areas statewide .... *D. spicata*

Danthonia compressa Austin, flattened oat grass

compressa, "flattened or compressed"

Illustration

A rare native of far northern Wisconsin, first collected by Steve Janke in 2006 in Florence County (on an old roadbed through a young fir, spruce, alder, white birch, and aspen forest) and then in 2010 by Matthew Wagner in Langlade County (in a red oak–sugar maple stand); probably overlooked.

Habit

Danthonia spicata (L.) P. Beauv. ex Roem. and Schult., poverty oat grass

spicata, "spiked"

Illustration

Abundant in dry, open, lightly vegetated sandy areas statewide, sometimes in lawns, where the low basal leaves escape the mower blades. Readily recognized even when sterile by its curling, dead basal leaves (the only Wisconsin grass with this character) and minute beards of white hairs at the leaf sheath summits. It is quite variable in size, and can be up to 1 m tall, with an open panicle.

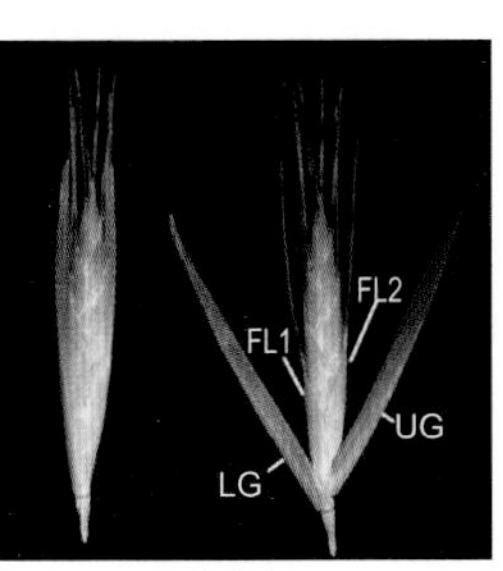

Spikelet

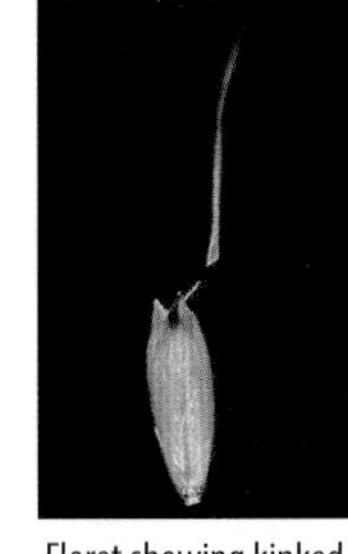

Floret showing kinked awn from between two lobes

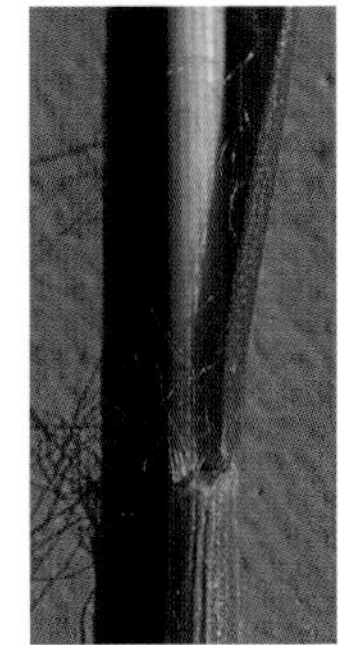

Leaf ligular area showing pilose hairs at sheath summit

Poverty oat grass, habit

31. *DESCHAMPSIA*, HAIR GRASS

for Louis-Auguste Deschamps (1774–1849), French botanist

Clump-forming perennials with abundant flat (not inrolled) leaves and panicles of small spikelets with two florets, the lemmas thin with a straight awn arising below the middle. Wavy hair grass (*Avenella flexuosa*) was previously placed in this genus (Valdes and Scholz 2006). Twenty-five to forty species; cosmopolitan.

Deschampsia cespitosa (L.) P. Beauv., tufted hair grass

cespitosa, *caespes*, "a turf, sod; field," and *osus*, "like"

SPECIAL CONCERN

Illustration

Uncommon; found in several habitats: wave-splashed sandstone (along Lake Superior) and dolomite (along Lake Michigan) ledges; fens (as in Waushara County); and cold, springy stream margins (as in Marquette and Taylor Counties), where this bunchgrass can be up to 1 m tall. It is one of the most naturally widely distributed grasses in the world and has been used in prairie restoration and mine site stabilization (USDA NRCS 2014).

Habit, young inflorescence

Portion of inflorescence

Mature inflorescences in springy fen habitat (with David Seils)

32. *DIARRHENA*, BEAK GRAIN OR BOTTLE GRASS

Greek, *dis*, "two," and *arren*, "man," from its two stamens

Woodland perennials with relatively broad, flat, shiny leaves, flowering in late summer; inflorescence a sparse panicle of several-flowered, awnless spikelets with small glumes, the grain bottle-shaped, protruding from between the lemma and palea at maturity. Six species of northern temperate areas.

Diarrhena obovata **(Gleason) Brandenb., obovate beak grain or bottle grass**

obovata, "inverted ovate," that is, egg-shaped with the broader end uppermost

ENDANGERED

Illustration

Rare in deciduous forests, often on mesic floodplain terraces and the lower slopes of rich forests near major rivers. Fassett (1951) and the Wisconsin Department of Natural Resources (2014) treat this as *Diarrhena americana*.

Inflorescence

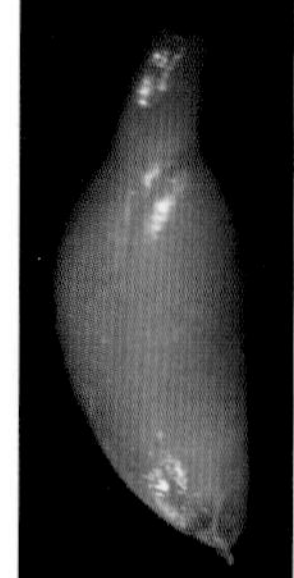

Grain, showing beak and bottle shape

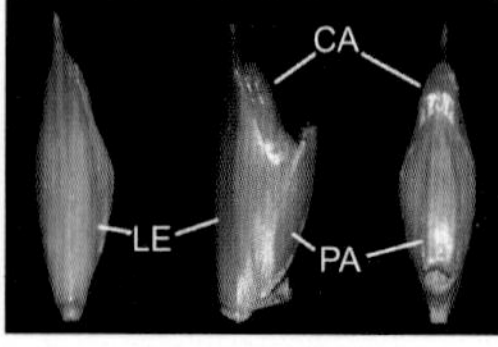

Spikelet

Habit showing broad, shiny leaves

33. *DICHANTHELIUM*, TWICE-BLOOMING PANIC GRASS

Greek, *dich*, "two," and *anthelium*, "flowering"—a fair translation is "twice-flowering"

Perennials with an overwintering rosette of short, wide blades; flowering at top of main culms in late spring, then from axillary branches until autumn (Shinners 1944). Seventy-two species, North and South America.

1. Basal blades narrow, erect (except *D. wilcoxianum*), grading into lower cauline leaves; lower internodes foreshortened, only upper two to four internodes elongated, the blades 1–5 mm wide, fifteen to sixty times as long; panicles produced after midseason partially hidden at base of plant
   2. Upper glumes and lower lemmas forming a beak extending 0.2–1 mm beyond the upper florets; spikelets 3.1–4.3 mm long, glabrous to sparsely pubescent; primary panicles with seven to twenty-five spikelets; open sandy areas .... *D. depauperatum*
   2. Upper glumes and lower lemmas equaling or exceeding upper florets by no more than 0.3 mm, not forming a beak; spikelets 2–3.4 mm long, pubescent or glabrous; primary panicles with twelve to seventy spikelets
      3. Cauline blades alike, 4–8 cm long; basal blades ascending to spreading; rare, dry hillside prairies, southern Wisconsin .... *D. wilcoxianum*
      3. Uppermost cauline blades 10–20 cm long, distinctly longer than the lower blades; basal blades erect to ascending
         4. Panicles 2–6 cm wide, with spreading branches and pedicels; plants pubescent or glabrous; spikelets not turgid, 2–3.2 mm long, 0.8–1.4 mm wide, the upper florets ellipsoid; open sandy areas .... *D. linearifolium*
         4. Panicles 1–3 cm wide, with ascending branches and appressed pedicels; plants pubescent; spikelets turgid, 2.6–3.4 mm long, 1–1.7 mm wide, the upper florets obovoid; prairies .... *D. perlongum*
1. Basal blades ovate to lanceolate, spreading, forming a rosette, well differentiated from cauline blades, or basal blades absent; four to fourteen internodes elongated, the cauline blades 2–40 mm wide, three to twenty times as long; culms usually branching from midculm nodes after midseason, often producing dense fascicles of branches partially enclosing small panicles
   5. Culms arising from rhizomes 3–5 mm thick, with five or seven to fourteen cauline blades; sheaths hispid, some sheaths in autumn swollen, containing hidden panicles; rare, Vilas County, where perhaps introduced .... *D. clandestinum*
   5. Culms arising from caudices or from rhizomes up to 2 mm thick, with three to seven or nine cauline blades; sheaths pubescent or glabrous, not hispid or swollen, without completely enclosed panicles
      6. Ligules with a membranous base, ciliate at apex; culms usually arising from short slender rhizomes; lower florets often male; cauline blades 5–40 mm wide, often slightly cordate at base
         7. Spikelets ellipsoid, not turgid, with pointed apices; cauline blades four to six, 15–40 mm wide, cordate at base; sheaths without papillose-based hairs; woods .... *D. latifolium*

7. Spikelets obovoid, turgid, with rounded apices; cauline blades three to four, 5–23 mm wide, their bases tapered, rounded, or truncate; sheaths with papillose-based hairs
    8. Panicles 3–5 cm wide, slightly longer than wide, with spreading to ascending branches; blades papillose-hispid, rounded to truncate at base; spikelets soft villous; prairies .... *D. leibergii*
    8. Panicles 1–2 cm wide, usually more than twice as long as wide, with ascending to nearly erect branches; blades glabrous, tapered at base; spikelets sparsely minutely pubescent; semiopen woods .... *D. xanthophysum*

6. Ligule of hairs only; culms arising from caudices; lower florets sterile; cauline blades 1–18 mm wide, their bases usually tapered, rounded, or truncate
    9. Spikelets 3.1–4.1 mm long, obovoid, turgid; upper glumes usually with an orange or purple spot at the base, the nerves prominent; sandy prairies and open areas .... *D. oligosanthes* var. *scribnerianum*
    9. Spikelets 1.3–2.8 mm long, ellipsoid to obovoid, not turgid; upper glumes lacking an orange or purple spot at base, purplish coloration, if present, concentrated near apex and margin; nerves not prominent
        10. Ligules 0.1–0.5 mm long, without adjacent pseudoligule of long hairs; internodes and axis of panicles glabrous; spikelets pubescent or glabrous
            11. Spikelets pubescent; larger blades more than 5 mm wide and 6 cm long, blade of flag leaf erect or ascending; thickets and moist woodlands .... *D. boreale*
            11. Spikelets glabrous; all blades less than 5 mm wide, 6 cm long and spreading; rare, Dane County .... *D. dichotomum*
        10. Ligules and adjacent pseudoligules of hairs 1–5 mm long, or, if shorter, culms puberulent; internodes, or at least the lower internodes, pubescent; spikelets pubescent
            12. Upper internodes glabrous; open areas .... *D. acuminatum* var. *lindheimeri*
            12. All internodes pubescent or puberulent
                13. Sheaths of main culms pubescent or pilose with hairs 0.5–5 mm long but not also puberulent (late season branches may be short-pubescent)
                    14. Sheaths densely pilose, with spreading to retrorse hairs 3–5 mm long; spikelets 1.8–2.4 mm long; axillary branches and panicles elongated, developing almost simultaneously with main panicles; prairies .... *D. villosissimum* var. *praecocius*
                    14. Sheaths pubescent to pilose with ascending to spreading hairs 2–4 mm long; spikelets 1.3–1.9 mm long; axillary branches forming dense fascicles after flowering of main panicles; various habitats .... *D. acuminatum* var. *fasciculatum*
                13. Sheaths of main culms puberulent with hairs 0.1–0.2 mm long, often also pubescent to hispid
                    15. Spikelets 2.1–2.8 mm long; panicle branches stiffly ascending, bearing spikelets only near tip; sheaths pilose to hirsute with stiffly ascending hairs; dry sandy sites .... *D. commonsianum* var. *euchlamydeum*

15. Spikelets 1.3–2.1 mm long; panicle branches spreading, not very stiff, bearing spikelets more than halfway to axis; sheaths puberulent only, or also sparsely pubescent
    16. Panicle axis puberulent only; ligule and pseudoligule hairs uniformly about 1 mm long; spikelets 1.6–2.1 mm long; sandy open areas .... *D. columbianum*
    16. Panicle axis pubescent to pilose, sometimes also puberulent; ligule and pseudoligule hairs 0.7–3 mm long; spikelets 1.3–1.7 mm long; sandy or wet areas .... *D. meridionale*

Dichanthelium acuminatum **(Sw.) Gould and C. A. Clark, twice-flowering panic grass**

acuminatum, "tapering gradually to a point"

The most abundant and widespread taxon of *Dichanthelium* in Wisconsin, found in a wide variety of weedy and native open habitats. Fassett (1951) treats this as *Panicum implicatum*, *P. lindheimeri*, and *P. subvillosum*. Hybrids with *D. boreale*, *D. oligosanthes* var. *scribnerianum*, and *D. xanthophysum* have been recorded in Wisconsin. There are two well-defined varieties.

Dichanthelium acuminatum **(Sw.) Gould and C. A. Clark var. *fasciculatum*** **(Torr.) Freckmann, hairy panic grass**

Common, found in a wide variety of weedy and native open habitats. Fassett (1951) treats this as *Panicum subvillosum* and *P. implicatum*.

Habit

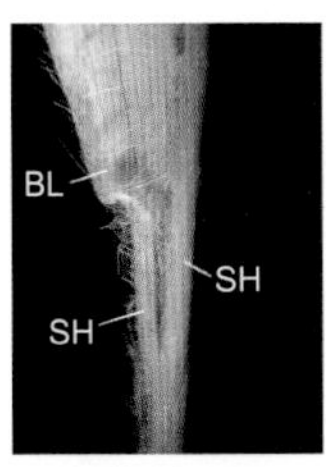

Leaf ligular area

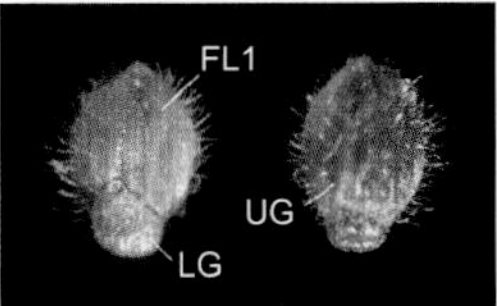

Spikelet

Habit

Dichanthelium acuminatum (Sw.) Gould and C. A. Clark var. *lindheimeri* (Nash) Gould and C. A. Clark, Lindheimer's panic grass

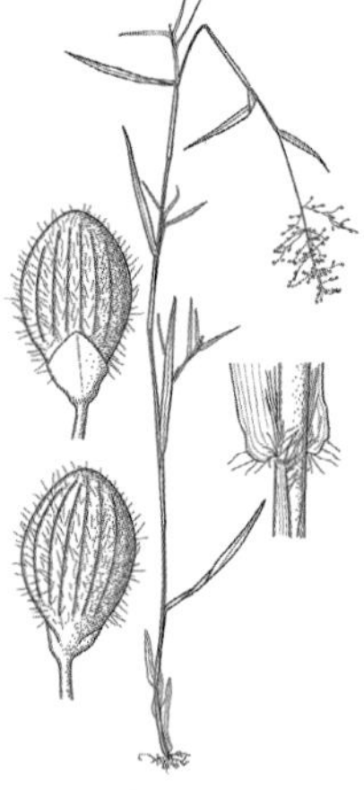

Illustration

Much less common than *Dichanthelium acuminatum* var. *fasciculatum* and restricted to moist sandy, gravelly, or peaty shores. Fassett (1951) treats this as *Panicum lindheimeri.*

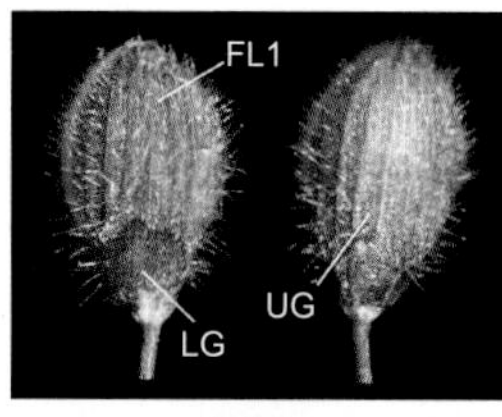

Spikelet

Habit

Dichanthelium boreale (Nash) Freckmann, northern panic grass

boreale, "northern"

Illustration

Wetland edges or occasionally in dry open woodlands; occasional statewide, most common in the bed of Glacial Lake Wisconsin in the central part of the state. Hybrids with *Dichanthelium acuminatum* have been recorded in Wisconsin. Fassett (1951) treats this as *Panicum boreale.*

Habit

Dichanthelium clandestinum **(L.) Gould, deer-tongue grass**

clandestinum, "hidden"

Rare, first collected in 2006 in Vilas County from a small grassy clearing in a sandy white pine–quaking aspen woods by Joshua D. Sulman, who speculated that the species was "presumably a recent introduction," since the species is used for revegetating sterile sandy sites and mine reclamation (USDA NRCS 2014). But it is also known from a 1919 collection made along a railroad in nearby Gogebic County, Michigan.

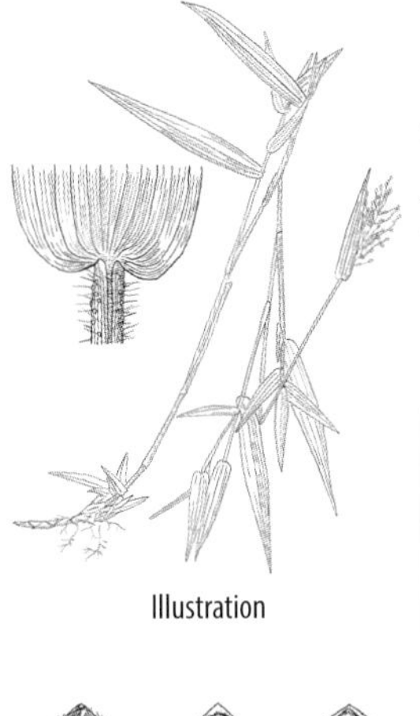

Illustration

Spikelet illustration

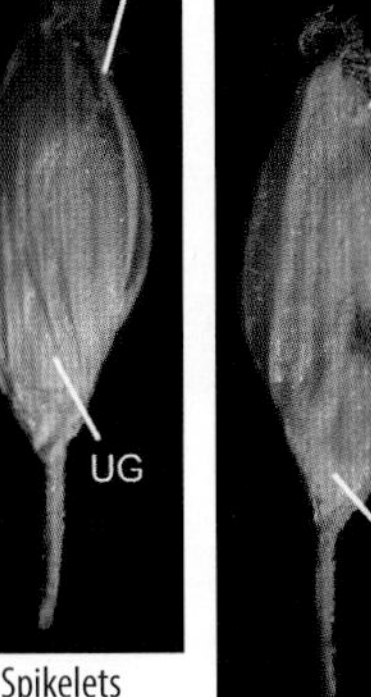

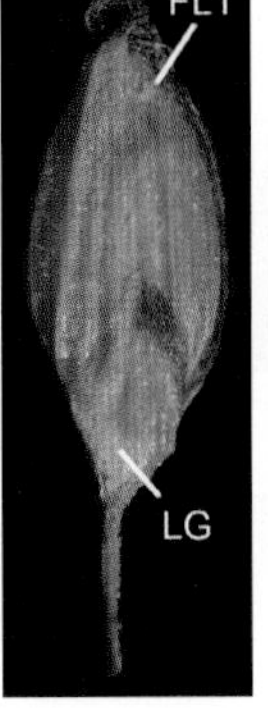

Spikelets

Leaf sheath showing papillose-pilose hairs

Dichanthelium columbianum **(Scribn.) Freckmann, puberulent panic grass**

columbianum, "from [District of] Columbia"

Locally common statewide in open weedy and native habitats. Hybrids with *Dichanthelium depauperatum* have been recorded in Wisconsin. Fassett (1951) treats this as *Panicum columbianum*; Barkworth et al. (2007) as *D. acuminatum* subsp. *columbianum*.

Habit

Dichanthelium commonsianum (Ashe) Freckmann var. *euchlamydeum* (Shinners) Freckmann, Shinners' panic grass

commonsianum, for Albert Commons (1829–1919), American botanist, its discoverer

Leaves and branching

Locally common in south-central Wisconsin, where it is characteristic of bare sandy blowouts (Shinners 1944). Fassett (1951) treats this as *Panicum commonsianum* var. *euchlamydeum*; Barkworth et al. (2007) as *Dichanthelium ovale* subsp. *pseudopubescens*.

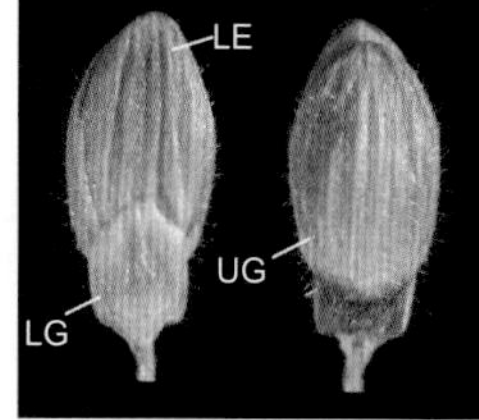

Spikelet

Spikelet illustration

Habit

Habit

Shinners' panic grass in sand barrens habitat

Dichanthelium depauperatum (Muhl.) Gould, poverty panic grass, starved panic grass
depauperatum, "starved, dwarf, depauperate"

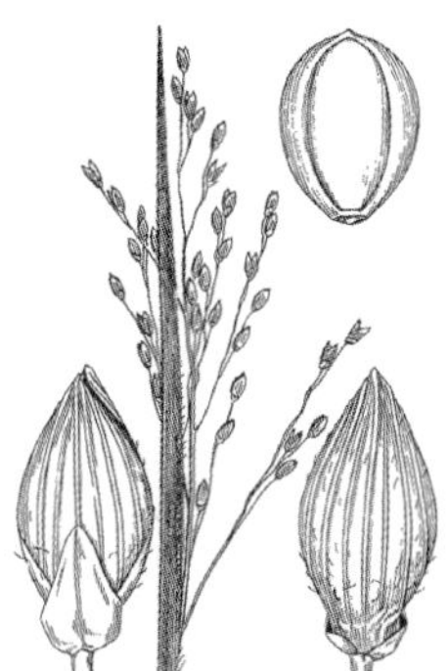
Illustration

Common statewide in areas with dry oak or pine forests, savannas, barrens, and woodlands. Hybrids with _Dichanthelium columbianum_ and with _D. perlongum_ have been recorded in Wisconsin. Fassett (1951) treats this as _Panicum depauperatum_.

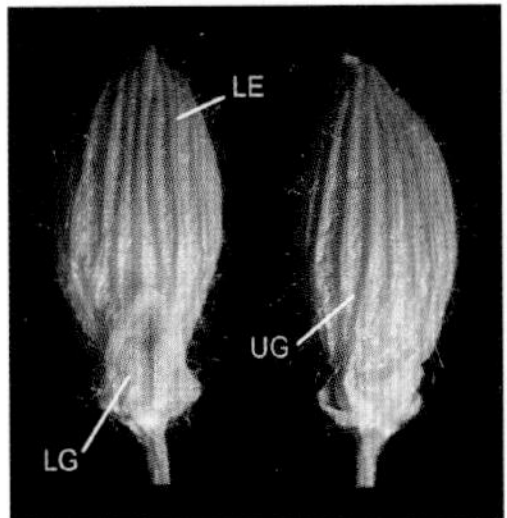

Spikelet

Habit

Dichanthelium dichotomum **(L.) Gould, forked panic grass**

dichotomum, "forked in pairs"

A rare, rather mysterious species, collected by John T. Curtis at Fox Point on Lake Mendota in Madison in 1941, where perhaps not native. It is the only species in the genus with completely smooth (glabrous) spikelets. Fassett (1951) treats this as *Panicum dichotomum*.

Illustration

Portion of inflorescence

Habit

Dichanthelium latifolium **(L.) Harvill, broad-leaved panic grass**

latifolium, Latin, "wide to broad leaves"

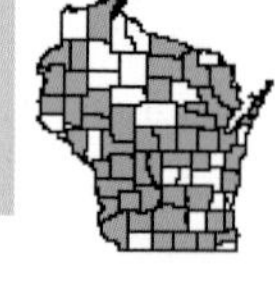

A common forest edge and understory grass of dry to mesic woods; much commoner in southern Wisconsin; easily recognized by its broad, bamboo-like leaf blades. Fassett (1951) treats this as *Panicum latifolium*.

Illustration

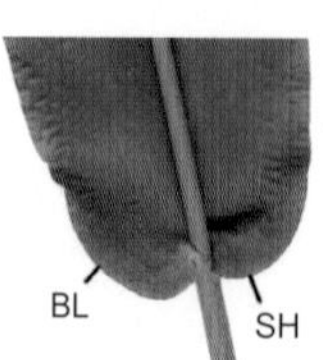

Cordate base of leaf blade

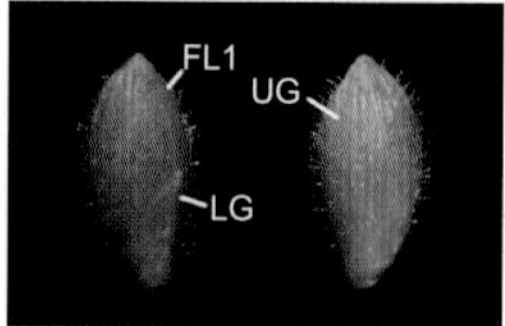

Spikelet

Habit

Dichanthelium leibergii (Vasey) Freckmann, Leiberg's panic grass, prairie panic grass

leibergii, for John Bernhard Leiberg (1853–1913), Swedish botanist, its discoverer

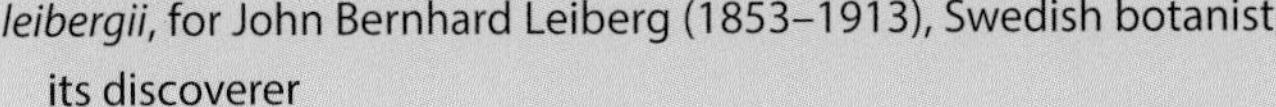

Illustration

Occasional in dry-mesic to wet-mesic prairies, sandy or gravelly hillsides, and railroad rights-of-way in southern Wisconsin. Fassett (1951) treats this as *Panicum leibergii.*

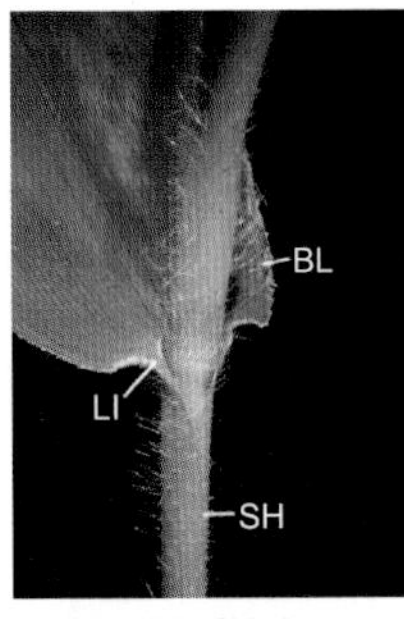

Base of leaf blade

Inflorescence emerging from uppermost sheath

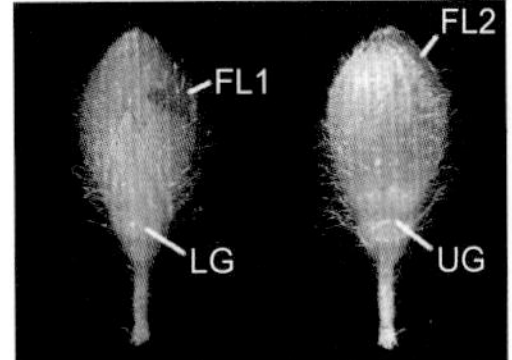

Spikelet

Habit

Dichanthelium linearifolium **(Scribn.) Gould, linear-leaved panic grass, slender-leaved panic grass**

linearifolium, "linear, parallel-sided leaves"

Habit

In the same habitats as *Dichanthelium depauperatum* (dry oak or pine forests, savannas, barrens, and woodlands), but less common than that species. Fassett (1951) treats this as *Panicum linearifolium*.

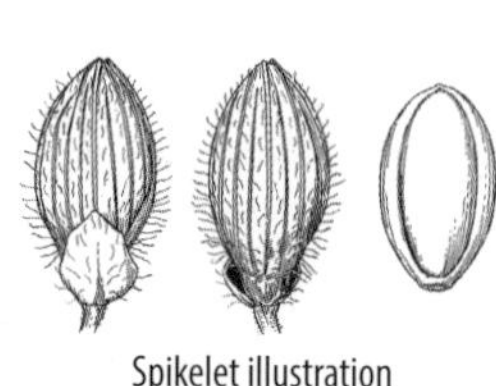

Spikelet illustration

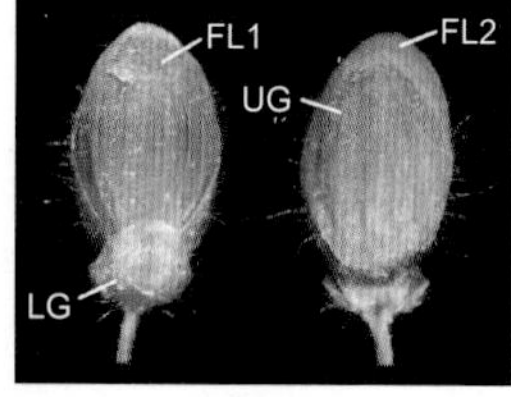

Spikelet

Habit

Dichanthelium meridionale (Ashe) Freckmann, slender panic grass

meridionale, "of noonday; blooming at noontime"

Habit

Fairly common statewide in dry open weedy and native habitats. Fassett (1951) treats this as *Panicum meridionale*, Barkworth et al. (2007) as *Dichanthelium acuminatum* subsp. *implicatum*.

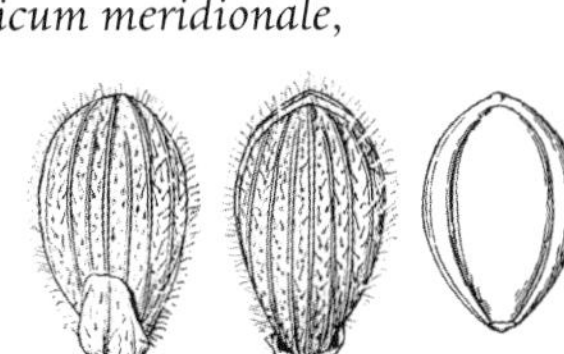

Spikelet illustration

Base of leaf blade

Spikelet

Habit

Dichanthelium oligosanthes (Schult.) Gould var. _scribnerianum_ (Nash) Gould, red dot panic grass, Scribner's panic grass

oligosanthes, *oligos*, "few," and *anthes*, "flower"

Illustration

Occasional in southern Wisconsin, where it is most typical of dry to mesic prairies and limey and sandy oak woodlands. Hybrids with *Dichanthelium acuminatum* and *D. villosissimum* var. *praecocius* have been recorded in Wisconsin. Fassett (1951) treats this as *Panicum oligosanthes* var. *scribnerianum*.

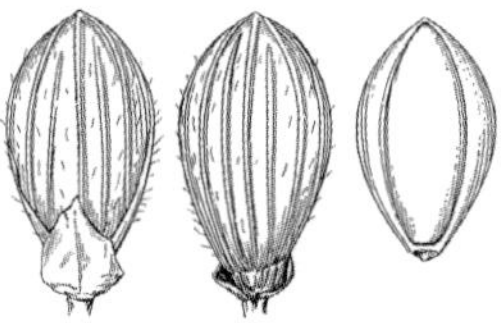

Spikelet illustration

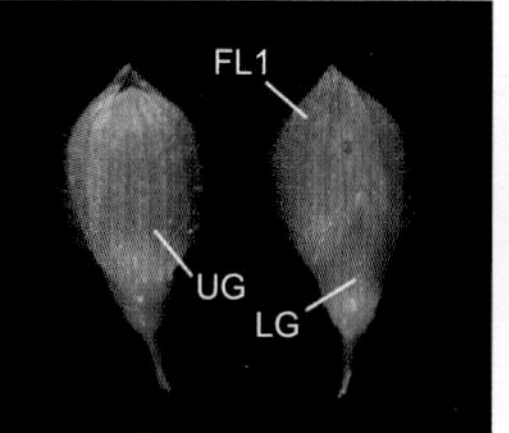

Spikelet

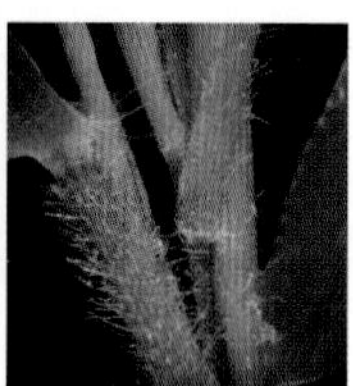

Base of leaf blade

Inflorescence emerging from uppermost sheath

Habit

Dichanthelium perlongum (Nash) Freckmann, long-stalked panic grass

perlongum, *per*, "through, extra, very," and *longus*, "long, extended"

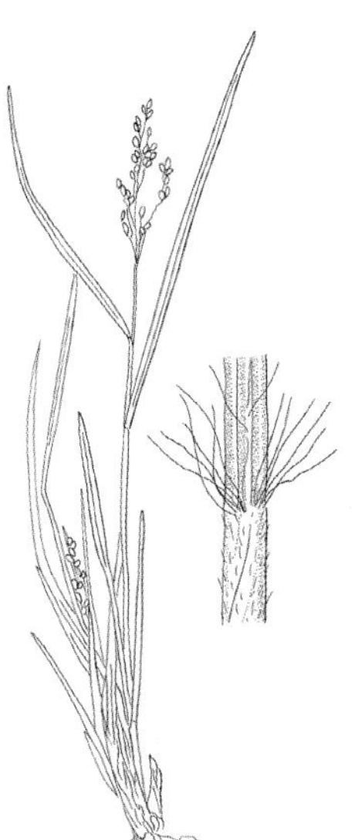

Illustration

Occasional, dry prairies in southern Wisconsin. Hybrids with *Dichanthelium columbianum* have been recorded in Wisconsin. Fassett (1951) treats this as *Panicum perlongum*.

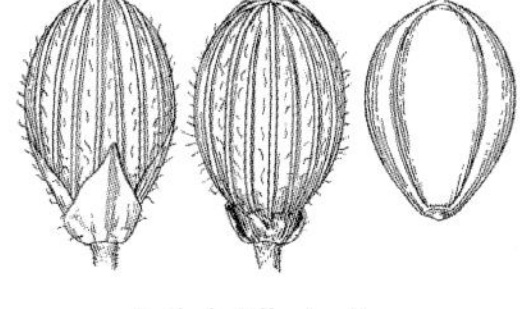

Spikelet illustration

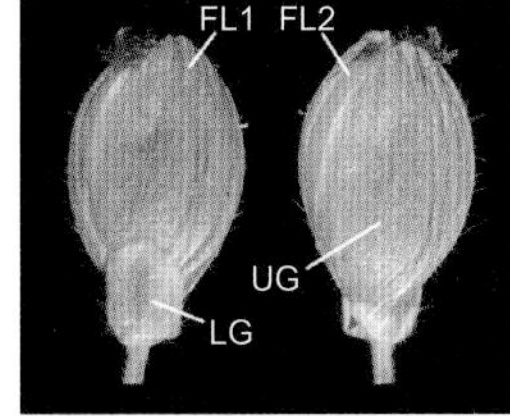

Spikelet

Inflorescence

Habit

Dichanthelium villosissimum (Nash) Freckmann var. _praecocius_ (Hitchc. and Chase) Freckmann, prairie panic grass

villosissimum, "hairiest"

Habit

Oak barrens and dry to mesic prairies in the southern two-thirds of the state; also old fields and roadsides. Hybrids with *Dichanthelium oligosanthes* var. *scribnerianum* have been recorded in Wisconsin. Fassett (1951) treats this as *Panicum praecocius*, Barkworth et al. (2007) as *Dichanthelium ovale* subsp. *praecocius*.

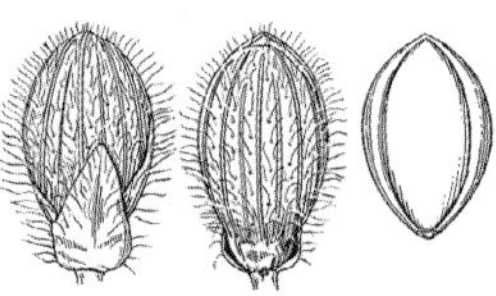

Spikelet illustration

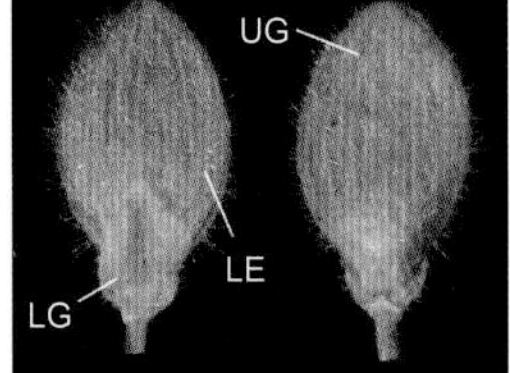

Spikelet

Dichanthelium wilcoxianum (Vasey) Freckmann, Wilcox's panic grass

wilcoxianum, for Gen. Timothy E. Wilcox, US Army, avid student of plants, ca. 1900

SPECIAL CONCERN

Illustration

Rare in sandy or gravelly dry hillside prairies in far southern Wisconsin; often associated with dolomite. Fassett (1951) and the Wisconsin Department of Natural Resources (2014) treat this as *Panicum wilcoxianum*.

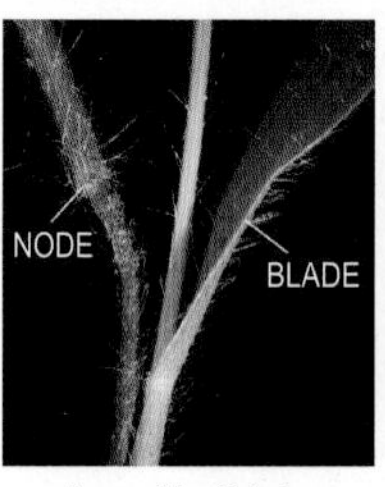

Base of leaf blade

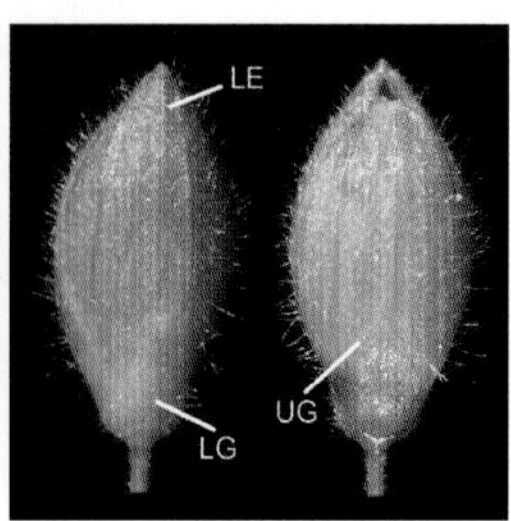

Spikelet

Wilcox's panic grass, habit

Wilcox's panic grass, habit

Wilcox's panic grass, habitat

Dichanthelium xanthophysum (A. Gray) Freckmann, pale panic grass, slender rosette grass
xanthophysum, xanthos, "yellow"

Dry acid oak or pine forests, barrens, and woodlands; locally common in northern Wisconsin, absent from the far south and the Lake Michigan counties. Hybrids with *Dichanthelium acuminatum* have been recorded in Wisconsin. Fassett (1951) treats this as *Panicum xanthophysum*.

Inflorescence

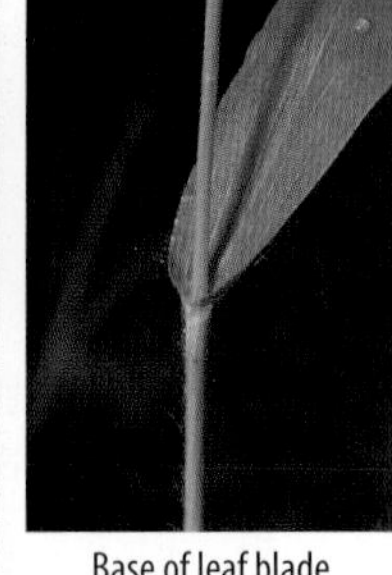

Base of leaf blade

Portion of inflorescence

Habit

34. *DIGITARIA*, CRABGRASS

Latin, *digitus*, "finger"

Usually annuals; inflorescence either an open panicle (*Digitaria cognata*) or more usually a series of digitately arranged one-sided spikes (racemes); spikelets pointed, the first glume minute or absent, the upper floret leathery. In addition to the following species, slender crabgrass (*D. filiformis* (L.) Koeler) might occur in southern Wisconsin. It is a southern US species with erect culms, digitate (finger-like) racemes, and a rachis less than 0.5 mm wide, and it has been found recently in the Chicago region. Two hundred tropical to warm temperate species; cosmopolitan.

1. Inflorescence an open panicle; native, sandy open ground .... *D. cognata*
1. Inflorescence a collection of three to five digitate, densely flowered spike-like racemes; exotic and weedy
   2. Spikelets 1.7–2.3 mm long, the second glume about as long as the spikelet; leaves more or less glabrous; common statewide .... *D. ischaemum*
   2. Spikelets 2.5–3.5 mm long, the second glume about half as long as the spikelet; leaves more or less hairy near the base; much commoner in southern Wisconsin .... *D. sanguinalis*

Digitaria cognata (Schult.) Pilg., fall witch grass

cognata, "closely related to"

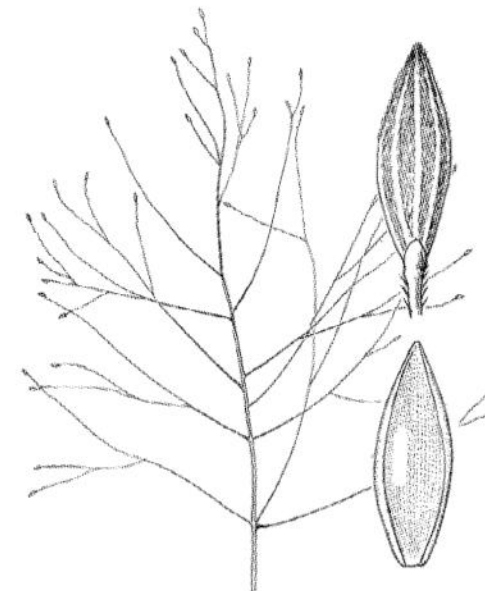

Illustration

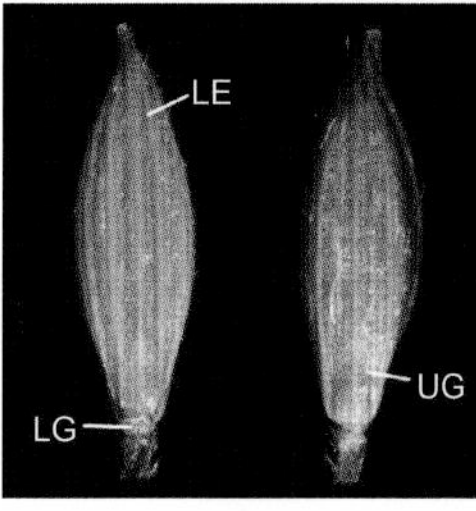

Spikelet

Much like poverty oat grass (*Danthonia spicata*), this weedy native has become locally abundant in dry, open, lightly vegetated sandy areas statewide; in late summer and early fall the purple inflorescences lend a "fog" to sandy lawns. There are only three collections before 1916, from sandstone outcrops and sand terraces along major rivers, but by 1944 it had become common in most of the southern half of the state (Shinners 1944). Like purple love grass (*Eragrostis spectabilis*) and witch grass (*Panicum capillare*), it is a purple-flowered tumbleweed. Fassett (1951) treats this as *Leptoloma cognata*.

Habit

Fall witch grass habitat, Nauke Lane, Oconto County

Digitaria ischaemum (Schreb.) Muhl., smooth crabgrass
ischaemum, ancient name, presumably from Greek, _ischaemos_, "styptic, blood restraining," from its supposed styptic properties

A Eurasian exotic, first collected in 1893.
One of the commonest sidewalk-crack, dooryard, driveway, and garden weeds, found statewide as a sprawling annual "plastered" close to the pavement.

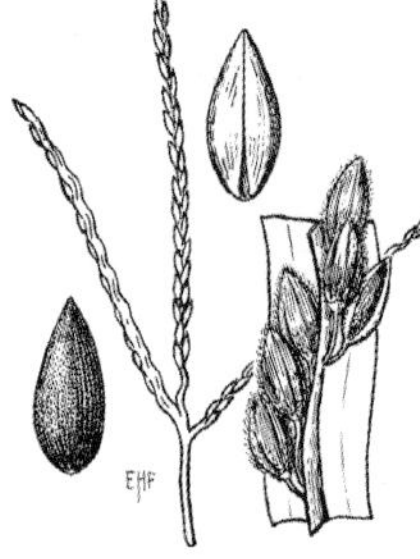

Illustration

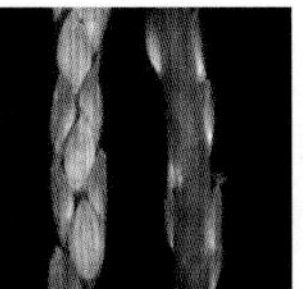

Portion of inflorescence; right view shows the rachis

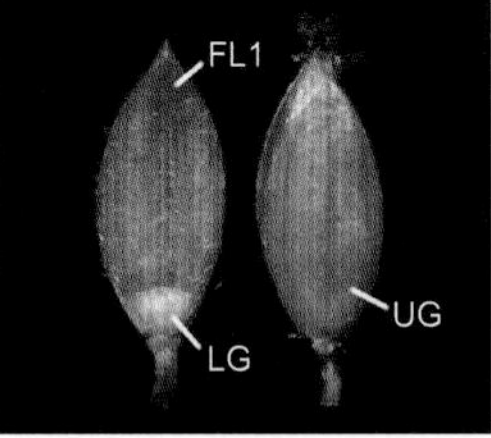

Spikelet

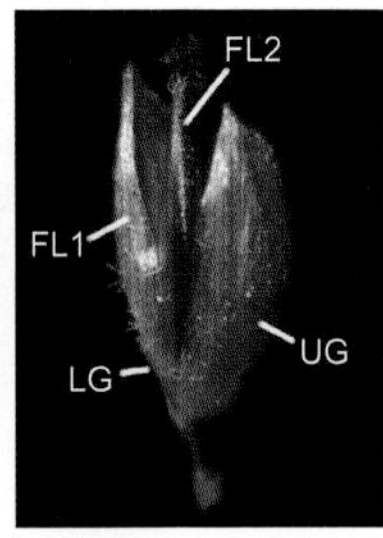

Spikelet, lateral view

Inflorescence

Habitat and habit

Digitaria sanguinalis (L.) Scop., hairy crabgrass, northern crabgrass

sanguinalis, Latin, "blood red"

Illustration

A Eurasian exotic, present early (Lapham 1853). Habitat and abundance as in the previous species, but absent as yet from far northern Wisconsin.

Spikelet

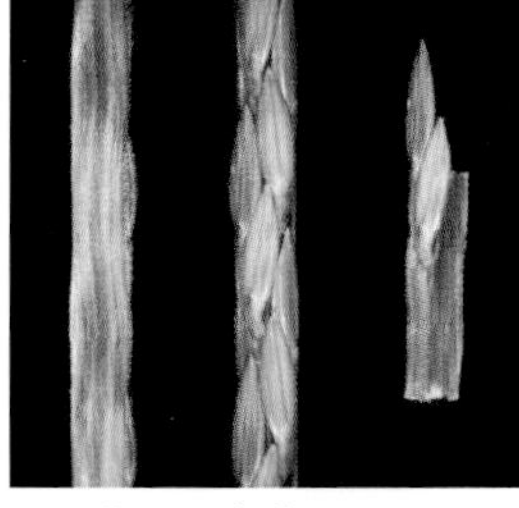

Portion of inflorescence, showing ventral and dorsal views and paired spikelets

Inflorescence from leaf sheath

Habitat and "crab-like" habit

35. *DIPLACHNE*, SPRANGLE-TOP

Greek, *diploos*, "double," and *achne*, "awn"

Sprawling annuals or perennials; panicles with spikelets appressed to lower side of the main branches, many-flowered, the lemmas short-awned from between two small teeth. Several dozen species; pantropical and warm temperate regions.

Diplachne fusca (L.) P. Beauv. ex Roem. and Schult. subsp. *fascicularis* (Lam.) P. M. Peterson and N. Snow, bearded sprangle-top

fusca, "dark or brown"

Illustration

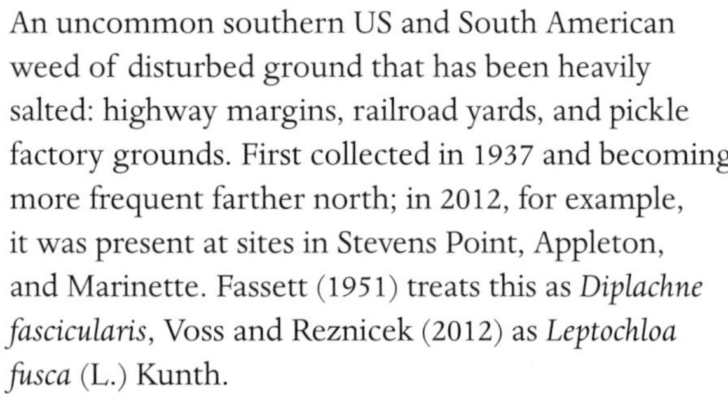

An uncommon southern US and South American weed of disturbed ground that has been heavily salted: highway margins, railroad yards, and pickle factory grounds. First collected in 1937 and becoming more frequent farther north; in 2012, for example, it was present at sites in Stevens Point, Appleton, and Marinette. Fassett (1951) treats this as *Diplachne fascicularis*, Voss and Reznicek (2012) as *Leptochloa fusca* (L.) Kunth.

Habit

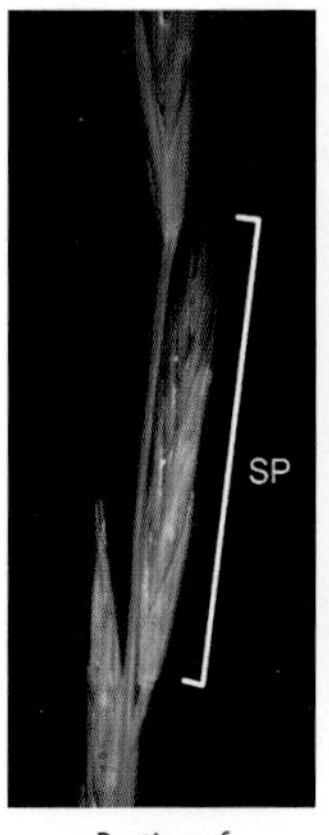

Portion of inflorescence

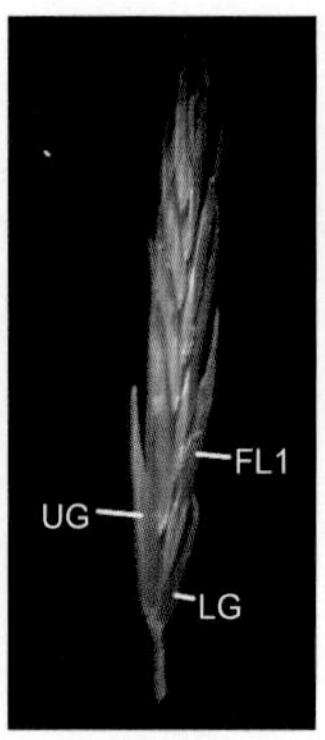

Spikelet

36. *DISTICHLIS*, SALT GRASS

Greek, *distichos*, "two-ranked," in reference to the arrangement of the leaves

Creeping perennials; foliage stiff, with numerous strongly two-ranked blades; panicles small, composed of rather large, awnless spikelets of two types, the male spikelets somewhat larger and with more numerous florets than the female spikelets, which are produced on separate plants. Five species of the Americas and Australia.

Distichlis spicata (L.) Greene, salt grass

spicata, "spiked"

A rare southern US and South American weed of saline, disturbed ground, collected in the state from 1956 to 1972, but not since.

Illustration

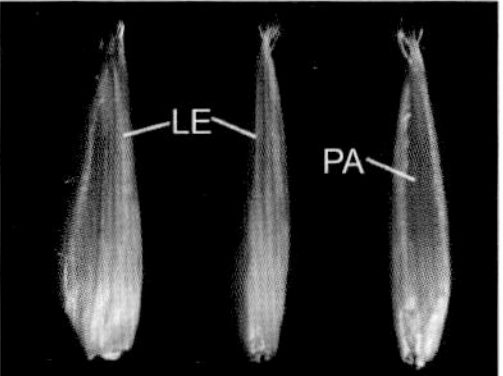

Florets

Inflorescence

Habit

37. *ECHINOCHLOA*, BARNYARD GRASS

Greek, *echinos*, "hedgehog," and *chloa*, "grass," referring to its bristling awns

Small to robust grasses of disturbed areas and wetland margins; inflorescence a panicle with spikelets tending to overlap along a few main branches, the spikelets rough-hairy and with the upper glume and lower (sterile) lemma short- to long-awned. Forty to fifty species; cosmopolitan.

1. Spikelets with both the upper glumes and lower lemma with awns 2–4 cm long; leaf sheaths rough-hairy; native, river floodplains and Green Bay shoreline .... *E. walteri*
1. Spikelets with upper glume awnless, the lower lemma with an awn less than 1 (rarely 2.5) cm long; leaf sheaths smooth; exotic and native but weedy
   2. Fertile lemma with a firm tip not sharply demarcated from the body; upper glume and sterile lemma with bulbous-based bristles .... *E. muricata*
   2. Fertile lemma with a withering green tip sharply demarcated from the body by a row of minute hairs; upper glume and sterile lemma lacking bulbous-based bristles
      3. Branches of inflorescence clearly visible throughout, not obscured by spikelets; abundant weed of fields, gardens, and roadsides .... *E. crus-galli*
      3. Branches of inflorescence obscured by densely congested spikelets; rare adventive .... *E. esculenta*

Echinochloa crus-galli (L.) P. Beauv., Eurasian barnyard grass

crus-galli, "cock's spur"

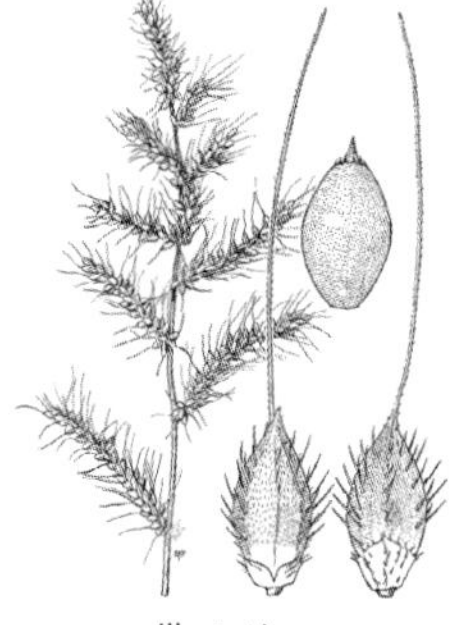

Illustration

A Eurasian exotic, first collected in 1887.
A common and familiar weed of fields, gardens, and roadsides, but also occurring in native plant communities.

Portion of inflorescence

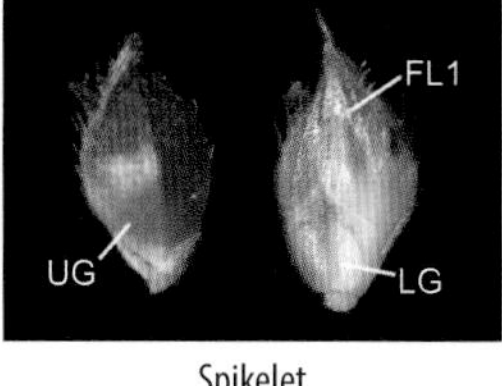

Spikelet

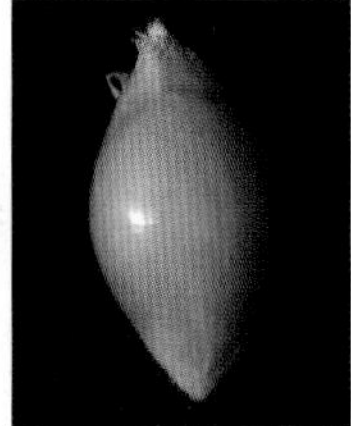

Upper floret, with minute hairs at summit

Habit

Echinochloa esculenta (A. Braun) H. Scholtz, Japanese millet

esculenta, "full of food; edible, esculent"

A rare Asian exotic, first collected in 1932.

Echinochloa muricata (P. Beauv.) Fernald, American barnyard grass

muricata, "muricate," roughened with hard points

Illustration

A common native species that is very similar to the exotic *Echinochloa crus-galli*, but often preferring wetter shoreline habitats. Fassett (1951) treats this as *E. pungens*.

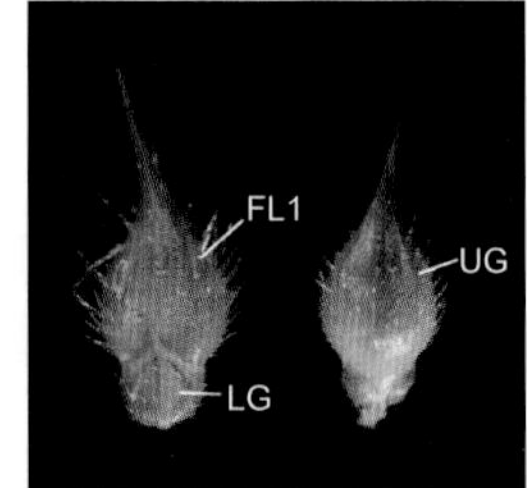

Spikelet

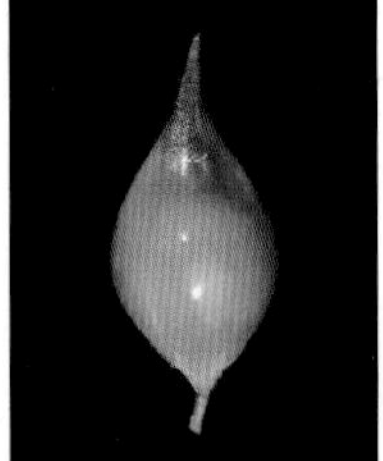

Upper floret, without minute hairs at summit

Portion of inflorescence

Habit

Echinochloa walteri (Pursh) A. Heller, bottlebrush barnyard grass

walteri, for Thomas Walter (1740–1789), British-born American botanist

The most robust, prickly looking (because of the long spikelet awns) member of the genus in Wisconsin; it could be mistaken at first glance for a species of ryegrass (*Elymus*), but differs in its branched panicle; locally common along the margins of major rivers (especially the Fox River system) and the swampy western shore of Green Bay.

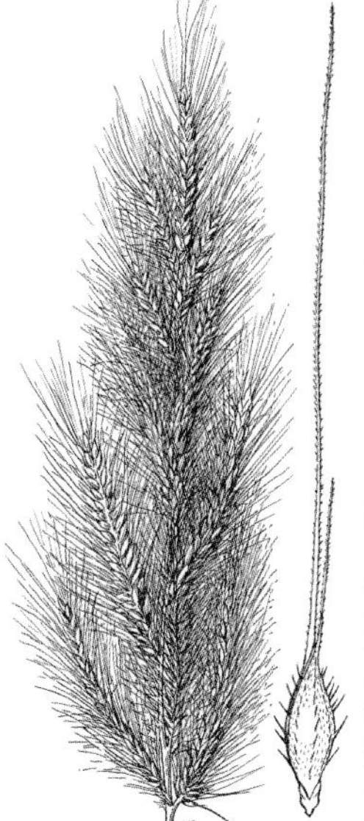

Illustration

Inflorescence

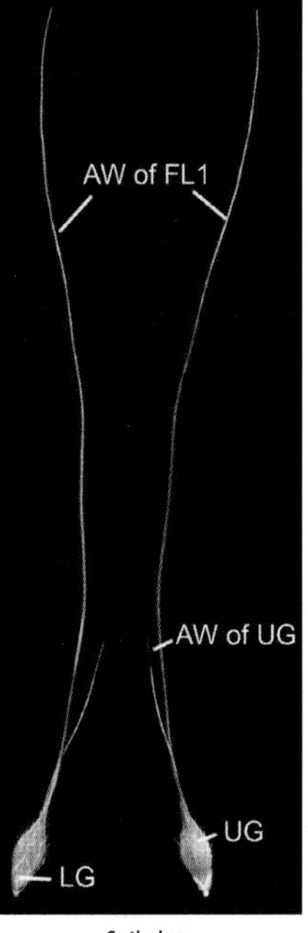

Spikelet

Habit

38. *ELEUSINE*, GOOSE GRASS

from the town where Ceres, goddess of harvests, was worshiped

A sprawling weed with several digitate (finger-like) slender spikes; spikelets several-flowered. Eight species of Africa and South America.

Eleusine indica (L.) Gaertn., goose grass

indica, "of India"

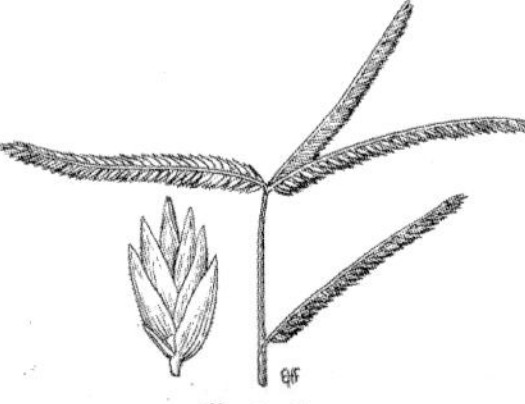

Illustration

An uncommon, sprawling, mostly urban Old World weed of compacted ground in southern Wisconsin, first collected in 1921; the leaf sheaths are broadly laterally compressed. Goose grass may be mistaken at first glance for crabgrass (*Digitaria*) or Bermuda grass (*Cynodon*), but those genera have single-flowered spikelets.

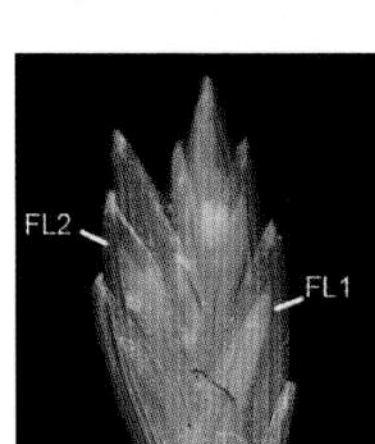

Spikelet

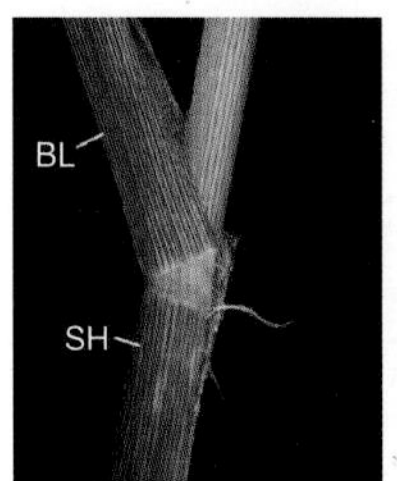

Leaf ligular area

Portion of inflorescence, dorsal and ventral views

Habit

39. *ELYMUS*, WILD RYE

Greek word for a type of grain, meaning generally "to roll up," referring to the lemma and palea, which are tightly rolled about the seed

Perennials of woodlands, streambanks; the inflorescence is a dense to open terminal spike with one to three spikelets per node; if there are three per node, then each lateral one is often reduced to a pair of awn-like glumes and an abortive floret. The solitary (or central) spikelet at each node is several-flowered and often has awned glumes, too. Specimens of *Elymus* are often difficult to key to species because of hybridization and polyploidy. The commonest species encountered in Wisconsin will be Canada wild rye (*E. canadensis*), the largest species and with robust, long-awned, often arching panicles; bottlebrush grass (*E. hystrix*), also long-awned, but with more erect, sparsely flowered panicles; Virginia wild rye (*E. virginicus*), with dense, short-awned panicles; and the weedy quack grass (*E. repens*), with narrow, sparsely flowered panicles with awns absent or at least not prominent. Several species have glaucous (waxy bluish white) foliage. 150 species of northern temperate areas.

Tall wheatgrass (*Thinopyrum ponticum* (Podp.) Z. W. Liu and R. R.-C. Wang) is a robust Mediterranean weed that has been collected several times recently in the Chicago area (Swink and Wilhelm 1994; vPlants 2014) and may eventually turn up in Wisconsin. It would key to *Elymus trachycaulus* below and resembles a quack grass on steroids, growing up to 2 m tall with spikes 20–40 cm long. The spikelets are distinctly curved and have truncate glumes, and the leaf sheath apices have auricles.

1. Spikelets solitary at each node of the rachis; glumes two
   2. Lemmas densely hairy; leaves inrolled, strongly blue-glaucous; rare, southern and eastern Wisconsin, including Lake Michigan dunes .... *E. lanceolatus*
   2. Lemmas glabrous; leaves flat, green- to sometimes green-blue-glaucous; common species statewide
      3. Anthers 3–6 mm long; spikelets with florets not easily disarticulating, even when dry; plants strongly rhizomatous; abundant weed .... *E. repens*
      3. Anthers 1–2.5 mm long; spikelets with florets easily disarticulating when mature; plants clump-forming, not strongly rhizomatous; common native .... *E. trachycaulus*
1. Spikelets two to three at each node of the rachis, including sterile spikelets, so that the number of glumes is four to six
   4. Spikelets horizontally spreading at maturity, giving the inflorescence the appearance of a bottlebrush, the rachis plainly visible between the nodes; glumes absent or awn-like; common, forests .... *E. hystrix*
   4. Spikelets ascending to erect at maturity, the rachis often not very evident between the nodes; glumes awn-like to lanceolate

5. Larger paleas 8–13 mm long; lemma awns spreading to recurved at maturity
   6. Leaves four to nine per culm, the blades 8–15 mm wide, glabrous above; abundant weedy native . . . . . . *E. canadensis*
   6. Leaves ten to twelve per culm, the blades 15–20 mm wide, hairy above; uncommon native . . . . . . *E. wiegandii*
5. Larger paleas 5–9 mm long; lemma awns more or less straight
   7. Glumes less than 1 mm wide, not widened above the base
      8. Palea (of lower floret) 7–9 mm long; leaves ca. ten, glabrous; occasional, mostly southern Wisconsin . . . . . . *E. riparius*
      8. Palea (of lower floret) 5–7 mm long; leaves five to seven, with hairy sheaths and blade undersurfaces; occasional except in northern Wisconsin . . . . . . *E. villosus*
   7. Glumes 1–2 mm wide, evidently widened and flattened above the base
      9. Spikelets spreading, the inflorescence 2.5–5 cm wide; glume awns 15–25 mm long; rare, southern Wisconsin . . . . . . *E. macgregorii*
      9. Spikelets ascending, the inflorescence 1–2.5 cm wide; glume awns usually 10 mm, up to 15 mm, long
         10. Glume awns absent or up to 3 mm long; uncommon . . . . . . *E. curvatus*
         10. Glume awns 3–10 (rarely 15) mm long; common . . . . . . *E. virginicus*

Elymus canadensis **L., Canada wild rye**
canadensis, of or referring to Canada

Abundant in a variety of dry to wet open habitats; most characteristic of savannas, prairies, and dunes; often weedy along roadsides, and also planted on roadside embankments, where its short-lived perennial habit makes it suitable for erosion control (USDA NRCS 2014). Highly variable; in some populations the foliage can be strongly glaucous (bluish whitened). *Elymus* ×*maltei* Bowden (Malte's wild rye) is a hybrid with *E. virginicus*.

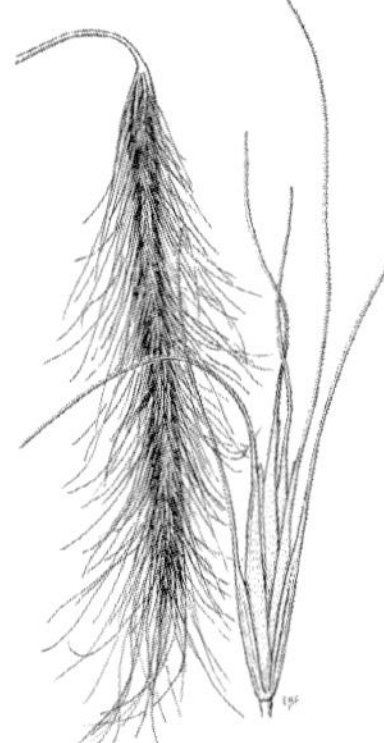
Illustration

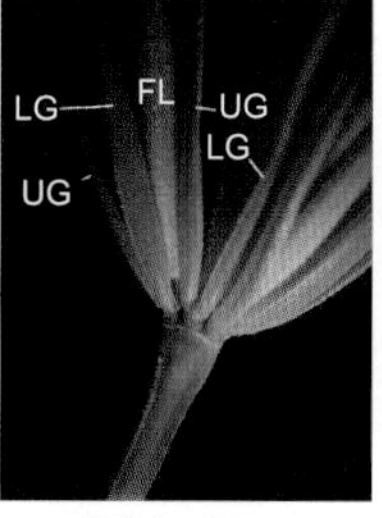

Spikelet cluster

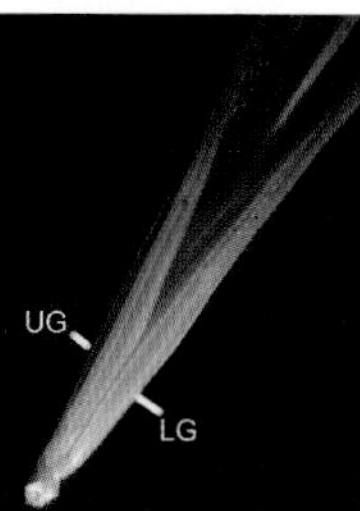

Glumes

Inflorescence at flowering time

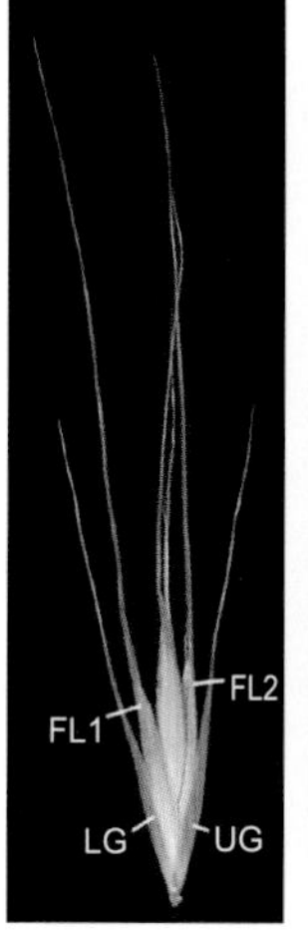

Spikelet

Emerging inflorescence (glaucous form)

Mature inflorescences

Canada wild rye, habit (glaucous form)

Elymus curvatus Piper, awnless wild rye

curvatus, "bent backward or curved"

Uncommon in a wide variety of habitats: streambanks, wetland edges, roadsides, old fields, and moist to dry forests. Fassett (1951) treats this as *Elymus virginicus* var. *submuticus*.

Inflorescence

Illustration

Portion of inflorescence

Elymus diversiglumis Scribn. and Ball, Minnesota wild rye

diversiglumis, "many-shaped hull or husk"

Rare, roadsides, shores, and power line and railroad rights-of-way. Fassett (1951) treats this as *Elymus interruptus.*

Habit

Illustration

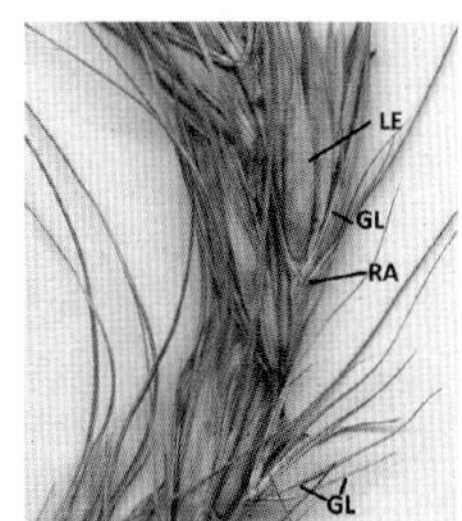

Portion of inflorescence

Elymus hystrix L., bottlebrush grass

hystrix, "bristly; porcupine-like"

Common, dry to wet woods, but most characteristic of moist to dry deciduous forests; also on roadsides and riverbanks. A distinctive and attractive grass with a perfect common name. *Elymus* × *ebingeri* G. C. Tucker, Ebinger's wild rye, is a hybrid with *E. virginicus*. Fassett (1951) treats this as *Hystrix patula*.

Illustration

Foliage

Inflorescence at flowering time

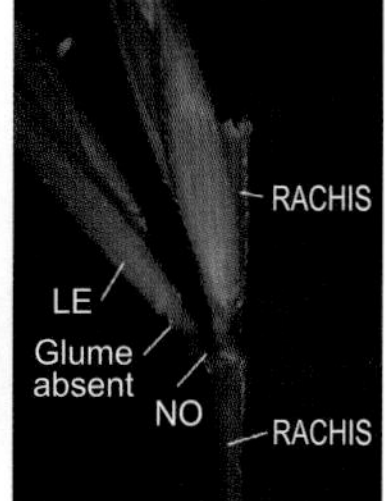

Spikelet cluster

Bottlebrush grass, habit

***Elymus lanceolatus* (Scribn. and J. G. Sm.) Gould, thick-spiked wheatgrass**

lanceolatus, "lance-shaped"

THREATENED (IN PART)

Illustration

Rare, sand dunes along Lake Michigan, where it has declined in abundance due to human disturbance. Our Lake Michigan shoreline plants, listed as threatened by the Wisconsin Department of Natural Resources, are subsp. *psammophilus* (J. M. Gillett and H. Senn) Á. Löve; inland populations are the typical subspecies and are not WDNR listed. Fassett (1951) treats this as *Agropyron dasystachyum*.

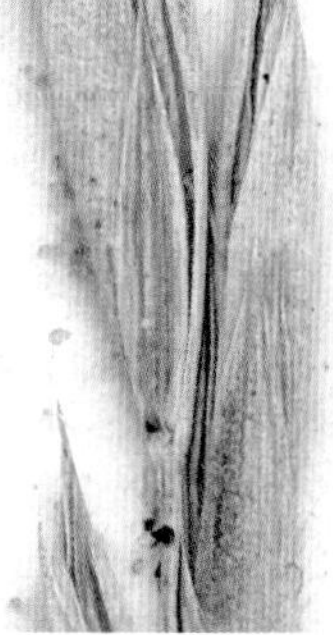

Portion of inflorescence

Thick-spiked wheatgrass,
habitat on Lake Michigan dunes

Elymus macgregorii R. E. Brooks and J. J. N. Campb., MacGregor's wild rye

macgregorii, in honor of the M(a)c "clan Gregor," represented by North American botanists Ronald Leighton McGregor (1919–2012) and John McGregor

Rare, wet prairies, floodplain forests, and old fields in southern Wisconsin.

Illustration

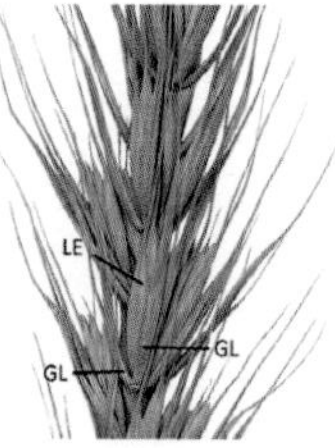

Portion of inflorescence

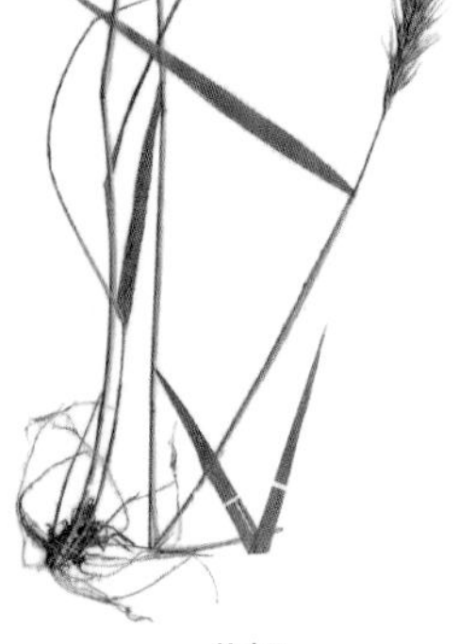

Habit

Elymus repens (L.) Gould, quackgrass

repens, "having creeping and rooting stems"

An abundant, strongly rhizomatous weed, familiar from gardens, lawns, old fields, and roadsides, but also well established in natural communities such as prairies and rock outcrops; known from 1853 (Lapham 1853). Fassett (1951) treats this as *Agropyron repens*.

Illustration

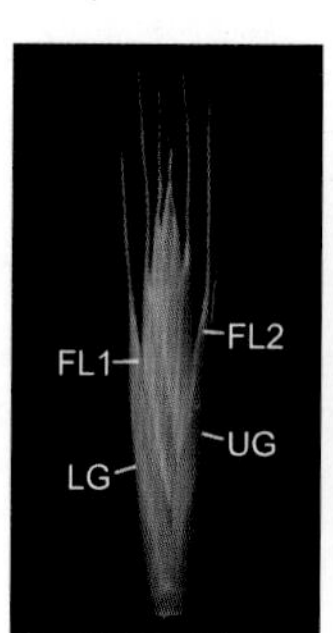

Spikelet

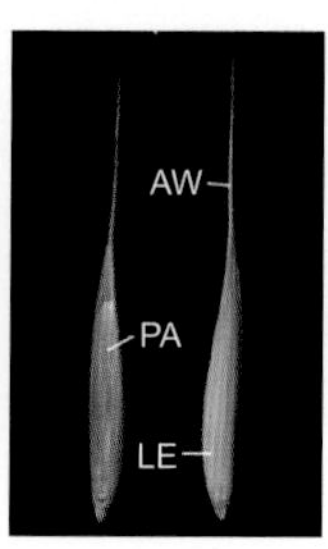

Florets

Inflorescence presenting broad side of spikelets

Inflorescence presenting narrow side of spikelets

Elymus riparius Wiegand, riverbank wild rye

riparius, Latin, generally meaning "growing near riverbanks or lakes"

Illustration

Uncommon in dry to more usually wet forests, streamsides, prairies, and roadsides; mostly in southwestern Wisconsin.

Inflorescences

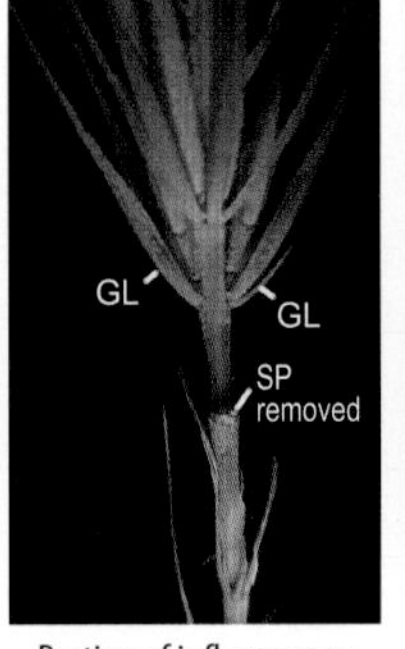

Portion of inflorescence

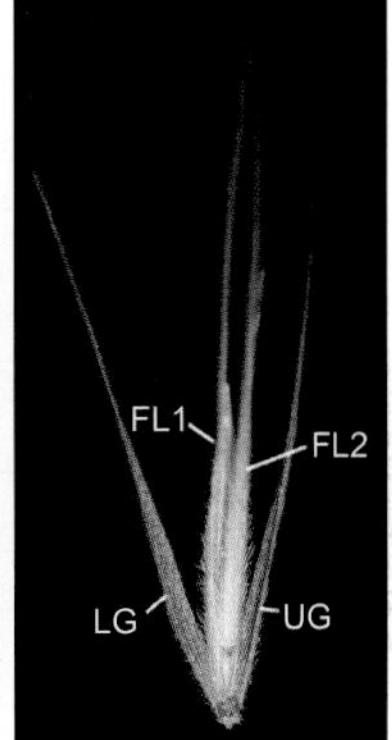

Spikelet

Elymus trachycaulus (Link) Gould ex Shinners, slender wheatgrass

trachycaulus, *trachelos*, "neck," and *caulos*, "stem of plant"

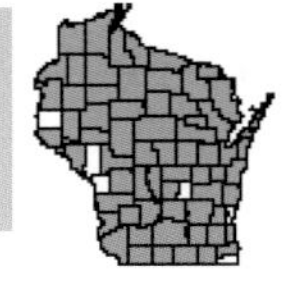

Illustration

A common, somewhat weedy native found in a variety of dry (less commonly wet) open or semiopen sites: bracken grasslands, sand barrens, dry prairies, and dunes. Sometimes used as a cover plant in erosion control (USDA NRCS 2014). Fassett (1951) treats this as *Agropyron trachycaulum*.

Inflorescence

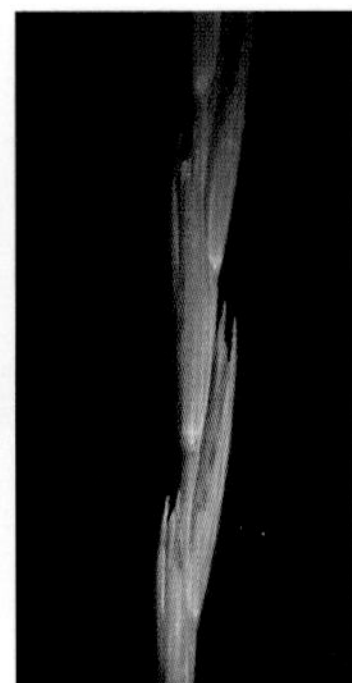

Portion of inflorescence

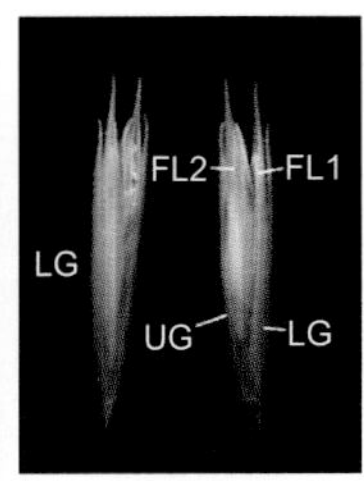

Spikelet

Slender wheatgrass, habit

Elymus villosus **Muhl. ex Willd., downy wild rye**

villosus, Latin, "with hairs"

Fairly common in southern Wisconsin in dry to wet forests and savannas; also on dry rock outcrops.

Illustration

Hairy upper surface of leaf blade

Portion of inflorescence

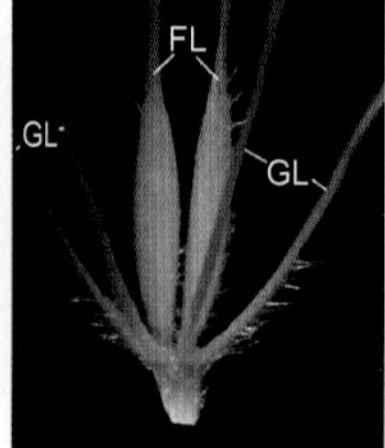

Spikelet

Inflorescence

Downy wild rye, habit

Elymus virginicus **L., Virginia wild rye**

virginicus, "of Virginia"

Illustration

Common in and characteristic of floodplain forests and the margins of open wetlands and shores. *Elymus* ×*ebingeri* G. C. Tucker (Ebinger's wild rye) is a hybrid with *E. hystrix*; *E.* ×*maltei* Bowden (Malte's wild rye) is a hybrid with *E. canadensis*.

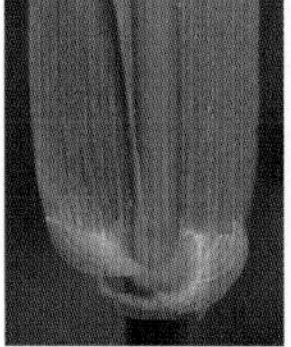

Leaf blade base with auricles

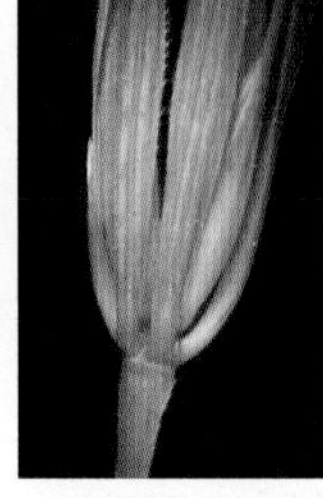

Portion of inflorescence

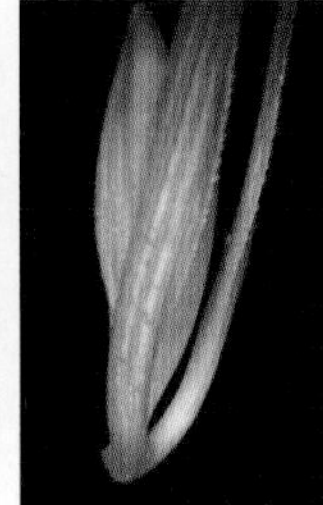

Portion of inflorescence

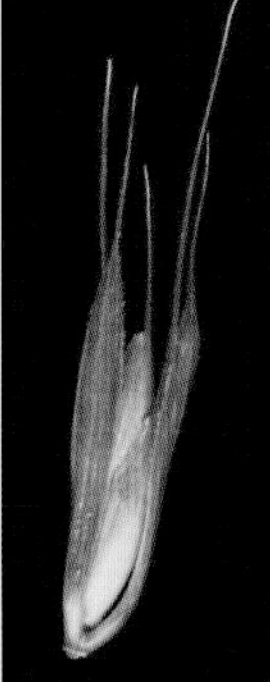

Spikelet

Portion of inflorescence

Habit

Elymus wiegandii **Fernald, Wiegand's wild rye**

wiegandii, for American botanist Karl McKay Wiegand (1873–1942)

Illustration

Uncommon, floodplain forests. Fassett (1951) treats this as part of *Elymus canadensis*.

Inflorescence

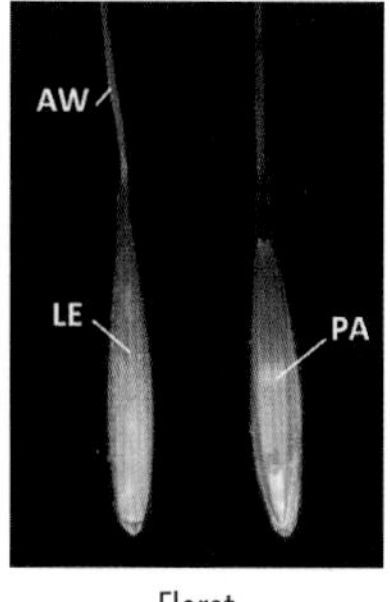

Floret

Spikelet

Habit

40. *ERAGROSTIS*, LOVE GRASS

Greek, *er*, "spring," and *agrotis*, "grass"

Summer- and fall-flowering grasses with hairy ligules and panicle inflorescences. The awnless spikelets are several- to many-flowered; the glumes are relatively short, while the lemmas have three prominent parallel nerves. A distinctive feature of love grass is that the ciliate-keeled paleas, as well as the rachilla that bears them, often persist for a while after the lemmas have fallen to reveal the grains. 350 species; cosmopolitan.

1. Creeping plants of wet muddy or sandy shores; culm nodes bearded; mostly southern Wisconsin .... *E. hypnoides*
1. Erect plants of terrestrial habitats; culm nodes bearded or glabrous
   2. Margins of culms and leaves (and often pedicels and spikelets) glandular warty; exotic
      3. Spikelets 2.5–4 mm wide; larger glume 1.8–2.5 mm long; common weed .... *E. cilianensis*
      3. Spikelets 1.5–2.4 mm wide; larger glume 1–2 mm long; uncommon weeds
         4. Larger glume 1.2–2.3 mm long; spikelets 4–11 mm long, five- to fifteen-flowered; grain ovoid, deeply grooved; rare .... *E. mexicana*
         4. Larger glume 1–1.8 mm long; spikelets 4–7 mm long, seven- to twelve-flowered; grain ellipsoid, not grooved; occasional .... *E. minor*
   2. Plants not glandular warty; mostly native species
      5. Spikelets red to purple; main panicle branches with a tuft of hairs at their bases where they join the inflorescence axis; perennials
         6. Plants 30–60 cm tall; larger glumes 1.5–2.5 mm long; lower branches of panicle reflexed at maturity; native, common in dry sandy ground, southern and western Wisconsin .... *E. spectabilis*
         6. Plants 50–120 cm tall; Larger glumes 2.5–5 mm long; lower branches of panicle ascending at maturity; rare adventive, southern Wisconsin .... *E. trichodes*
      5. Spikelets green to dark gray; main panicle branches lacking a tuft of hairs at their bases; annuals
         7. Well-developed spikelets five- to ten-flowered, rarely fifteen-flowered, on pedicels appressed to the panicle branches; well-developed lemmas 1.5–2 mm long, their lateral nerves distinct; common weed .... *E. pectinacea*
         7. Well-developed spikelets two- to four-flowered, rarely five-flowered, on pedicels that spread from the panicle branches; well-developed lemmas 1.2–1.5 mm long, their lateral nerves obscure
            8. Leaf sheaths long-hairy; uncommon .... *E. capillaris*
            8. Leaf sheaths smooth
               9. Axils of inflorescence branches smooth; tip of the upper glume opposite the tip of the lowest lemma; grain grooved; native, occasional, southern Wisconsin ..... *E. frankii*
               9. Axils of inflorescence branches long-hairy; the tip of the upper glume shorter than the tip of the lowest lemma; grain not grooved; rare adventive .... *E. pilosa*

Eragrostis capillaris **(L.) Nees, lace grass**
capillaris, "hair-like"

Illustration

An elegant southeastern US species, rarely found in Wisconsin, where it was first collected in 1900. There are evidently native populations such as those on granite and rhyolite outcrops in Marquette, Waupaca, and Green Lakes Counties, but more typically the species occurs in waste, often saline ground such as near factories or along railroad tracks (as in downtown Appleton in 2012).

Habitat and habit

Hairy surfaces of leaf sheaths

Inflorescence

Spikelet

Habitat and habit

Eragrostis cilianensis **(All.) Vignolo ex Janch., stink grass**

cilianensis, for Cigliano, Italy

A European exotic, first collected in 1874. A common garden, agricultural, and railroad and roadside weed of rich ground, more frequent in southern Wisconsin; when wet, it has an unpleasant, fetid odor.

Illustration

Inflorescence

Spikelet

Spikelet base

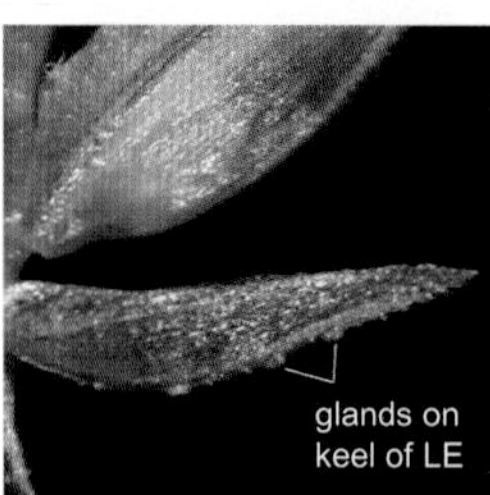

Glume, showing warty keel

Habit

Eragrostis frankii C. A. Mey. ex Steud., sandbar love grass
frankii, for German botanist Joseph C. Frank (1782–1835), its discoverer

Illustration

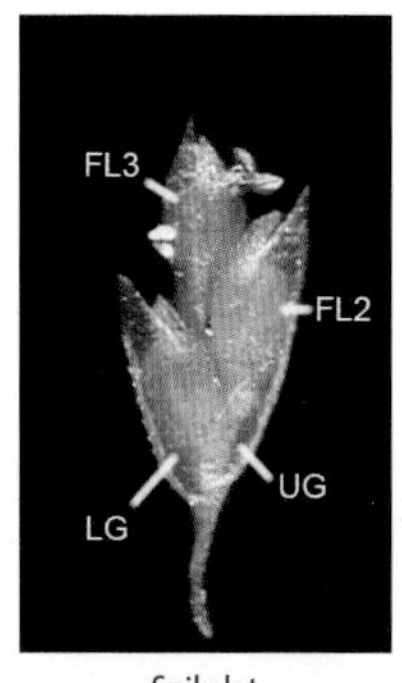

Spikelet

Muddy or, less commonly, sandy streamsides, lakeshores, and ditches; also weedy in gardens and on roadsides. An uncommon, purportedly native species of southern Wisconsin, collected now with much less frequency than in the early twentieth century.

Habit

Eragrostis hypnoides (Lam.) Britton, Sterns and Poggenb., creeping love grass

hypnoides, "moss-like"

Illustration

A small, creeping, mat-forming grass on the muddy or sandy shores of streams and lakes, mostly in southern Wisconsin. The closely related *Eragrostis reptans* (Michx.) Nees, with larger lemmas (2–4 mm long), unisexual spikelets, and larger anthers 1.5–2 mm long, occurs in an Illinois border county and could occur in Wisconsin in the same habitat as *E. hypnoides*.

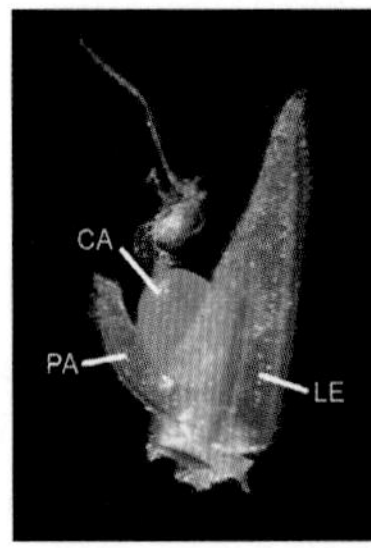

Floret

Inflorescence

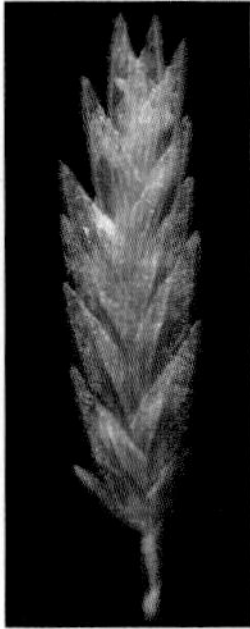

Spikelet

Eragrostis mexicana (Hornem.) Link, Mexican love grass

mexicana, "of Mexico"

A rare adventive of the southwestern United States and south, collected once in a railroad yard in Milwaukee by W. W. Oppel in 1937. Fassett (1951) treats this as *Eragrostis neomexicana*.

Inflorescence

Illustration

Eragrostis minor Host, little love grass

minor, "smaller, lesser"

Illustration

A European exotic, first collected in 1938 by Lloyd H. Shinners, who notes it as a common railroad weed in Milwaukee; an occasional weed, mostly along or near railroads. Fassett (1951) treats this as *Eragrostis poaeoides.*

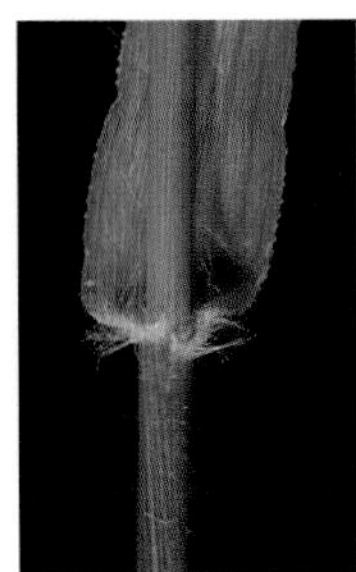

Leaf ligular area

Portion of inflorescence

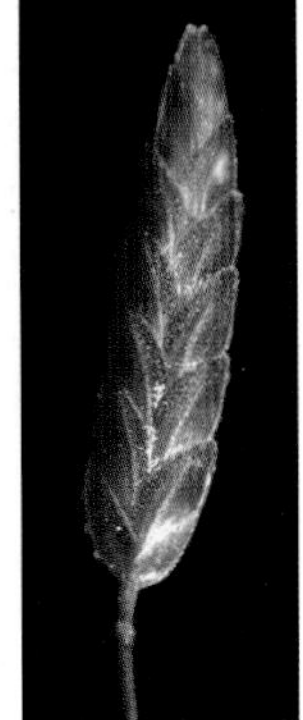

Spikelet

Eragrostis pectinacea (Michx.) Nees ex Steud., tufted love grass

pectinacea, Latin, "comb-like"

Illustration

Our most common species in the genus, abundant along roads and railroads, in sidewalk cracks, and in gardens, pastures, and old fields.

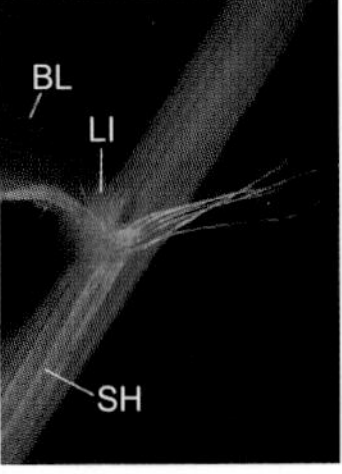

Leaf ligular area showing elongate cilia

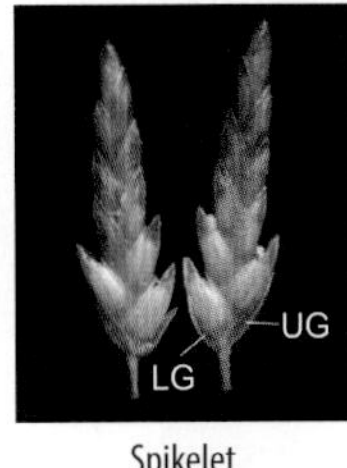

Spikelet

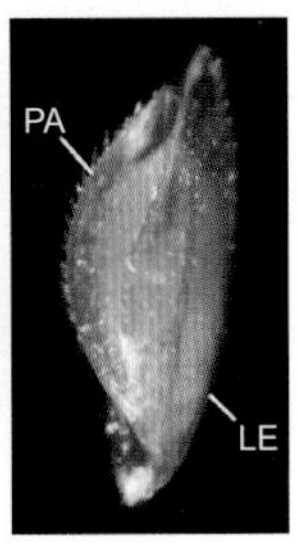

Floret

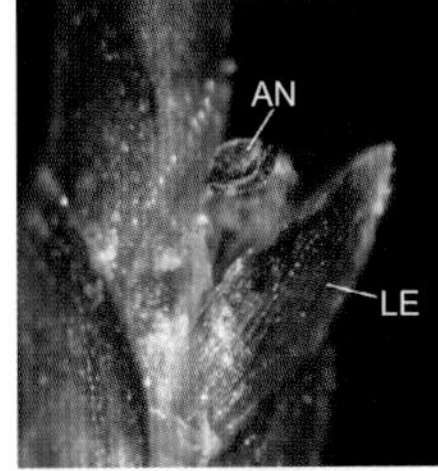

Floret

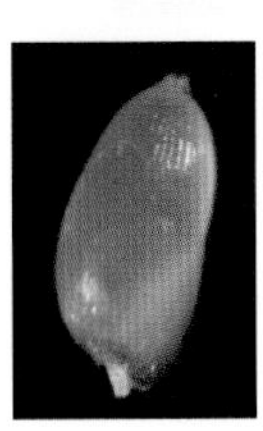

Grain

Eragrostis pilosa (L.) P. Beauv., Indian love grass

pilosa, "soft, hairy"

A rare, supposedly European exotic, collected from 1938 to 1975; Shinners (1940) notes it as a common street weed in Milwaukee. Fassett (1951) treats this as *Eragrostis multicaulis.*

Habit

Illustration

Spikelet

Eragrostis spectabilis (Pursh) Steud., purple love grass

spectabilis, "spectacular or showy"

Portion of inflorescence

Common in dry sandy old fields, barrens, and prairies. This is the largest of our three "purple tumbleweed grasses," which flower and fruit in the late summer and early fall; the others are fall witch grass (*Digitaria cognata*) and witch grass (*Panicum capillare*). It has prominent tufts of hairs at the base of the main panicle branches.

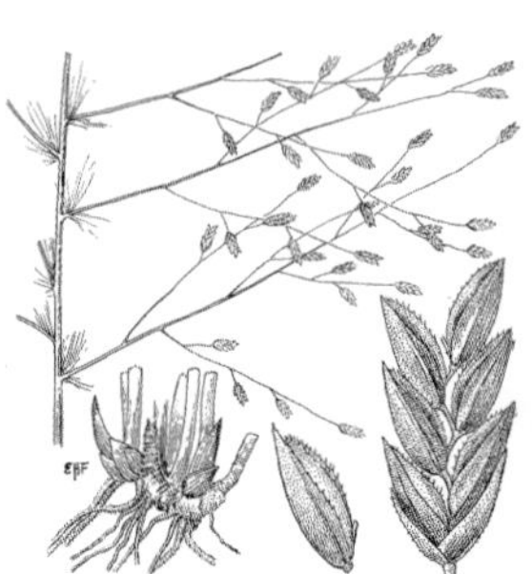
Illustration

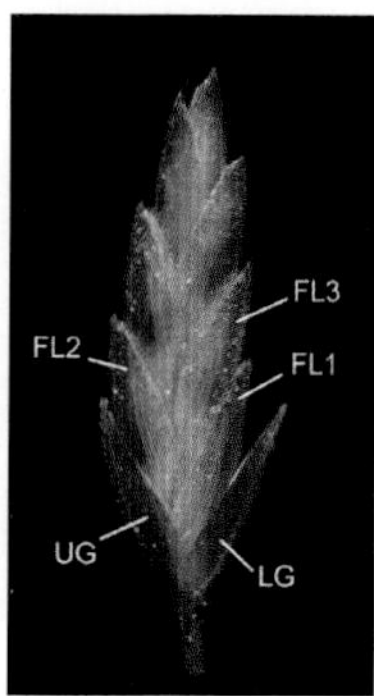

Spikelet

Eragrostis trichodes **(Nutt.) A. W. Wood, sand love grass**
trichodes, thrix, "hairy"

Illustration

Portion of inflorescence

Our most robust species of love grass, this native of the southern Great Plains was first collected in 1974 by Charlene Bennett. A yet-uncommon, recent introduction, used in at least one site by the Wisconsin Department of Transportation in prairie restoration efforts (T. S. Cochrane, 1990 collection note). It is sometimes used as a cover plant in erosion control on sandy sites, where it "greens up" early in the season and remains so well into the fall; very palatable to livestock, an alternate common name is "ice cream grass" (USDA NRCS 2014).

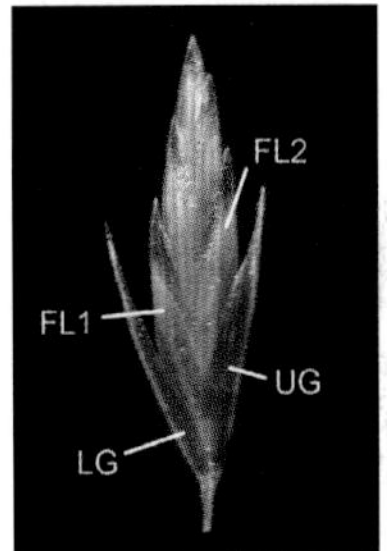

Spikelet

41. *ERIOCHLOA*, CUP GRASS
Greek, *erion,* "wool," and *chloa,* "grass"

A weedy annual grass whose spikelets resemble those of *Panicum*, *Dichanthelium*, or *Paspalum*, but with one distinguishing difference: each spikelet has a small swollen knob at the base of the fertile floret. Twenty to thirty species; cosmopolitan, tropical to warm temperate areas.

Eriochloa villosa **(Thunb.) Kunth, Chinese cup grass**

villosa, Latin, "with hairs"

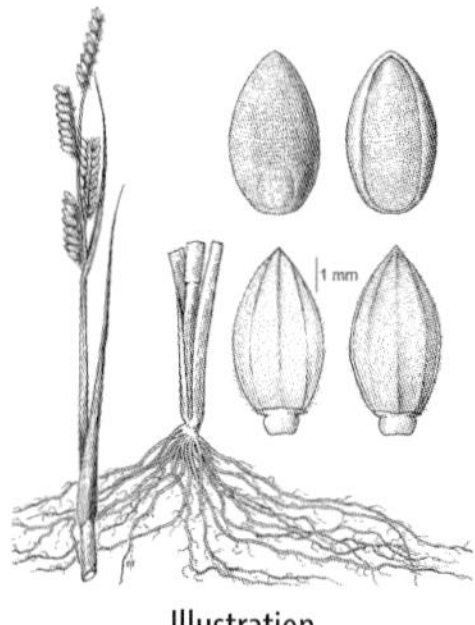

Illustration

A locally common Asian roadside and agricultural weed in southern Wisconsin, first collected in 1966 by R. E. Doersch. Prairie cup grass (*Eriochloa contracta* Hitchc.) is a Great Plains species with shorter pedicel hairs (1 mm vs. 2–3 mm long) and awned (vs. awnless) upper lemmas; it has been collected numerous times since 1990 along interstate highways in the Chicago area and will probably eventually be found in Wisconsin.

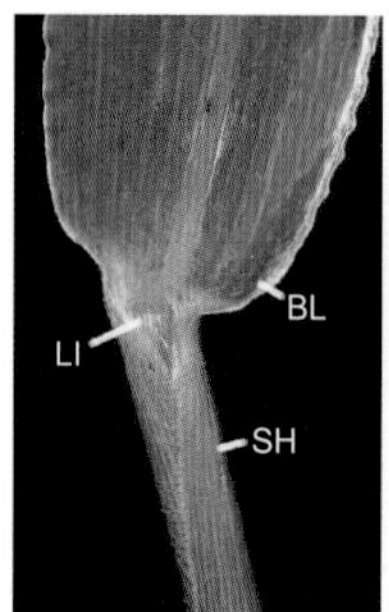

Leaf ligular area

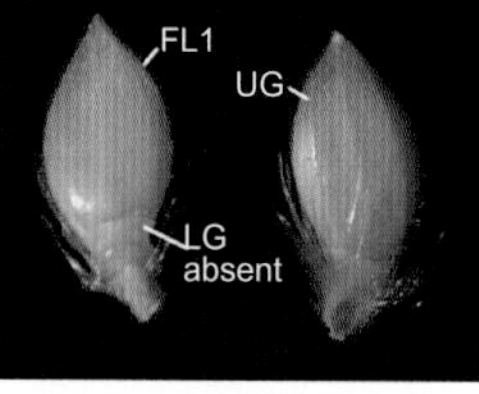

Spikelet, showing (left) cup-like pedicel summit and absence of lower glume

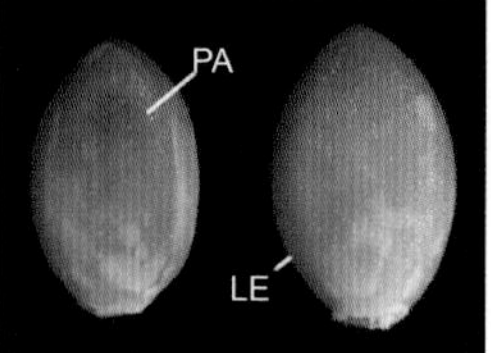

Florets

Portion of inflorescence, dorsal and ventral views

Inflorescence with four spikes

42. *FESTUCA*, FESCUE

ancient Latin word for some type of grass, obscure

Perennial bunchgrasses (less commonly annuals) with panicle inflorescences, the spikelets several-flowered, the lemmas faintly nerved and awned (if at all) from their tips. The spikelets quickly shatter at maturity between the florets and above the persistent glumes. The following genera may resemble fescues: bromes (*Bromus*), with fused leaf sheaths and strongly nerved spikelets, those species that have lemma awns produced from between two tiny apical teeth; and the meadow fescues (*Lolium arundinaceum* and *L. pratense*), with flat (not inrolled) leaves. Five hundred species, cosmopolitan.

1. Plants annual; anther one
 2. Lower glume less than half the length of the upper glume; rare adventive .... *F. myuros*
 2. Lower glume more than half the length of the upper glume; occasional native ..... *F. octoflora*
1. Plants perennial; anthers three
 3. Leaf blades 3–8 mm wide, flat, not inrolled; lemmas awnless or with awns less than 1 mm long; native, forests
  4. Spikelets 4–6 mm wide, four- to five-flowered, in clusters of ten to twenty at ends of lower panicle branches; rare, floodplain forests, southern Wisconsin .... *F. paradoxa*
  4. Spikelets 2–4 mm wide, two- to four-flowered, not clustered at ends of lower panicle branches; common, mesic forests statewide .... *F. subverticillata*
 3. Leaf blades 1–3 mm wide, inrolled; lemmas with awns 1–10 mm long
  5. Lemma margins delicate and membranous; awns 3–10 mm long; panicle open, with lax branches; native, rare on beaches and dunes of Lake Michigan, in Door County .... *F. occidentalis*
  5. Lemma margins firm; awns less than 3 mm long; panicle narrow, with erect to spreading branches
   6. Leaf sheaths closed when plants immature, shredding into reddish-brown fibers at maturity; culms curved or bent upward at the very base; common weed .... *F. rubra*
   6. Leaf sheaths open even when plants immature, tan or pale brown, not shredding into fibers at maturity; culms erect, not curved at the base
    7. Panicle branches (at least lower) ascending; anthers 1–2 mm long; native, uncommon in sandy areas .... *F. saximontana*
    7. Panicle branches (at least lower) somewhat spreading; anthers 2–3 mm long; exotic; common in sandy areas .... *F. trachyphylla*

Festuca myuros L., rat-tail fescue

myuros, "mouse's tail"

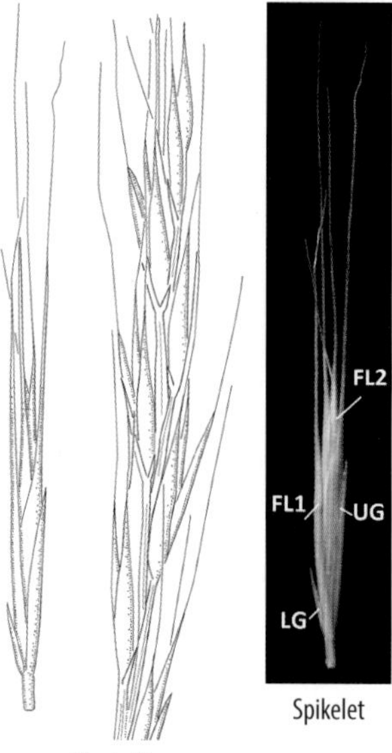

Illustration

Spikelet

Rarely adventive in disturbed ground; annual; collected from 1914 to 1970. Barkworth et al. (2007) treat this as *Vulpia myuros.*

Festuca occidentalis Hook., western fescue

occidentalis, Latin, "western"

THREATENED

Illustration

Spikelets

Rare on dry beaches and dunes near Lake Michigan in Door County.

Festuca octoflora **Walter, six-weeks fescue**

octoflora, "eight-flowered"

A weedy native annual of dry, disturbed, usually sandy ground, but also found in dry prairies; less commonly collected now than it was fifty years ago. Barkworth et al. (2007) treat this as *Vulpia octoflora*.

Illustration

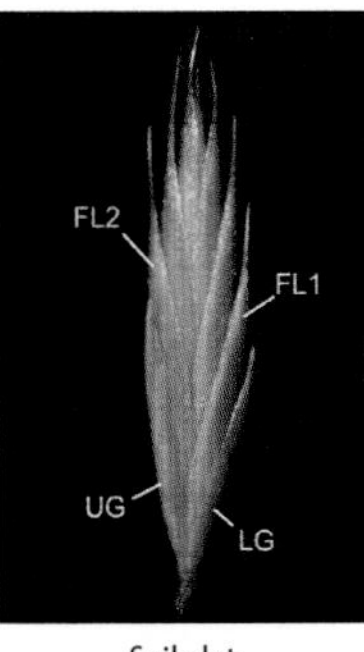

Spikelet

Inflorescence

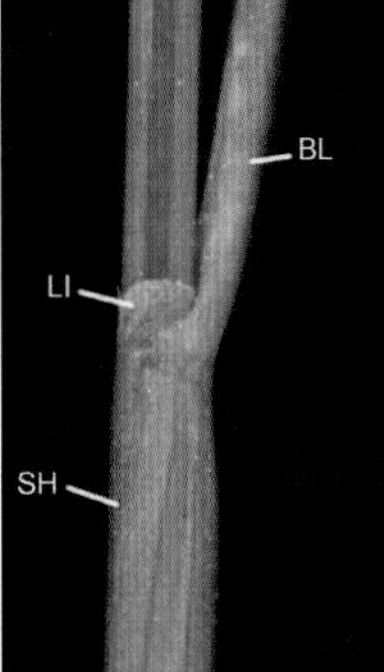

Leaf ligular area

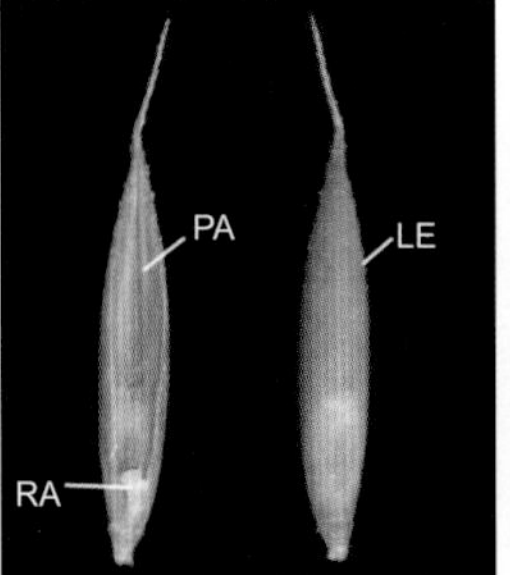

Florets

Inflorescence

Festuca paradoxa Desv., cluster fescue

paradoxa, "unusual, paradoxical"

SPECIAL CONCERN

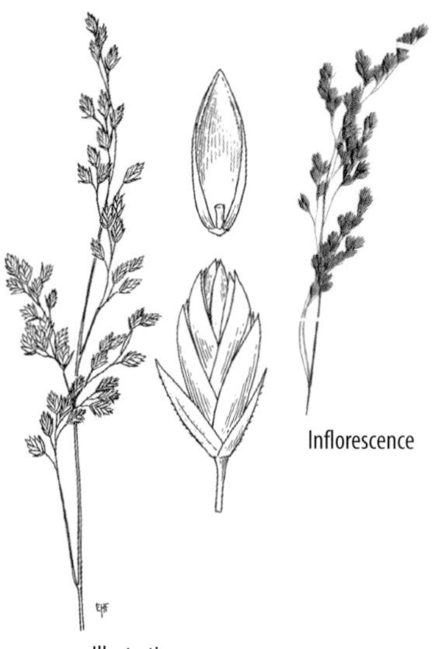

Inflorescence

Illustration

Rare in southwestern Wisconsin in peaty and marshy meadows near rivers; also in open sand along roadsides. Last collected in Adams County in 1962.

Habit

Festuca rubra L., red fescue

rubra, Latin, "red"

Illustration

Portion of inflorescence

A Eurasian exotic, first collected in 1935; now a fairly common weed of roadsides, gravel pits, and trails and clearings in forests; often planted in lawns, parks, and on golf courses, where it forms a dense turf. This species includes plants annotated as *Festuca diffusa* Dumort. by the late Leon Pavlick in 1977.

Festuca saximontana Rydb., Rocky Mountain fescue

saximontana, *saxum*, "rock," and *montanus*, "of mountains"

Illustration

A somewhat weedy native species of dry sandy dunes and roadsides; most common in sandy habitats along Lake Michigan (Door County) and on sand spits in the Apostle Islands (Lake Superior).

Portion of inflorescence

Festuca subverticillata (Pers.) E. B. Alexeev, nodding fescue

subverticillata, sub, "below, almost, less so than a similar plant," and *verticillus*, "whorled"

Our most shade-loving fescue, commonly in moist deciduous or mixed forests; has characteristic elongate, drooping panicle branches. The lemma often has a subtle, slightly hooked tip reminiscent of the upper jaw of a Chinook salmon. Fassett (1951) treats this as *Festuca obtusa*.

Illustration

Habit

FL3
FL2
PA
LE
UG
LG

Spikelet

Festuca trachyphylla (Hack.) Krajina, hard fescue, sheep fescue

trachyphylla, *trachelos*, "neck," and *phyllon*, "leaf"

A European exotic, first collected in 1915, and now a common weed of sandy disturbed ground: roadsides, old fields, and dunes along the Great Lakes. It is widely planted as a roadside stabilizing species (USDA NRCS 2014). Fassett (1951) treats this as *Festuca ovina*.

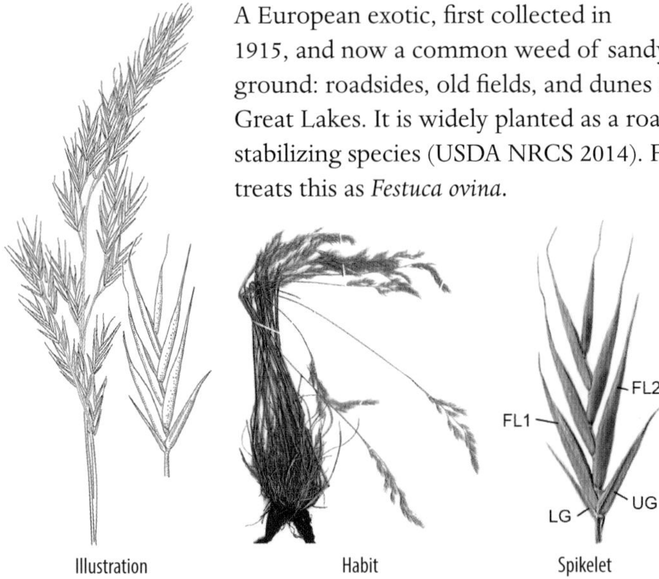

Illustration Habit Spikelet

43. *GLYCERIA*, MANNA GRASS

Greek, "sweet," from the taste of the lower part of the culm

Medium- to large-sized, mainly native panicle-producing grasses with completely fused leaf sheaths and a wetland habitat preference; several species are able to produce rafts of floating leaves. The spikelets are short-glumed and several- to many-flowered, and they readily fall apart between the florets at maturity. The lemmas have prominent, parallel nerves. Other genera that might be mistaken for *Glyceria* include the bluegrasses, with open leaf sheaths, blades with boat-shaped tips, and spikelet lemmas with less prominent, convergent nerves; and the smaller-statured, less common wetland alkali grasses (*Puccinellia*) and false manna grasses (*Torreyochloa*), both with leaf sheaths fused only near the base. Thirty-five species; mostly in the temperate Northern Hemisphere.

1. Spikelets 10–25 mm long, linear, with eight to fifteen florets; ligules 5–15 mm long
   2. Leaf blades 2–5 (rarely 10) mm wide; lemmas smooth on the back between the nerves; anthers 0.5–1 mm long; fruit 1–1.5 mm long; common statewide . . . . *G. borealis*

2. Leaf blades 5–10 mm wide; lemmas scabrous on back; anthers 1–2 mm long; fruit 1.5–2.5 mm long; uncommon, mostly in vernal ponds in southeastern Wisconsin .... *G. septentrionalis*

1. Spikelets 3–12 mm long, lanceolate to ovoid, with two to ten florets; ligules 0.5–6 mm long
   3. Lemmas smooth, the nerves not raised much, the apex acute, distinctly exceeding the bluntly rounded palea; anthers two; native statewide, but less common in the southeast .... *G. canadensis*
   3. Lemmas with prominently raised nerves, the apex short-acute, only slightly exceeding the palea; anthers two or three
      4. Upper glumes 0.5–1.5 mm long; lemmas strongly corrugated; culms slender, less than 4 mm thick; anthers two .... *G. striata*
      4. Upper glumes 1.5–4 mm long; lemmas not strongly corrugated; culms 5–10 mm thick; anthers three
         5. Lower glumes 1–2.3 mm long; upper glumes 1.5–2.7 mm long; lemmas less than 3 mm long; common native statewide .... *G. grandis*
         5. Lower glumes 2–3 mm long; upper glumes 3–4 mm long; lemmas more than 3 mm long; locally aggressive exotic, most common in southeastern Wisconsin .... *G. maxima*

Glyceria borealis (Nash) Batch., northern manna grass

borealis, "northern"

Common, wetlands and their edges, except in far southwestern Wisconsin. This species grows in water up to 2 feet deep and often develops rafts of floating leaves; *Glyceria borealis*, *G. grandis*, and the much less common *G. septentrionalis* are the aquatic species in this genus.

Illustration

Spikelet

Portion of inflorescence

Spikelet

Floating leaf

Floating leaves

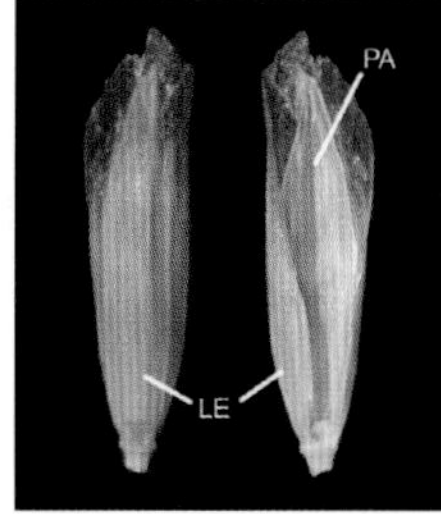

Florets

Northern manna grass, floating leaves

Northern manna grass, floating leaves

Northern manna grass, portion of inflorescence

Glyceria canadensis (Michx.) Trin., rattlesnake manna grass

canadensis, of or referring to Canada

Illustration

An abundant wetland grass of central and northern Wisconsin, rare or absent in the far south. Characteristic of sedge meadows, bogs, and fens, it may occur in shallow water but does not produce floating leaves.

Portion of inflorescence

Young inflorescence

Mature inflorescence

Glyceria grandis S. Watson, reed manna grass

grandis, "large"

Illustration

Common in wetlands statewide, often in shallow water, and producing rafts of floating leaves.

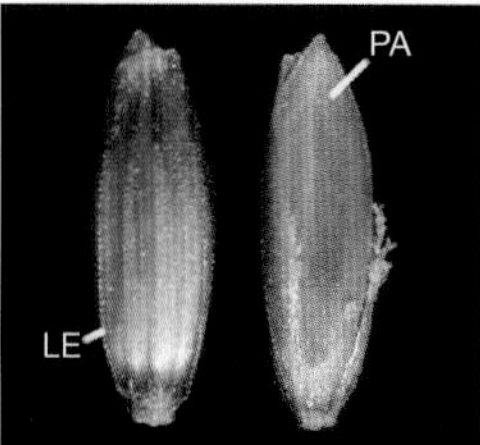

Florets

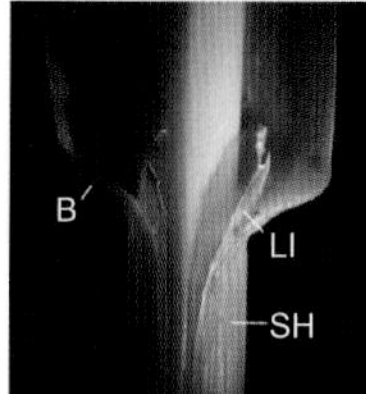

Leaf ligular area

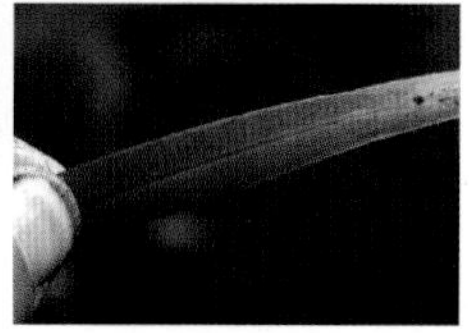

Floating leaf

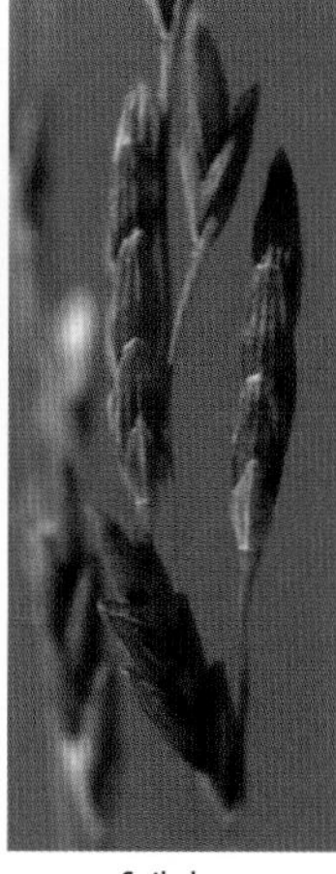

Spikelet

Spikelet

Reed manna grass, habit

Reed manna grass, floating leaves

Glyceria maxima (Hartm.) Holmb., tall manna grass

maxima, "largest"

A robust Eurasian exotic, first collected in 1975 by Gerould Wilhelm and Raymond Schulenberg (Freckmann and Reed 1979). Sedge meadows, river floodplains, and lakeshores in southern Wisconsin; an aggressive, uncommon weed that should be eradicated whenever encountered.

Portion of inflorescence

Inflorescence

Glyceria septentrionalis **Hitchc., eastern manna grass**

septentrionalis, "northern," from the Latin word septentrio, "Big or Little Dipper"

Illustration

An uncommon wetland species, frequent in vernal woodland ponds in southeastern Wisconsin; like its much commoner, more northern relative *G. borealis*, it will produce rafts of floating leaves in shallow water.

Inflorescence

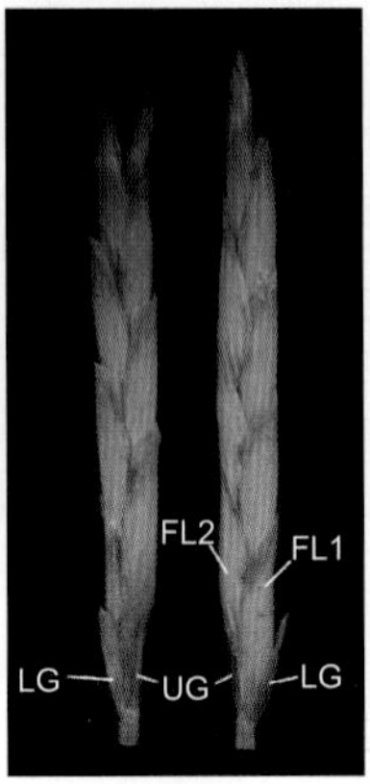

Spikelet

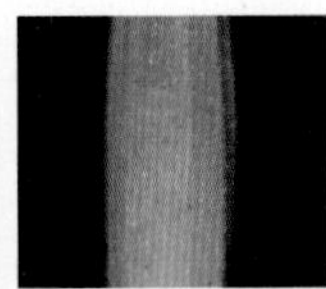

Floret, showing very short stiff hairs between the nerves

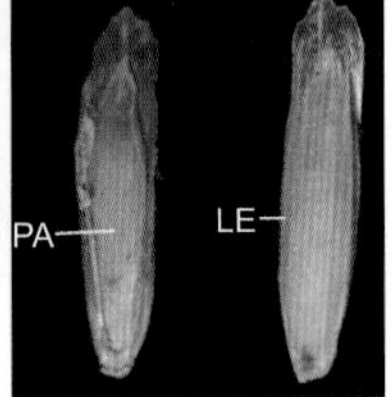

Florets

Glyceria striata **(Lam.) Hitchc., fowl manna grass**

striata, "striped"

An abundant wetland grass statewide, but usually not growing in shallow water. It is also frequent in wet to moist forests, often in small depressions or in the ruts of trails or old logging roads.

Illustration

Portion of inflorescence

Spikelet

Fowl manna grass, habit

44. *GRAPHEPHORUM*, MELIC-OATS

possibly from *grapheiron*, "pencil, stylus, or paintbrush," and *phorus*, "bearing, carrying," referring to the hairy appendage formed by the extension of the rachilla of the spikelet

Perennials with panicles of awnless, two-flowered spikelets (Finot et al. 2005). The relatively long glumes and the pilose rachilla distinguish this genus from bluegrass (*Poa*), manna grass (*Glyceria*), or false manna grass (*Torreyochloa*). Twenty species of temperate areas of the Northern Hemisphere.

Graphephorum melicoides (Michx.) P. Beauv. melic-oats

melicoides, Greek, meli, "honey," and *deum*, "like"

ENDANGERED

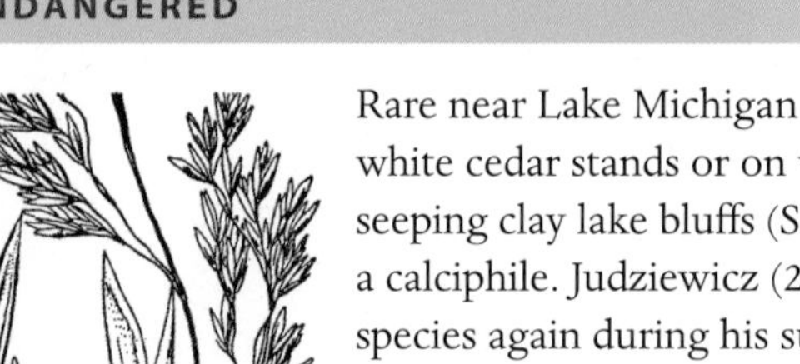

Illustration

Rare near Lake Michigan in sandy white cedar stands or on wooded dunes and moist, seeping clay lake bluffs (Shinners 1940); apparently a calciphile. Judziewicz (2002) did not find this species again during his surveys of the Grand Traverse Islands (Door County) in 1998 and 1999. According to Gary Fewless (pers. comm.), it still survives in sunny wet swales and along a shaded roadside at the Ridges Sanctuary in Door County (2010 sighting); look for it in August. Fassett (1951), Barkworth et al. (2007), and the Wisconsin Department of Natural Resources (2014) all treat this as *Trisetum melicoides.*

Portion of inflorescence

Inflorescence

45. *HESPEROSTIPA*, NEEDLE GRASS

Greek, *hesperos*, "western" or "evening," and *stipa*, "fiber"

Perennials with panicles of single-flowered spikelets, the latter with slender glumes and a distinctive hard, terete floret with a sharp, bearded base and a robust, elongate, once-bent terminal awn; green needle grass (*Nassella*) has more slender inflorescences and more delicately awned spikelets. Five North American species.

1. Lemma body 10–15 mm long, pale in color, its awn slender and flexible; glumes 20–30 mm long; uncommon introduction . . . . *H. comata*
1. Lemma body 18–25 mm long, dark in color, its awn stiff and stout; glumes 30–45 mm long; fairly common native . . . . *H. spartea*

Hesperostipa comata (Trin. and Rupr.) Barkworth, needle-and-thread

comata, "hairy"

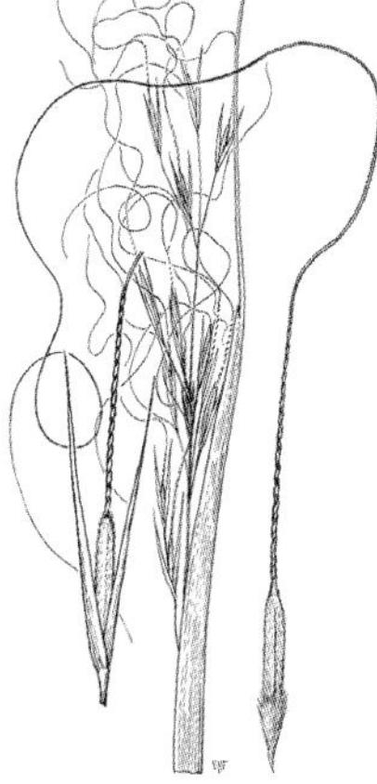

Illustration

A native of the Great Plains, coming east to western Minnesota as a native plant; in Wisconsin a rare weedy introduction from farther west, first collected in 1915. Most commonly encountered along railroad rights-of-way in southern Wisconsin. Fassett (1951) treats this as *Stipa comata*.

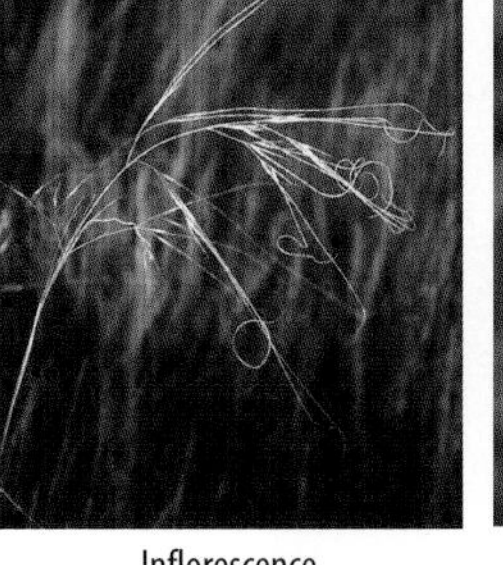

Inflorescence

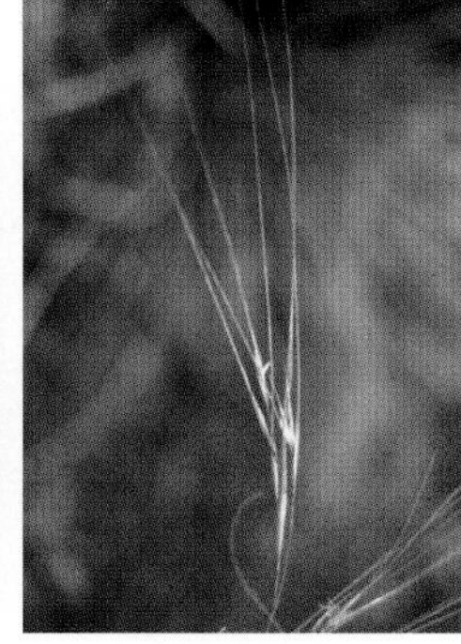

Inflorescence

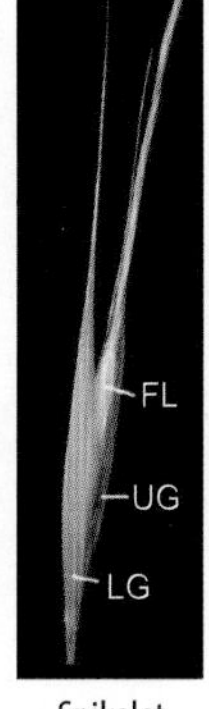

Spikelet

Floret

Hesperostipa spartea (Trin.) Barkworth, needle grass

spartea, Latin, "broom-like"

Illustration

Floret

Fairly common in dry to mesic prairies, barrens, and savannas in southern and central Wisconsin; slightly weedy, and sometimes found along roadsides and railroad rights-of-way. Fassett (1951) treats this as *Stipa spartea*.

Spikelet

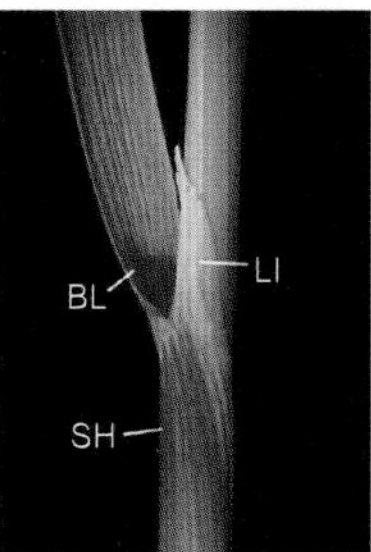

Leaf ligular area

Needle grass, inflorescence

Needle grass, inflorescence

46. *HOLCUS*, VELVET GRASS

word used by Pliny for a type of grass, from *holcus*, "attractive"

Perennial pasture grasses with soft, hairy foliage and panicle inflorescences, the spikelets two-flowered, falling intact, the lower floret perfect and awnless, the upper floret male and with a "cute," tiny hooked awn. Eight Old World species.

Holcus lanatus L., velvet grass or Yorkshire fog

lanatus, Latin, "wooly"

A European exotic, first collected in 1966 by C. H. Porter; rarely adventive in lawns and on sandy roadsides, perhaps introduced as a seed contaminant; last collected in 1975, it is perhaps not currently present in Wisconsin.

Illustration

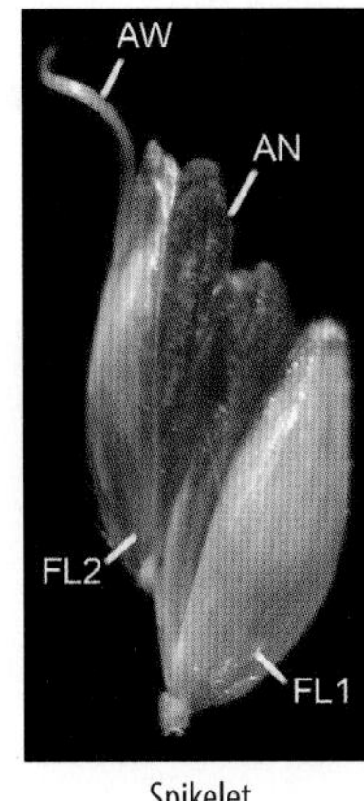

Spikelet

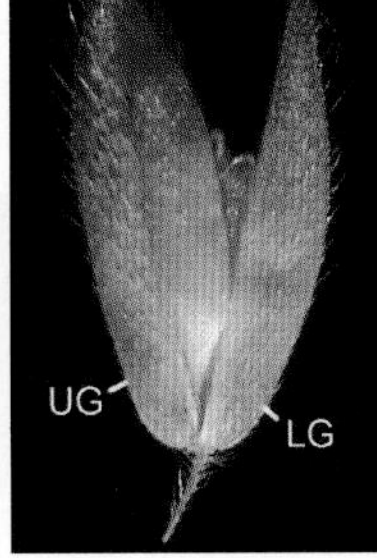

Floret

47. *HORDEUM*, BARLEY

ancient Latin word for barley

Annuals or tufted perennials; three spikelets per node, the lateral pair often stalked and sterile; one floret per spikelet; glumes and lemma often long-awned; rachis usually breaking at each node. Thirty-two species of Eurasia, Africa, and North America.

1. Bodies of lemmas 8–10 mm long; leaf with auricles at base of blade, glabrous; rachis not disarticulating at maturity; occasional escape from cultivation .... *H. vulgare*
1. Bodies of lemmas 3–6 mm long; leaf lacking auricles at base of blade, hairy; rachis disarticulating at maturity
   2. Awns 3–5 cm long; common weed .... *H. jubatum*
   2. Awns up to 1 cm long; rare adventive .... *H. pusillum*

Hordeum jubatum **L., squirrel-tail barley**

jubatum, "crested"

Illustration

A native of the western United States, purportedly exotic in Wisconsin, but early noted on "damp level prairies" (Lapham 1853). First collected in 1878. A fairly common roadside weed; the distinctive, silky-looking "squirrel-tail" inflorescences quickly disintegrate at maturity.

Portion of inflorescence

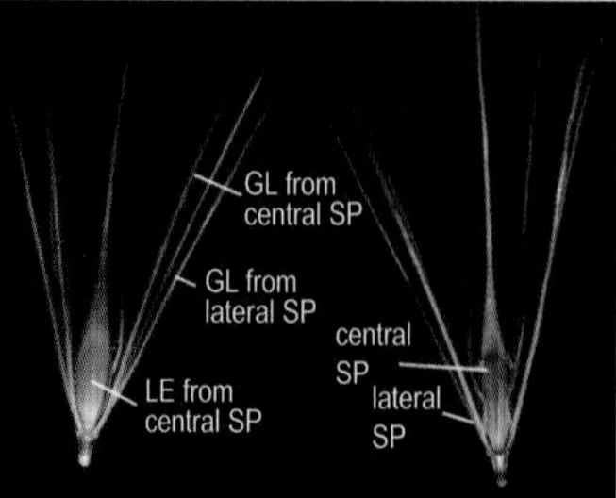

Spikelet cluster; only the central spikelet produces a floret, the lateral ones are represented only by needle-like glumes

Young inflorescence showing "squirrel-tail" aspect

Squirrel-tail barley, mature disarticulating inflorescence with "spidery" aspect

Hordeum pusillum Nutt., little barley
pusillum, "little"

A western US exotic, first collected in 1983 by Hugh Iltis; this rare adventive was collected in a sidewalk crack in Madison, Dane County.

Inflorescence and spikelet cluster

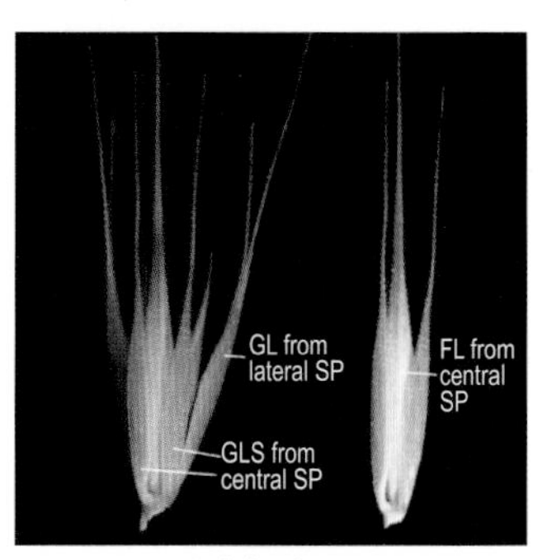

Spikelet cluster

Inflorescence and spikelet cluster

Habit as cover crop

Hordeum vulgare **L., barley**

vulgare, Latin, "common"

Illustration

A European exotic, first collected in 1908; a rare escape from cultivation, not collected since 1976.

Inflorescence Inflorescence

48. *KOELERIA*, JUNE GRASS

for Georg Ludwig Koeler (1765–1807), German professor at Mainz and student of grasses

Small perennials with contracted panicles, the rachis characteristically finely hairy (use a hand lens) near the base. Spikelets two-flowered, awnless, the subequal glumes about as long as the florets, which fall from them. Thirty-five species; cosmopolitan except in the lowland tropics.

Koeleria macrantha (Ledeb.) Schult., June grass

macrantha, "large-flowered or anthered"

Illustration

Common and characteristic of dry and sand prairies, barrens, and oak savannas. When flowering and the panicle is open, it is sometimes mistaken for wedge grass (*Sphenopholis*), but that genus has smooth inflorescence rachises, glumes of distinctly unlike shapes, and a usually wetter habitat preference. June grass is often used in prairie restoration mixes and has been used to revegetate mine sites (USDA NRCS 2014). Fassett (1951) treats this as *Koeleria cristata*.

Mature inflorescence

Young inflorescence

Habit

49. *LEERSIA*, CUT GRASS

for Johann Daniel Leers (1727–1772), German botanist

Perennials with panicles of glumeless spikelets comprised of a single, strongly laterally flattened spikelet with a five-nerved lemma and a barely separable three-nerved palea. Seventeen species; cosmopolitan.

1. Spikelets nearly round; leaf blades 10–20 mm wide; uncommon, floodplains of large rivers, southwestern Wisconsin . . . . *L. lenticularis*
1. Spikelets elliptical, much longer than wide; leaf blades 3–15 mm wide
   2. Spikelets 4–7 mm long; culms and leaves harshly scabrous (sandpapery); panicle branches two to three per node; abundant, wetlands . . . . *L. oryzoides*
   2. Spikelets 3–4 mm long; culms and leaves smooth, never harshly scabrous; panicle branch just one per node; occasional, upland forests and riverbanks in central and southern Wisconsin . . . . *L. virginica*

Leersia lenticularis Michx., catchfly grass

lenticularis, "shaped like a lens"

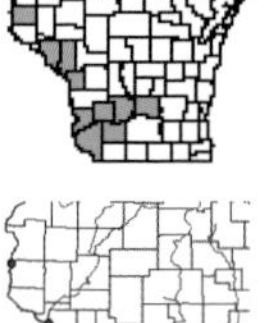

Forested floodplains of the Mississippi, lower Wisconsin, St. Croix, and Baraboo Rivers, where it can form robust pure stands in the forest understory.

Illustration

Spikelets along inflorescence branch

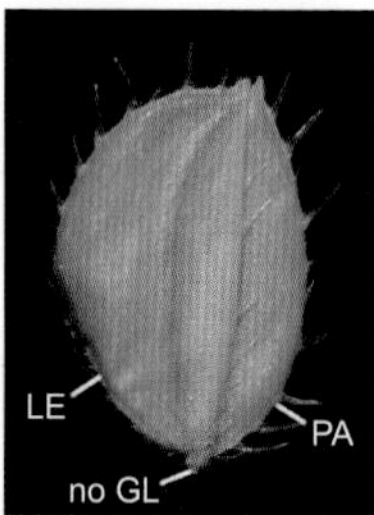

Spikelet; glumes are absent

Habitat along Mississippi River

Leersia oryzoides **(L.) Sw., rice cut-grass**

oryzoides, like Greek and Latin, *oryza*, "rice"

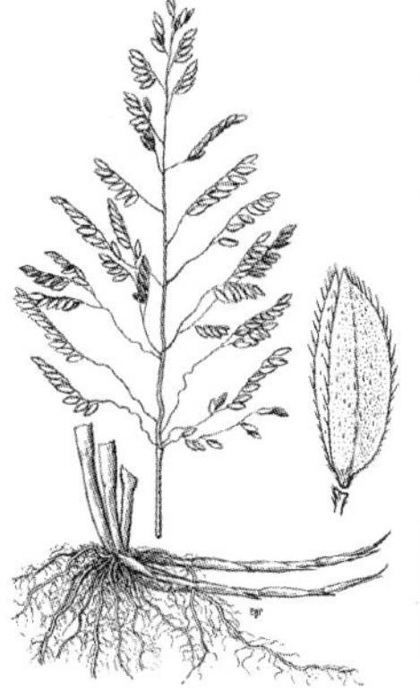
Illustration

Common statewide in marshes and on the margins of lakes, ponds, and rivers. It has been used to revegetate watercourses and can become weedy in cranberry bogs (USDA NRCS 2014). It flowers in late summer and fall and often produces cleistogamous (hidden) flowers within the upper leaf sheaths. The culms and leaves are harshly scabrous and will slightly cut human skin.

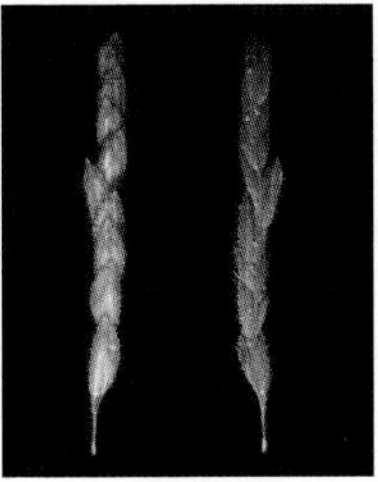
Spikelets on branch of inflorescence

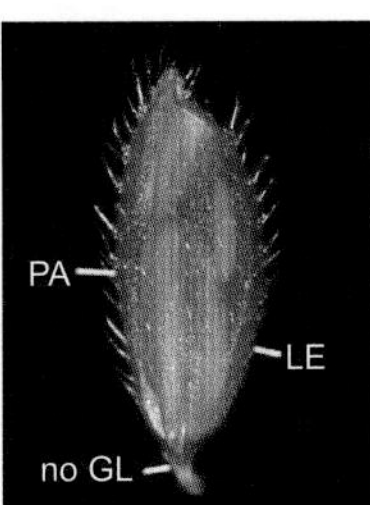

Spikelet; glumes absent

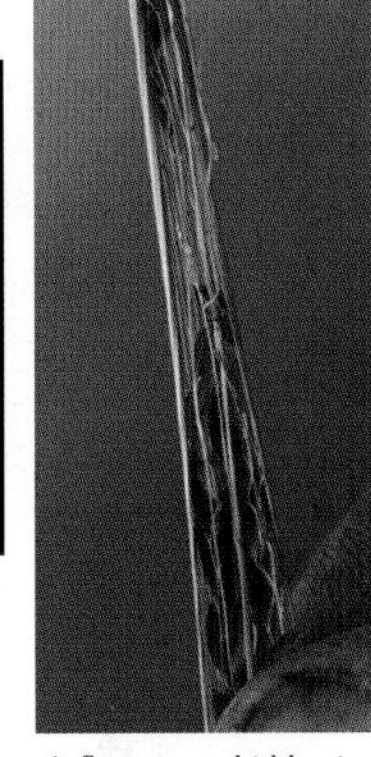
Inflorescence hidden in leaf sheath

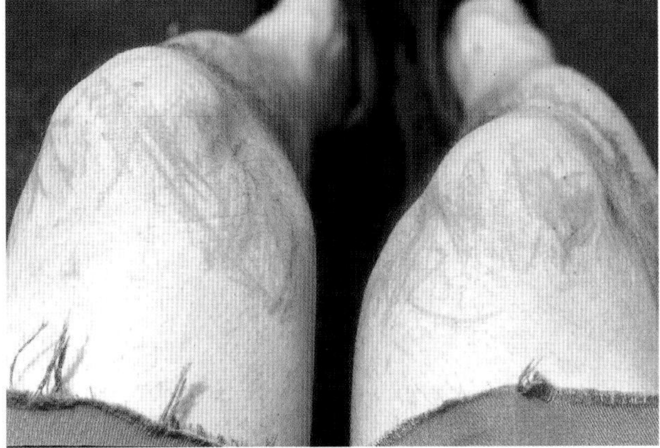
Cuts caused by leaf margins

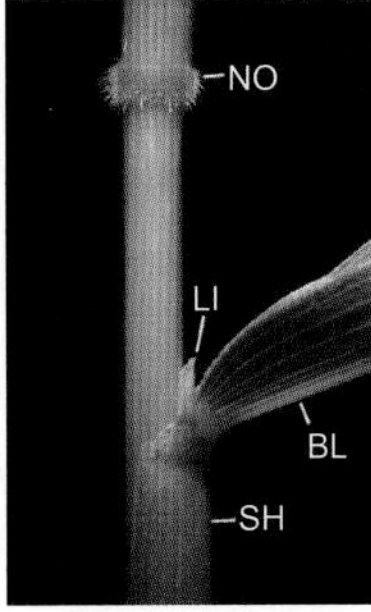

Leaf ligular area

Rice cut-grass, habit

Rice cut-grass, young inflorescence

Leersia virginica **Willd., white grass**
virginica, "of Virginia"

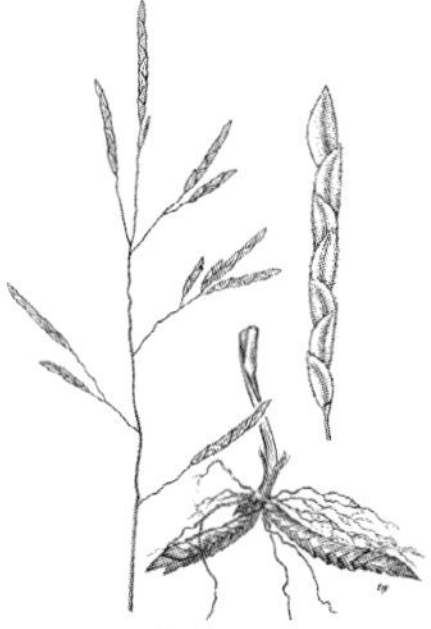

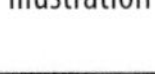
Illustration

Occasional in moist to dry upland woods, floodplains, and trail margins in central and southern Wisconsin. Flowering in late summer; the foliage is not harshly scabrous, as in rice cutgrass. It appears to be spreading northward along the western shore of Green Bay (Gary Fewless, pers. comm.) and in inland Oconto County.

Habit

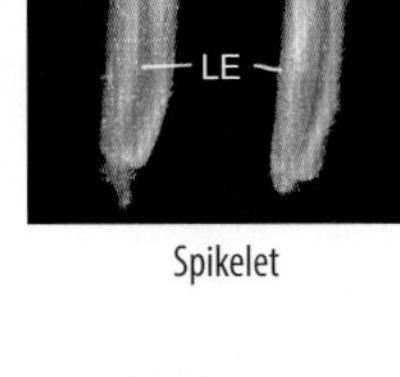

Spikelet

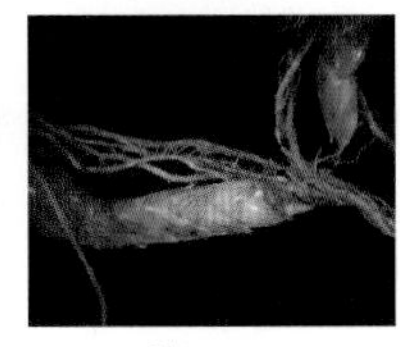
Rhizomes

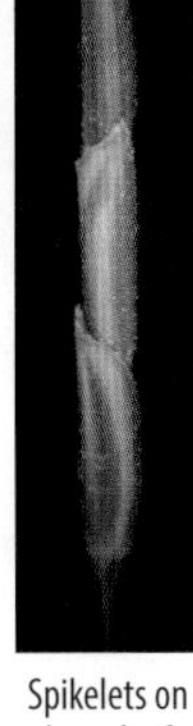
Spikelets on branch of inflorescence

50. *LEYMUS*, LYME GRASS

a modification of Elymus, which all Leymus spp. were formerly classified as

A robust perennial that resembles species of wild rye (*Elymus*), differing in its much broader (3.5–4.5 mm wide) glumes and always awnless spikelets. Fifty species of north temperate areas.

***Leymus arenarius* (L.) Hochst., lyme grass**
arenarius, "relating to sand"

Portion of inflorescence

A European exotic, first collected in 1959 by J. A. Reed at Point Beach, Manitowoc County. This robust, locally common dune grass of the Lake Michigan shoreline has attractive bluish foliage and is occasionally cultivated as an ornamental; nevertheless, this is an invasive species that should not be planted anywhere near Great Lakes dunes. It is sometimes mistaken for beach grass (*Ammophila breviligulata*), but it is bluer, with several-flowered spikelets.

Portion of inflorescence

Habitat on Lake Michigan dunes

51. *LOLIUM*, RYEGRASS AND MEADOW FESCUE

ancient Latin name

Ryegrass (*Lolium perenne*) resembles quackgrass (*Elymus repens*), but its spikelets are turned thin edgewise (rather than sidewise) to the inflorescence axis; the meadow fescues (*L. arundinacea* and *L. pratensis*) are tussock-forming grasses with leaf blades that persist over winter in a green state and narrow panicles of several-flowered spikelets (Darbyshire 1993). They differ from the "true" fescues (*Festuca*) most obviously in their stiff, flat-bladed leaves with a prominent pair of sheath auricles ("ears") at the base of the blade; bromes (*Bromus*) differ in their fused leaf sheaths and more open panicles of less strongly nerved spikelets, these awned in most species.

1. Inflorescence a spike; leaf auricles absent; upper glume absent .... *L. perenne*
1. Inflorescence a panicle; leaf auricles present; both glumes present
   2. Leaf auricles ciliate; lemmas 7–9 mm long; panicle branches two per node .... *L. arundinaceum*
   2. Leaf auricles glabrous; lemmas 5.5–7 mm long; panicle branch one per node ..... *L. pratense*

Lolium arundinaceum (Schreb.) Darbysh., tall meadow fescue

arundinaceum, "reed-like"

Illustration

A European exotic, first collected in 1940, and now planted as a turf or pasture grass and becoming an occasional weed of sandy, gravelly, or clayey roadsides and old fields. Most plants harbor an endophytic fungus that is damaging to domestic and native grazers and to birds and rodents that eat the fruits, leading to a diminishment of biodiversity on several levels, so this species is no longer recommended for planting (USDA NRCS 2014). It can hybridize with *Lolium pratense*. Fassett (1951) treats this as *Festuca elatior* var. *arundinacea*, Barkworth et al. (2007) as *Schedonorus arundinaceus*.

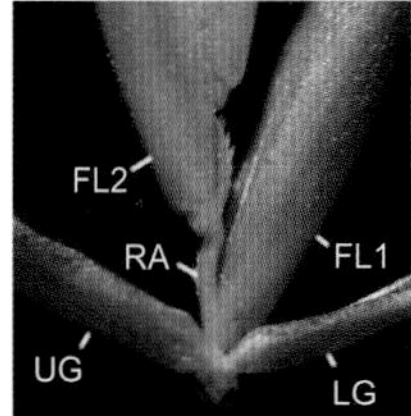

Base of spikelet

Portion of inflorescence

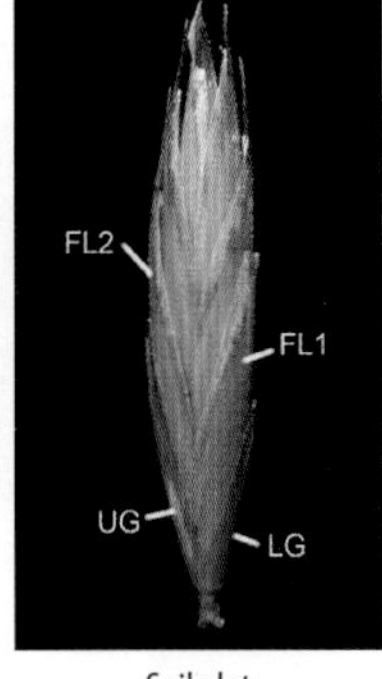

Spikelet

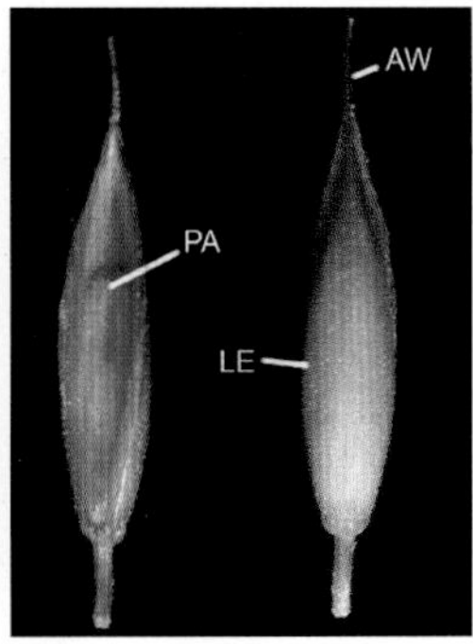

Florets

Tall meadow fescue, habit

Lolium perenne **L., perennial ryegrass**

perenne, "perennial"

Illustration

A variable annual to perennial Eurasian exotic, first collected in 1887, and now a fairly common weed of roadsides, lawns, gardens, and fields, sometimes invading native habitats. It is widely used in lawns and as a cover plant. The spike consists of alternating (two-ranked), many-flowered, strongly laterally compressed spikelets with only the thin edge of their "profile" touching the rachis (the inflorescence axis); glume single, opposite the rachis. This species includes darnel (*Lolium temulentum* L.), a variant with a longer glume, as treated by Barkworth et al. (2007). Another variant, with long-awned, many-flowered spikelets, has been called *L. multiflorum* Lam.

Portion of inflorescence

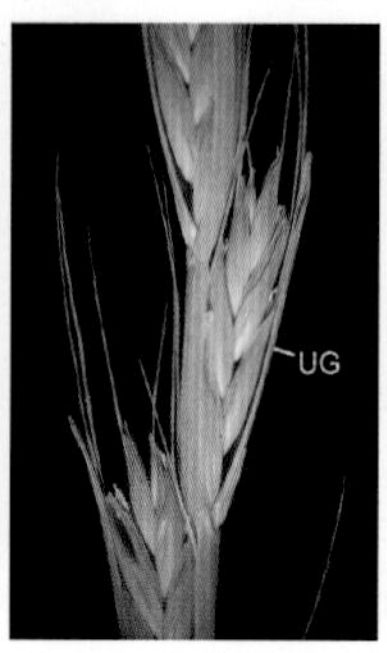

Portion of inflorescence

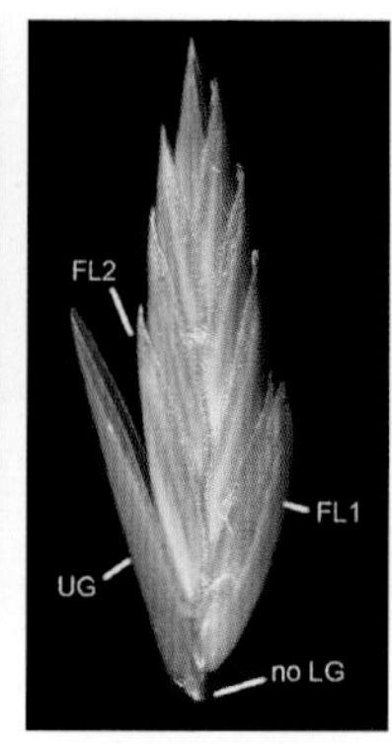

Spikelet; lower glume absent

Lolium pratense (Huds.) Darbysh., clustered meadow fescue

pratense, "of meadows"

A European exotic, first collected in 1879. A common weed in a variety of disturbed sites, including old fields and roadsides. Fassett (1951) treats this as *Festuca elatior* var. *elatior*, Barkworth et al. (2007) as *Schedonorus pratensis*.

Illustration

Spikelet

Habit

52. *MELICA*, MELIC GRASS

Greek, *melike*, deriving from *mel*, "honey," and applied to a kind of sorghum or other plant with sweet sap

Perennials with fused leaf sheaths and panicles of several-flowered spikelets; florets falling separately from above the persistent, rather large glumes. Our two species are quite different from each other. *Melica nitens* has awnless spikelets with broad, top-heavy-appearing glumes and lemmas, and the uppermost floret is a club-shaped rudiment, while *M. smithii* has slender, awned lemmas with the uppermost rudimentary floret slender, not club-shaped. Twinflower melic grass (*M. mutica* Walter) has been recorded from the Illinois counties bordering Walworth and Kenosha Counties and so could eventually be found in Wisconsin. It differs from *M. nitens* in its strongly perpendicular uppermost rudimentary spikelet and in its more sparsely flowered panicles with only two to five spikelets per branch; it prefers calcareous woods. Eighty species, cosmopolitan except in Australia.

1. Lemmas awnless; spikelets 9–12 mm long; rare, dry forest edges, southwestern Wisconsin . . . . . . *M. nitens*
1. Lemmas with awns 3–5 mm long; spikelets 12–20 mm long; rare, mesic forests, far northern Wisconsin . . . . *M. smithii*

Melica nitens (Scribn.) Nutt. ex Piper, tall melic grass

nitens, "shining"

SPECIAL CONCERN

A robust grass that is rare in dry prairies and on outcrops and railroad embankments in far southwestern Wisconsin, where it was collected as recently as 2012 by Judziewicz.

Illustration

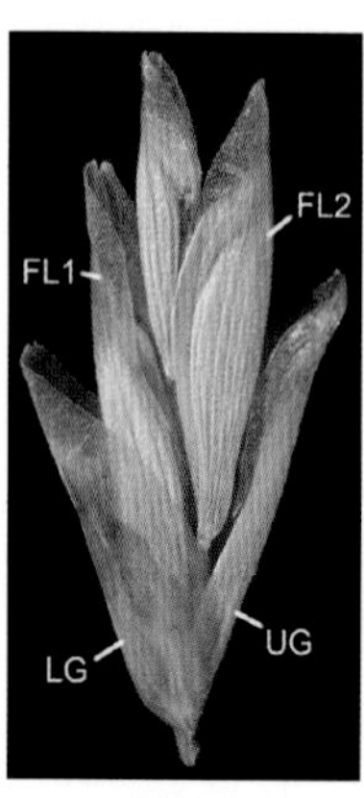

Spikelet

Habit

Melica smithii **(Porter ex A. Gray) Vasey, Smith's melic grass**

smithii, for Charles Eastwick Smith (1820–1900), American engineer and amateur botanist

ENDANGERED

Inflorescence

First collected in Florence by Hugh Iltis in 1964, but probably present in each of the northernmost tier of counties; rare in the understory of mesic hardwoods (sugar maple, yellow birch, basswood). Easily confused with false melic grass (*Schizachne purpurascens*); see discussion under that species. Also resembling some of the woodland bromes (*Bromus*), but differing in the completely glabrous spikelets.

Foliage

Portion of inflorescence

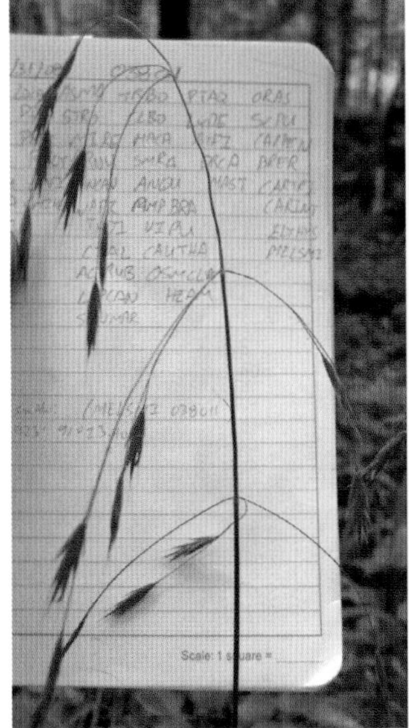

Inflorescence

Habit

53. *MILIUM*, WOOD MILLET

ancient Latin word for "millet," which belongs to another genus; therefore, of uncertain source

Large, broad-leaved, somewhat blue-green forest understory perennials with pyramidal panicles of awnless spikelets with large, equal, ovate glumes and a single, hardened, whitish, nerveless floret. Four species; circumboreal.

Milium effusum L., wood millet

effusum, "scattered, spread out, loose"

Fairly common in the understory of mesic hardwood or mixed hardwood hemlock forests; occasionally in white cedar swamps and boreal forests. A robust, handsome grass with broad leaves and somewhat glaucous foliage.

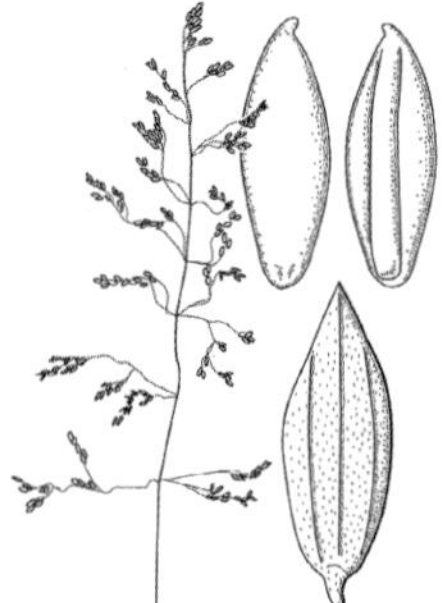

Illustration

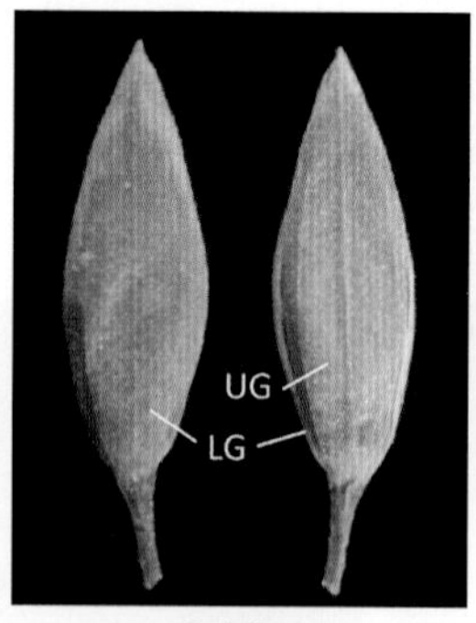

Spikelets

Florets

Inflorescence

Inflorescence

54. *MISCANTHUS*, SILVER GRASS

Greek, an allusion to the stalked spikelets

Large, clump-forming, often cultivated grasses, in late summer producing digitate (finger-like) spike-like racemes of fuzzy white spikelets. Twenty-five species, mostly from Southeast Asia.

Miscanthus sacchariflorus (Maxim.) Benth., Amur silver grass

sacchariflorus, "sweet flowers"

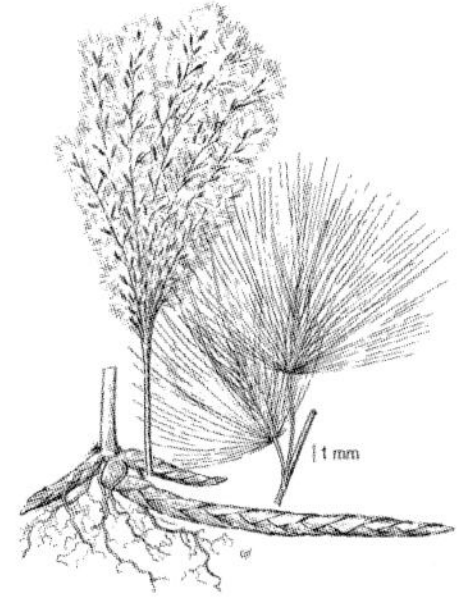

Illustration

An eastern Asian exotic, first collected in 1959 by Bill Collien at Potosi, Grant County. This is a locally common, conspicuous, rhizomatous, clonal weed found in roadside ditches and along railroads and shorelines. Perhaps introduced as an ornamental by troops returning from the Korean War. Chinese silver grass or eulalia (*Miscanthus sinensis* Andersson) is a common, robust (to over 1 m tall) cultivar that is clump-forming rather than rhizomatous. The spikelet hairs are much longer than those of *M. sacchariflorus*, and the entire plant has a more yellowish cast in the winter.

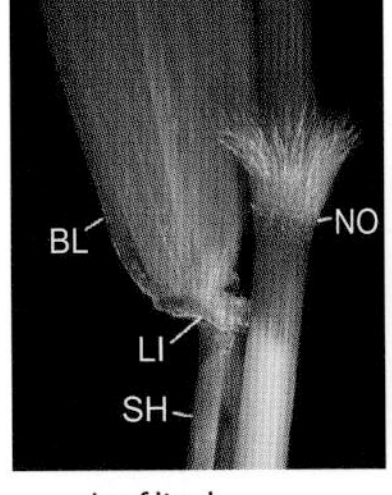

Leaf ligular area

Portion of inflorescence

Leaf ligular area

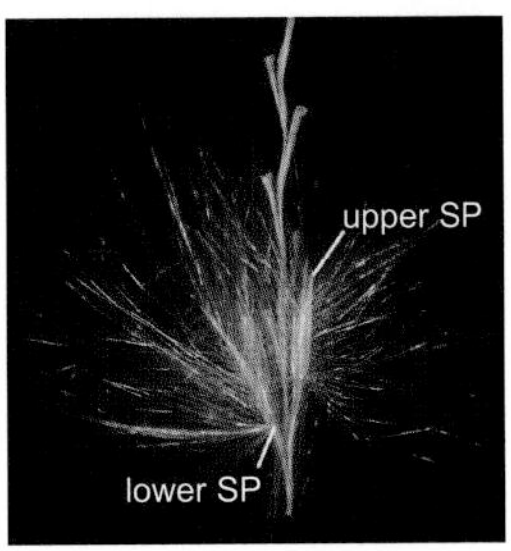

Spikelets

Amur silver grass, habit

55. *MOLINIA*, MOOR GRASS

for Juan Ignacio Molina (1740–1829), Jesuit Chilean author of civil and natural history

Robust clump-forming grasses with narrow panicles of two- to four-flowered awnless spikelets, the florets rather distant. At arm's length it could resemble switchgrass (*Panicum virgatum*), but that species has only one functional floret per spikelet. Manna grasses (*Glyceria*) have more densely flowered spikelets and fused leaf sheaths. Several European species.

Molinia caerulea (L.) Moench, purple moor grass

caerulea, "blue"

An invasive European exotic that was first collected in 1986 in a sphagnum bog harvested for peat in Monroe County by Frank Bowers and Lyman Echola (Freckmann, Bowers, and Echola 1989); then in 1989 on a sandy roadside in nearby Wood County by Alvin Bogdansky. In 1996 it was collected by Steven Spickerman, who found it to be frequent in an old field near Glidden, Ashland County; in 2012 it was still spreading at this site.

Illustration

Habit

Portion of inflorescence

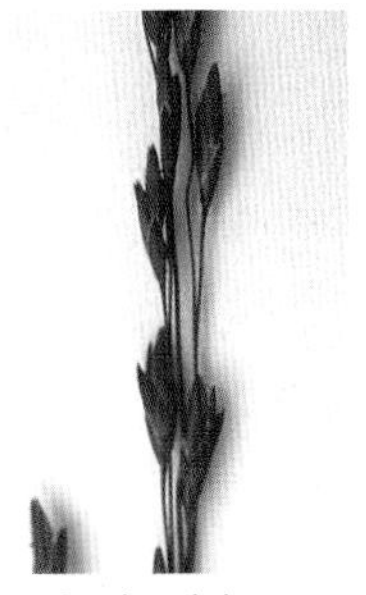

Purple spikelets

Habitat along Highway 13, Ashland County

56. *MUHLENBERGIA*, MUHLY GRASS

for Gotthilf Heinrich Ernst Muhlenberg (1753–1815), German-educated American botanist

Perennials from scaly rhizomes, often densely leafy and branching aboveground, the ligules membranous; panicles narrow (or in *Muhlenbergia asperifolia* and *M. uniflora* with open panicles); spikelets single-flowered, the floret falling from the glumes, the bracts all slender and long-acuminate to awned, the glumes of many species subtly but characteristically S-shaped (Shinners 1941). 155 species of the Western Hemisphere.

1. Lemmas smooth at the base, awnless; plants tufted to weakly rhizomatous; uncommon to rare
   2. Panicle open, the spikelets 1–2 mm long
      3. Panicles 10–20 cm wide; uncommon weed of dry areas, eastern Wisconsin .... *M. asperifolia*
      3. Panicles 1–5 cm wide; uncommon native of wet sandy-peaty shores, central and northern Wisconsin .... *M. uniflora*
   2. Panicle contracted, the spikelets 2.5–3.5 mm long
      4. Ligules less than 0.5 mm long, or absent; uncommon, dry dolomitic bluffs in southwestern Wisconsin .... *M. cuspidata*
      4. Ligules 1–2.5 mm long; rare, fens .... *M. richardsonis*
1. Lemmas long-hairy at the base, awned or awnless; plants strongly rhizomatous; common to uncommon
      5. Glumes less than 0.5 mm long; plants somewhat running (stoloniferous) but lacking rhizomes .... *M. schreberi*
      5. Glumes 1.5–8.5 mm long; plants rhizomatous
         6. Glumes (awns included) 3.5–8 mm long, evidently longer than the lemma body; lemmas awnless or short-awned
            7. Culm internodes roughened or minutely hairy (use a strong lens); ligules ca. 0.5 mm long; common, wetlands .... *M. glomerata*
            7. Culm internodes smooth and glabrous; ligules 0.5–1 mm long; occasional, dry areas .... *M. racemosa*
         6. Glumes 1.3–3.5 mm long, equal to or shorter than the lemma body; lemmas awnless to long-awned
            8. Glumes narrowly ovate, 0.5–1 mm wide
               9. Lemmas awnless or with an awn up to 1 mm long; rare, dry forests and cliffs, mostly southwestern Wisconsin .... *M. sobolifera*
               9. Lemmas with awn 4–18 mm long
                  10. Culm internodes and leaf sheaths smooth and glabrous; leaf blades 4–6 mm wide; ligules 1.5–2 mm long; uncommon, dry areas, mostly in southwestern Wisconsin .... *M. sylvatica*

10. Culm internodes and leaf sheaths minutely hairy; leaf blades 6–15 mm wide; ligules 0.5–1 mm long; occasional, dry areas, southern Wisconsin .... *M. tenuiflora*

8. Glumes narrowly lanceolate, ca. 0.5 mm wide
   11. Culm internodes and leaf sheaths smooth and glabrous; panicles with bases usually included in the upper leaf sheath; dry areas, mostly southern Wisconsin .... *M. frondosa*
   11. Culm internodes and leaf sheaths minutely hairy; panicles all exserted from the uppermost leaf sheath
      12. Ligules 0.5–1 (rarely 1.3) mm long; some spikelets nearly sessile (without pedicels); common statewide .... *M. mexicana*
      12. Ligules 1.5–2 mm long; all spikelets on distinct pedicels; uncommon, dry areas, mostly in southwestern Wisconsin .... *M. sylvatica*

Muhlenbergia asperifolia (Nees and Meyen ex Trin.) Parodi, scratch grass

asperifolia, "rough-leaved"

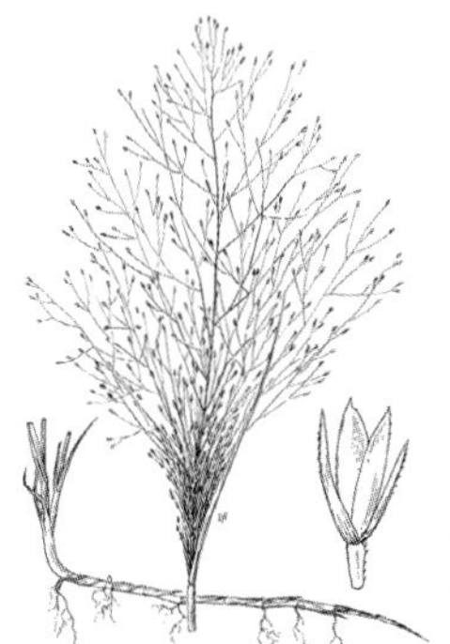

A western US species, native as far east as western Minnesota and Iowa; in Wisconsin, a rare weed of dry ground along railroads, but also collected in 2001 on the sandy shoreline of Lake Michigan on Sea Gull Bar, Marinette County. First collected in 1937 by W. W. Oppel.

Illustration

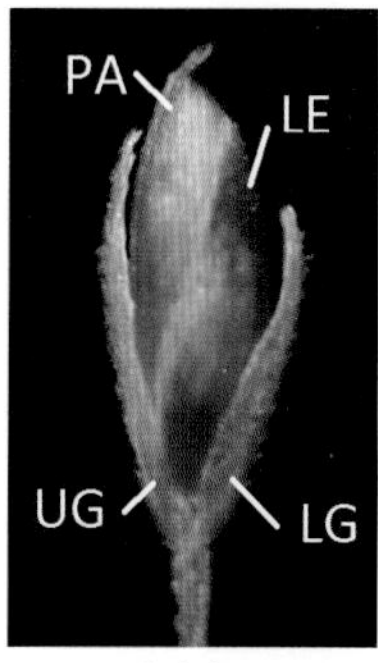

Spikelet

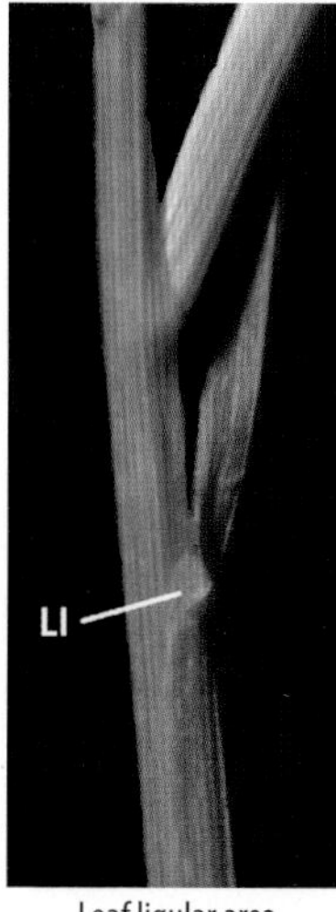

Leaf ligular area

Inflorescence

Muhlenbergia cuspidata **(Torr. ex Hook.) Rydb., plains muhly**

cuspidata, Latin, "with a cusp," a sharp and rigid point

A western dry prairie species mainly restricted in Wisconsin to the summits of dolomite bluffs above the Mississippi, St. Croix, and Pecatonica Rivers.

Illustration

Inflorescence

FL
UG
LG

Spikelet

Habit

Illustration

Muhlenbergia frondosa (Poir.) Fernald, wire-stemmed muhly

frondosa, "leafy"

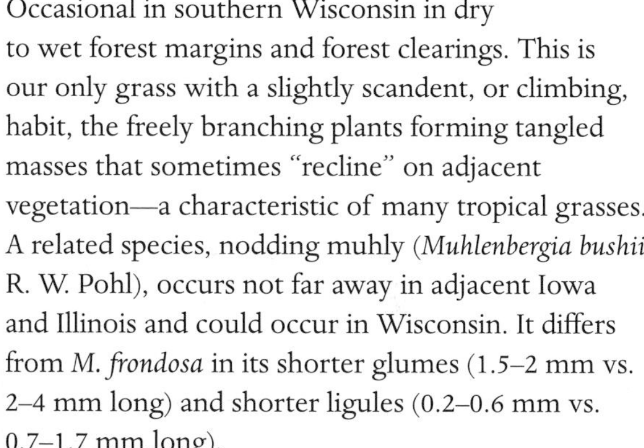

Occasional in southern Wisconsin in dry to wet forest margins and forest clearings. This is our only grass with a slightly scandent, or climbing, habit, the freely branching plants forming tangled masses that sometimes "recline" on adjacent vegetation—a characteristic of many tropical grasses. A related species, nodding muhly (*Muhlenbergia bushii* R. W. Pohl), occurs not far away in adjacent Iowa and Illinois and could occur in Wisconsin. It differs from *M. frondosa* in its shorter glumes (1.5–2 mm vs. 2–4 mm long) and shorter ligules (0.2–0.6 mm vs. 0.7–1.7 mm long).

Inflorescence

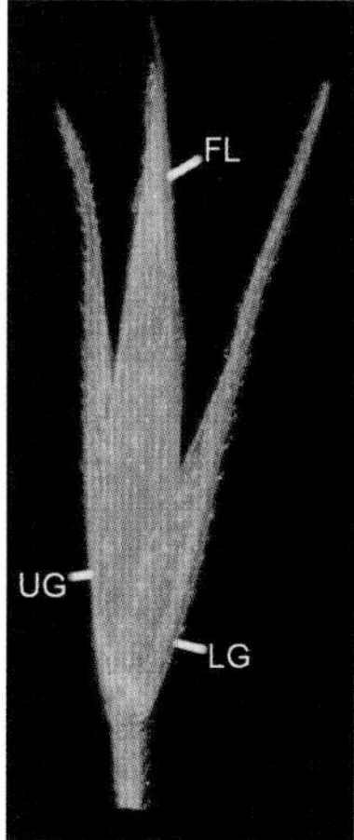

Spikelet

Habit

***Muhlenbergia glomerata* (Willd.) Trin., marsh wild-timothy**
glomerata, "clustered"

Common statewide; characteristic of sedge meadows, fens, and bogs; closely related to and intergrading with *M. racemosa*.

Illustration

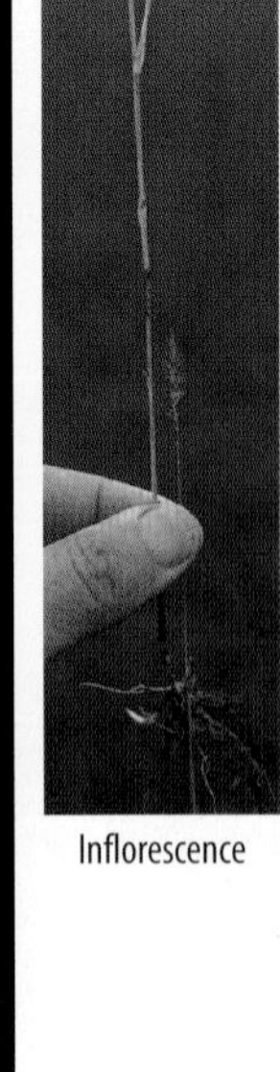

Inflorescence

Spikelet

Habit

Muhlenbergia mexicana **(L.) Trin., Mexican muhly**

mexicana, "of Mexico"

By far the commonest species of *Muhlenbergia*, found in a variety of wet to dry habitats, most typically on forest margins or along roads, trails, and shores. Awn lengths are variable.

Illustration

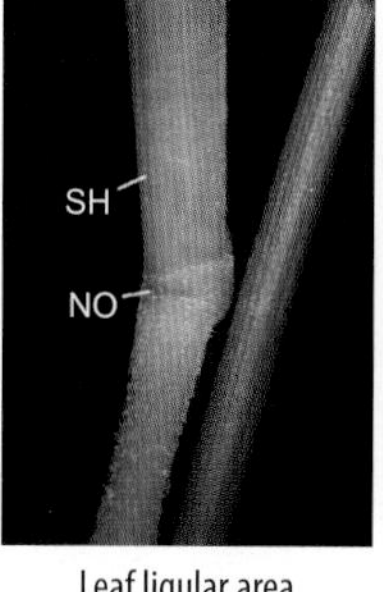

Leaf ligular area

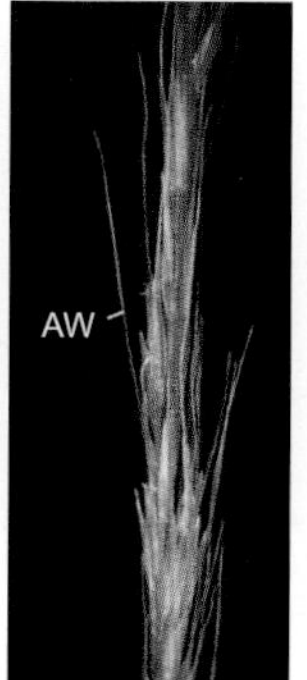

Inflorescence

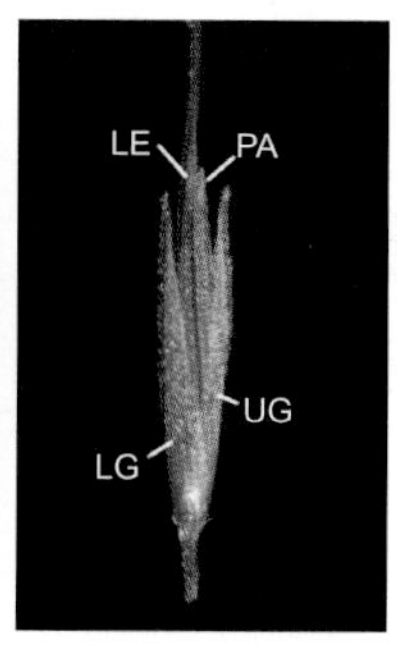

Spikelet, showing long awn

Habit

Spikelet, showing long awn

Muhlenbergia racemosa (Michx.) Britton, Sterns and Poggenb., upland wild-timothy

racemosa, Latin, "having a raceme"

Occasional in mesic prairies, on rock outcrops, along roadsides, and in other disturbed places; more common in the southwestern half of Wisconsin; closely related to *Muhlenbergia glomerata*.

Illustration

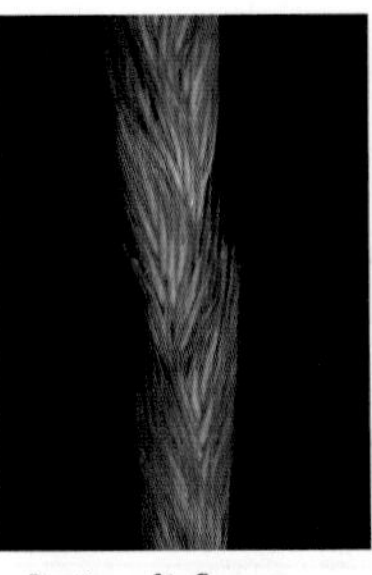

Portion of inflorescence

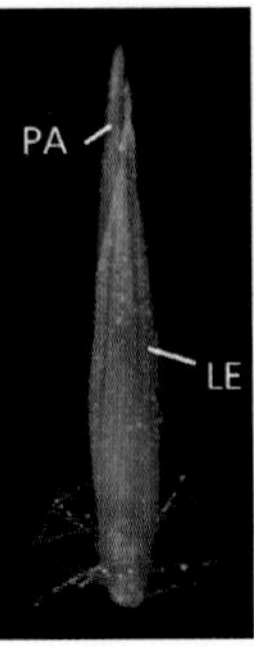

Floret

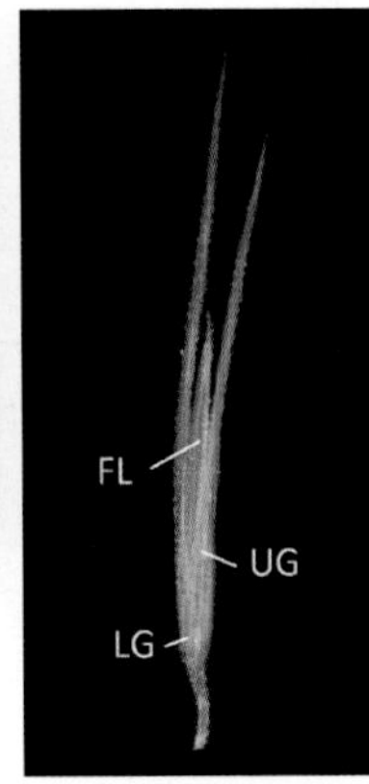

Spikelet

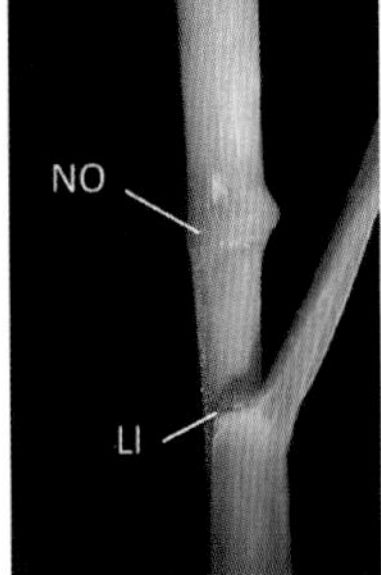

Leaf ligular area

Habit

Muhlenbergia richardsonis (Trin.) Rydb., mat muhly

richardsonis, for Sir John Richardson (1787–1865), its discoverer, Scottish naturalist and boreal and arctic North American explorer

Illustration

Rare in calcareous fens in central and southeastern Wisconsin, where first collected in 1989 by Neil Harriman and Thomas Underwood; mat-forming, with small inflorescences that are easy to overlook. In the nearby Upper Peninsula of Michigan, it is also found on seasonally dry dolomitic alvar.

Portion of inflorescence

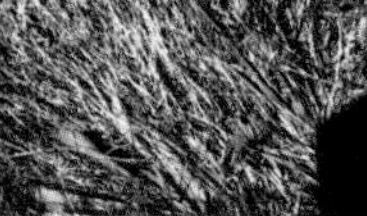

Habit

Inflorescence

Muhlenbergia schreberi J. F. Gmel., nimble-will

schreberi, for Johann Christian Daniel Schreber (1736–1810), German botanist

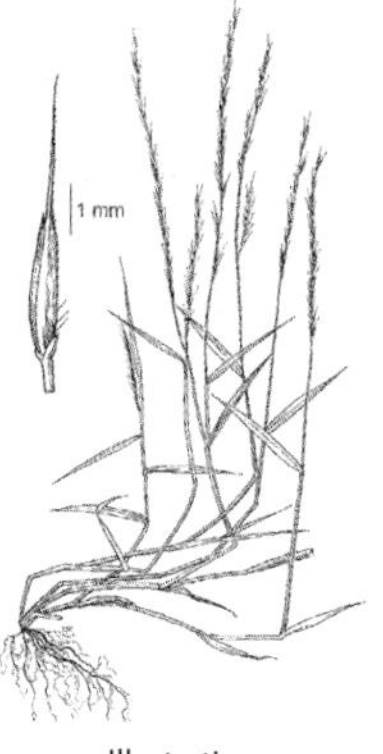

Illustration

Occasional in oak woods, pastures, and prairies; also weedy in urban areas and along trails; mostly in southern Wisconsin.

Habit

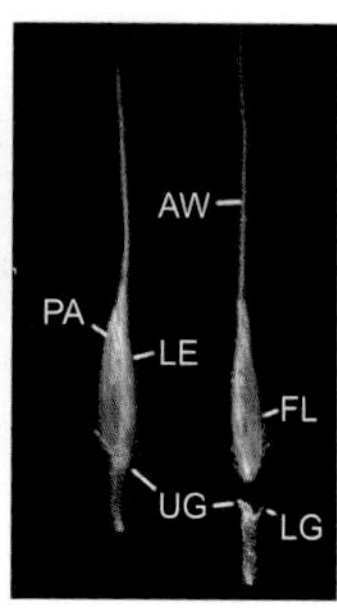

Spikelet

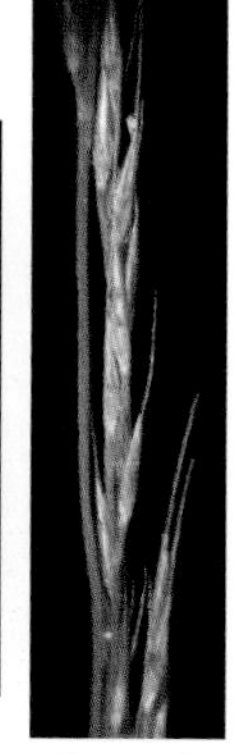

Portion of inflorescence

Muhlenbergia sobolifera (Muhl. ex Willd.) Trin., creeping muhly

sobolifera, "bearing offshoots"

Illustration

Dolomite cliffs and slopes, mainly in southwestern Wisconsin, where rare; the most recent collections have been made in Grant County by Michael Nee in 1972 and Judziewicz in 2012.

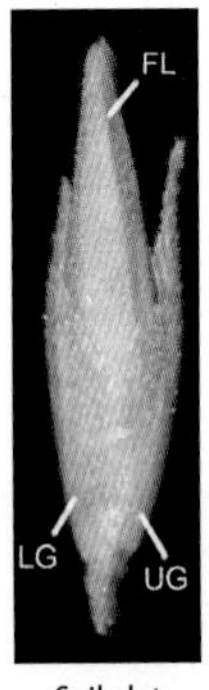

Spikelet

Habit

Muhlenbergia sylvatica (Torr.) Torr. ex A. Gray, forest muhly

sylvatica, "of or growing in woods, sylvan"

Illustration

Rare, mainly on dolomitic bluffs above the Mississippi and Lower Wisconsin Rivers in southwestern Wisconsin, but also a 1998 collection by Douglas Fields from a rich mesic forest in Taylor County.

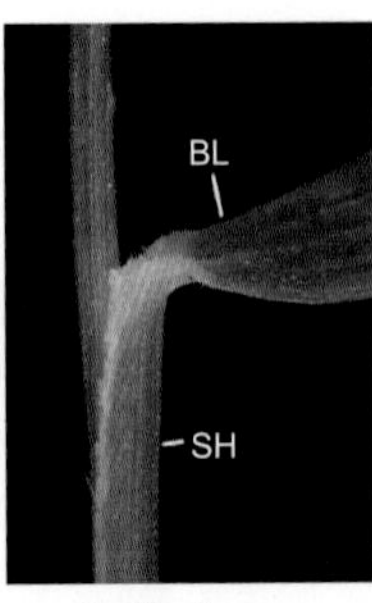

Leaf ligular area

Portion of inflorescence

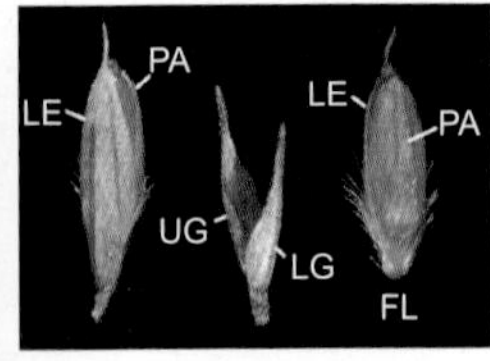

Spikelet

Muhlenbergia tenuiflora (Willd.) Britton, Sterns and Poggenb., slender muhly

tenuiflora, "with fine or delicate flowers"

Illustration

Uncommon, dry to dry-to-mesic forests, woodlands, and bluffs in southern Wisconsin.

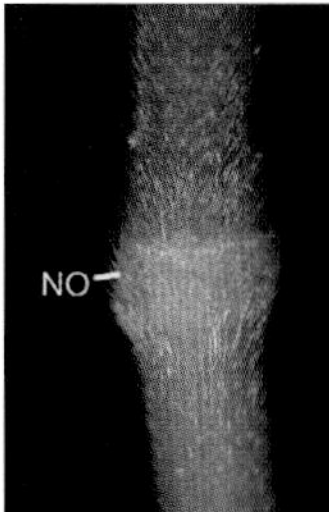

Leaf ligular area

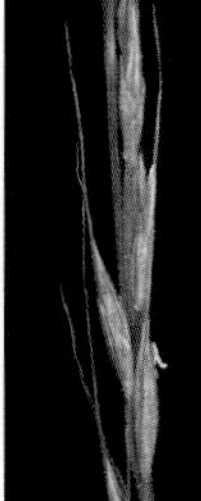

Portion of inflorescence

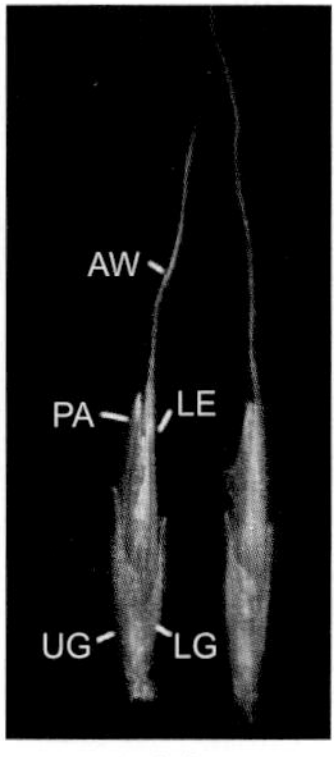

Spikelet

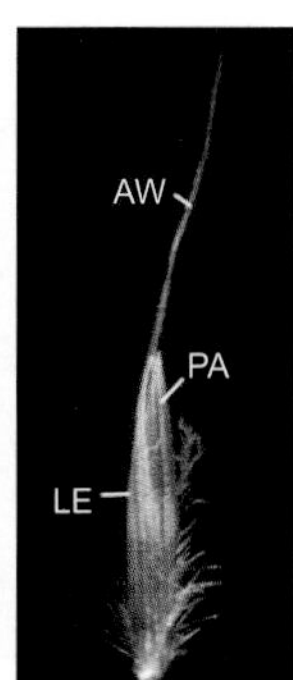

Floret

Muhlenbergia uniflora (Muhl.) Fernald, bog muhly

uniflora, Latin, "single-flowered"

Illustration

Moist, sandy-peaty shores and ditches in the bed of Glacial Lake Wisconsin and sandy outwash plains in northwestern Wisconsin (mainly the barrens country of Douglas, Bayfield, and Washburn Counties); also Vilas County. Local, but often abundant where found. Bog muhly could be mistaken for a species of *Dichanthelium* or *Panicum*, but its panicle is narrower, and the spikelets have small, subequal glumes with just one floret; no rudimentary lemma is present.

Inflorescence

Bog muhly, portion of inflorescence

Bog muhly, habit

57. *NASSELLA*, GREEN NEEDLE GRASS

Latin, *nassa*, a fishing basket with a narrow neck, and *ella*, "small"

Slender perennials with panicles of single-flowered spikelets, the floret with a sharp, bearded base and an elongate, twice-bent terminal awn; needle grasses (*Hesperostipa*) have more robust inflorescences and spikelets with awns that are only once-bent. 116 species; South America and western North America.

Nassella viridula (Trin.) Barkworth, green needle grass

viridula, Latin, *viridis*, "green"

A western US species, native as far east as western Minnesota and Iowa; in Wisconsin, a rare weed of dry ground along railroads, where it was collected from 1915 to 1965 but not since, and so perhaps no longer a part of our grass flora. Fassett (1951) treats this as *Stipa viridula*.

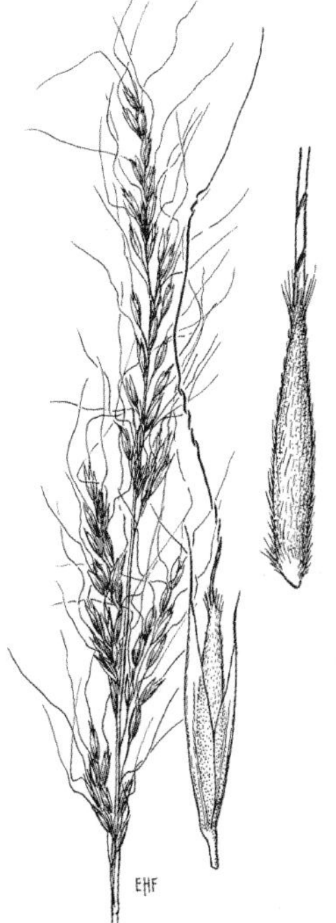

Illustration

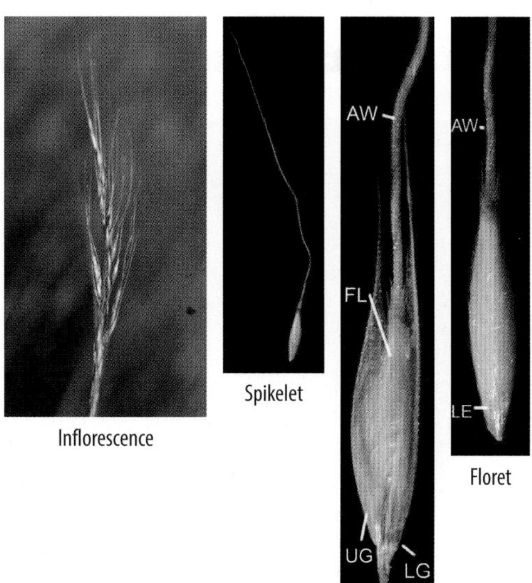

Inflorescence

Spikelet

Spikelet

Floret

58. *ORYZOPSIS*, ROUGH-LEAVED RICE GRASS

orysa, "rice," and *opsis*, "appearance," referring to its similar appearance to rice

Woodland perennials with spike-like, few-flowered panicles; spikelets single-flowered, the glumes fairly wide, the lemma hard, cylindrical, with obscure nerves, and tipped by a weak awn. Several other species traditionally placed in this genus are now treated as species of *Patis* and *Piptatheropsis* (Romaschenko et al. 2011). A single North American species.

Oryzopsis asperifolia Michx., rough-leaved rice grass

asperifolia, "rough-leaved"

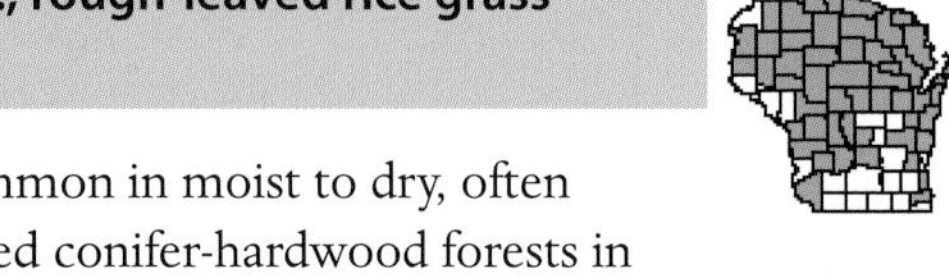

Illustration

Common in moist to dry, often mixed conifer-hardwood forests in central and northern Wisconsin; early flowering (April–May) and with relatively wide, coarse all-basal leaves (turned upside down!) that remain green all winter.

Florets

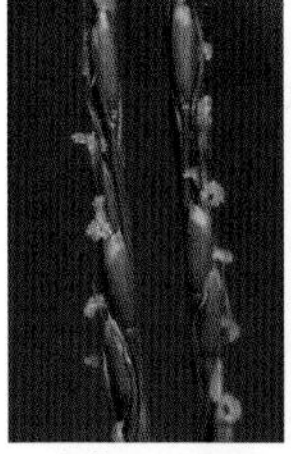

Inflorescence

Spikelets

Habit

59. *PANICUM*, PANIC GRASS

Latin, *panus*, probably meaning "an ear of millet"

Small annuals to robust perennials lacking an overwintering rosette of short, wide leaves, flowering occurring during summer and early fall in terminal panicles; spikelets awnless, mostly slenderly to widely ovate, consisting of a short lower glume, an upper glume as long as the spikelet, a sterile lower lemma resembling the upper glume, and a slightly smaller, disarticulating, smooth, leathery to hard, fertile floret (Shinners 1944). Several genera resemble *Panicum*, most notably the commoner and generally smaller-of-stature genus *Dichanthelium*, which differs in the presence of an overwintering rosette of leaves, a tendency to begin flowering in the spring, and spikelets that are usually plumper and hairier. Other similar grasses include cup grass (*Eriochloa*), but with a distinctive knob-like floret base; bead grass (*Paspalum*), with hemispherical spikelets and the lower glume essentially absent; and the rare Munro grass (*Coleataenia*), with tiny prickle-like hairs present at the summit of the fertile floret. Common witch grass (*Panicum capillare*) and switchgrass (*P. virgatum*) are the two common native species, and fall panic grass (*P. dichotomiflorum*) is the most common introduction. About one hundred species; cosmopolitan.

1. Spikelets 3–6 mm long
   2. Lower glume at least half as long as the spikelet; lower floret male; plants usually over 0.75 m tall with a terminal panicle at last 20 cm tall; plants nearly glabrous .... *P. virgatum*
   2. Lower glume about one-quarter to one-third as long as the spikelet; lower floret completely sterile; plants usually less than 0.75 m tall, with a terminal and/or lateral panicles less than 15 cm tall; plants glabrous to pubescent, weedy
      3. Leaves more or less glabrous; common, sprawling .... *P. dichotomiflorum* (in part)
      3. Leaves harshly pubescent on the backs of the leaf sheaths; uncommon, often near bird feeders .... *P. miliaceum*
1. Spikelets 1.5–3 mm long
   4. Backs of leaf sheaths glabrous
      5. Spikelets 1.5–2.3 mm long; ligules membranous, 0.5–1 mm long; tufted perennial; rare, sandbars and slough margins along the lower Wisconsin River, perhaps extirpated .... *Coleataenia rigidula*
      5. Spikelets at least 2.4 (to 3.8) mm long; ligules membranous with a ciliate apex, 1–2.5 mm long; common sprawling annual weed .... *P. dichotomiflorum* (in part)
   4. Backs of leaf sheaths long-hairy, often with harsh, papillose-based hairs
      6. Spikelets 1.5–2 mm long, ovate; uncommon weedy native, mostly central and northern Wisconsin .... *P. philadelphicum*
      6. Spikelets over 2 (to 4) mm long, lanceolate

7. Mature panicle several times as tall as wide, with ascending branches; rare, fens and dry dolomite, eastern Wisconsin . . . . *P. flexile*
7. Mature panicle about as long as wide, with spreading branches; common weedy native statewide . . . . *P. capillare* (in part)

Panicum capillare **L., common witch grass**

capillare, "hair-like"

Illustration

Abundant weedy native found in many open habitats; most characteristic of roadsides, gardens, agricultural fields, and pastures, but also found in prairies, barrens, and wetland margins. One of our three "purple tumbleweed grasses"; the other two, fall witch grass (*Digitaria cognata*) and purple love grass (*Eragrostis spectabilis*), generally flower a bit later in the season.

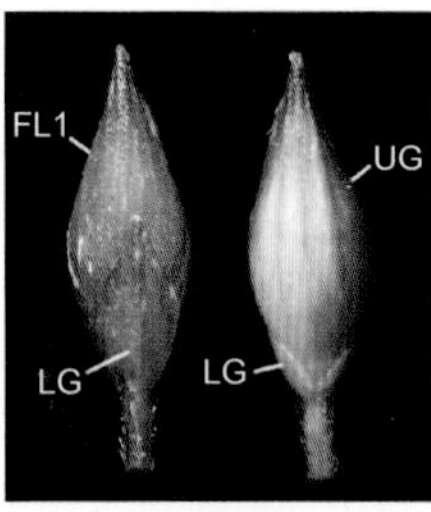

Spikelet

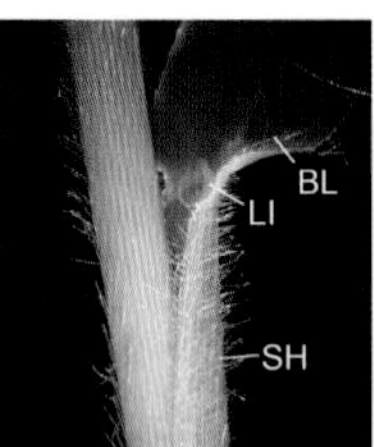

Leaf ligular area

Habit, mature inflorescence

Habit, emerging inflorescence

Panicum dichotomiflorum Michx., elbow grass, fall panic grass

dichotomiflorum, "flowers forked in pairs"

Illustration

Perhaps native to the eastern United States, but probably introduced to Wisconsin, where first collected in 1914, and widespread by 1940 (Shinners 1940). It is most frequent as a cornfield and roadside weed, but is also found in natural habitats such as marshes and shorelines.

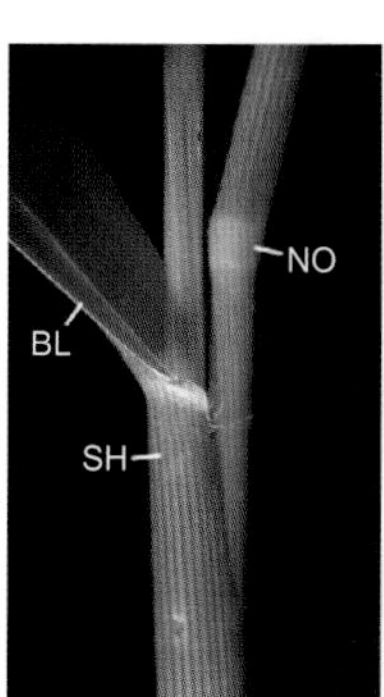

Leaf ligular area

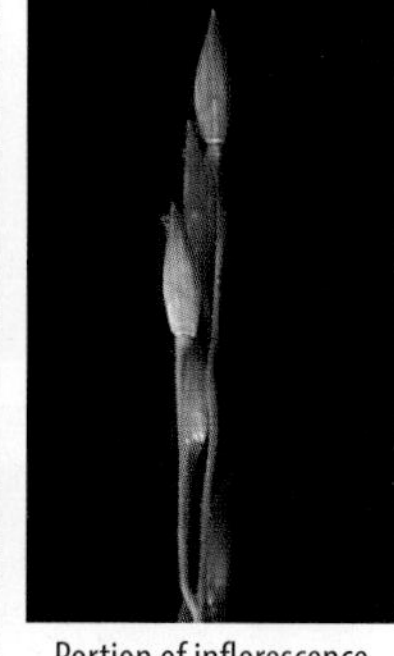
Portion of inflorescence

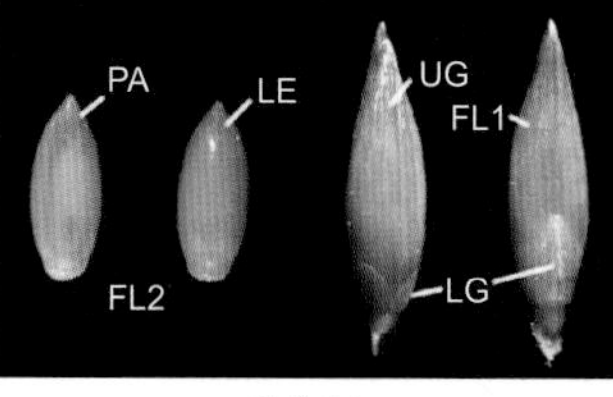

Spikelet

Panicum flexile (Gatt.) Scribn., fen panic grass

flexile, "flexible"

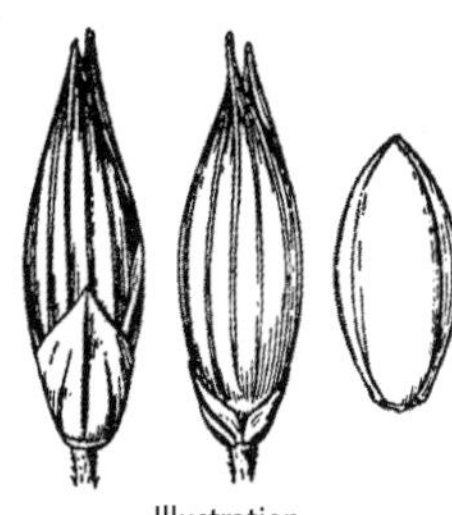
Illustration

Rare; usually in fens and other wetland margins in southeastern Wisconsin, but also a recent collection by Gary Fewless from seasonally dry dolomitic alvar in Brown County; first collected in 1940 by Lloyd Shinners.

Portion of inflorescence

Panicum miliaceum **L., broomcorn millet, proso millet**

miliaceum, pertaining to millet, or "millet-like"

A Eurasian exotic, first collected in 1916, and now an occasional weed in railroad rights-of-way, roadsides, waste ground, and urban areas; sometimes becoming established from birdseed provided at feeders. The typical subspecies has yellow upper florets, while the uncommon invasive subspecies *ruderale* (Kitag.) Tzvelev has dark-colored upper florets.

Leaf ligular area showing papillose-pilose hairs

Spikelet

Upper floret

Spikelets

Habit

Portion of inflorescence

Panicum philadelphicum Bernh. ex Trin., Philadelphia panic grass

philadelphicum, "of Philadelphia"

Habit

Uncommon; sandy roadsides, ditches, shores, and prairies. Voss and Reznicek (2012) consider the correct name for midwestern plants to be *Panicum tuckermanii* Fernald.

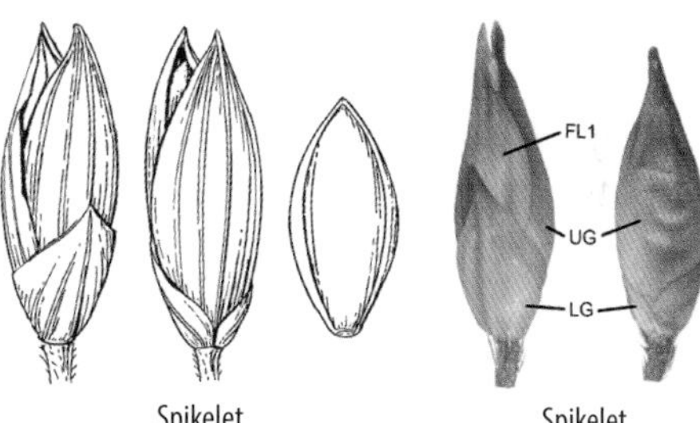

Spikelet Spikelet

Panicum virgatum L., switchgrass

virgatum, Latin, "wand-like"

Illustration

Common in prairies and sand barrens and on riverbanks in the southern two-thirds of Wisconsin; often planted north of its natural range and sometimes aggressively spreading from plantings and along roadsides. A large, robust bunchgrass, but one that was not a dominant in Wisconsin prairies in presettlement times. It is very productive and has been used as a biofuels source (USDA NRCS 2014).

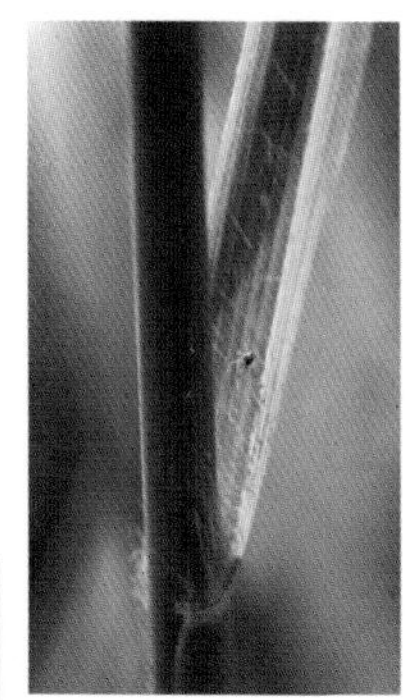
Leaf ligular area

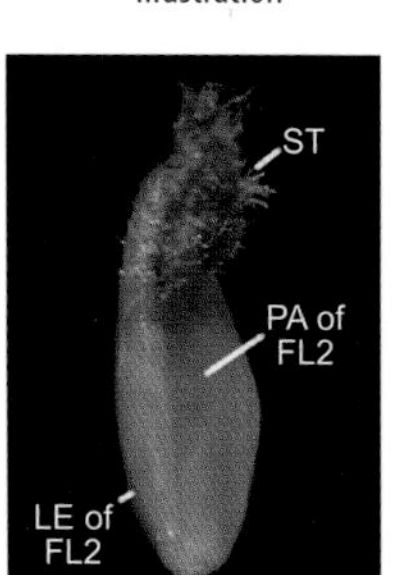

Upper floret

Spikelet

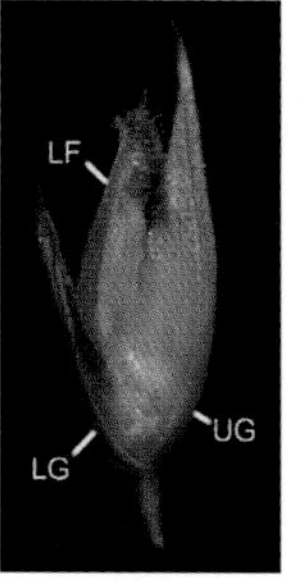

Spikelet

Switchgrass, portion of inflorescence

60. *PASCOPYRUM*, WESTERN WHEATGRASS

Latin, *pasco*, "pasture," and Greek, *pyros*, "grain"

This rare introduction resembles quack grass (*Elymus repens*), differing in that the glumes are ciliate (not smooth) and that the upper glume is longer than the lower one; the leaf blades tend to be bluish and stiff, with a few coarse veins. A single western North American species.

***Pascopyrum smithii* (Rydb.) Á. Löve, western wheatgrass**
smithii, for Jared Gage Smith (1866–1925), botanist and agrostologist for the United States Department of Agriculture

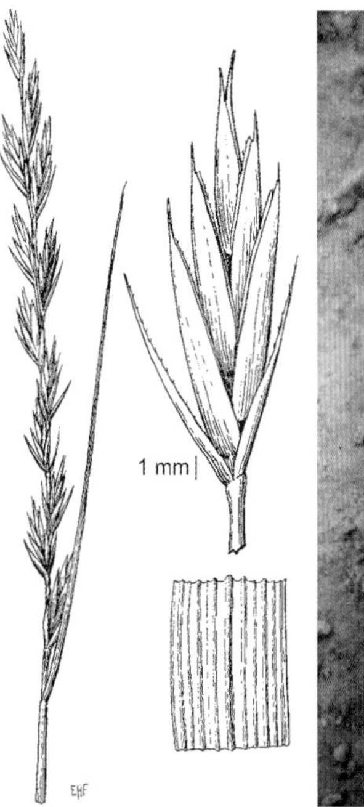

Illustration

Portion of inflorescence

A western species that was formerly frequently collected along railroads from 1889 until 1958; since then the only collection is one made in 1993 in Washington County by Daniel Spuhler. The foliage tends to be glaucous (bluish whitened). Fassett (1951) treats this as *Agropyron smithii.*

61. *PASPALUM*, BEAD GRASS

Greek word for "millet"

A late summer–flowering relative of the panic grasses *Panicum* and *Dichanthelium*, but easily distinguished from them by the awnless, nearly glumeless bead-like spikelets, which are round in outline and plano-convex in shape, like a split pea. Three hundred to four hundred species of tropical and warm temperate areas, mostly in the Western Hemisphere.

Paspalum setaceum **Michx., bead grass**

setaceum, "bristled"

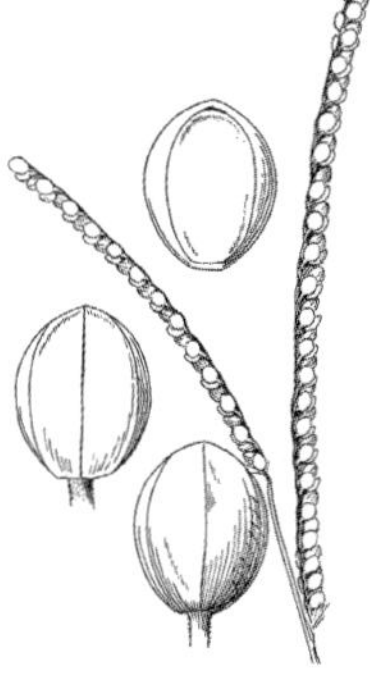
Portion of inflorescence

Occasional to fairly common in southern and central Wisconsin in sand prairies and oak woodlands and on dunes along the lower Wisconsin River; also found along roadsides and in old fields; it may be spreading northward. Fassett (1951) treats this as *Paspalum ciliatifolium*.

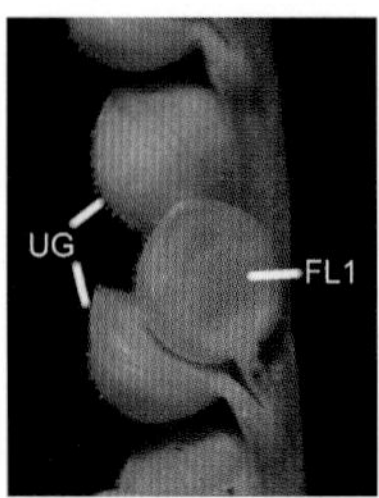

Portion of inflorescence

Leaf ligular area

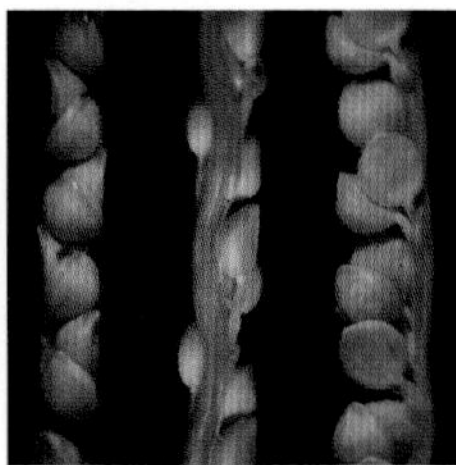
Portion of inflorescence

Habit

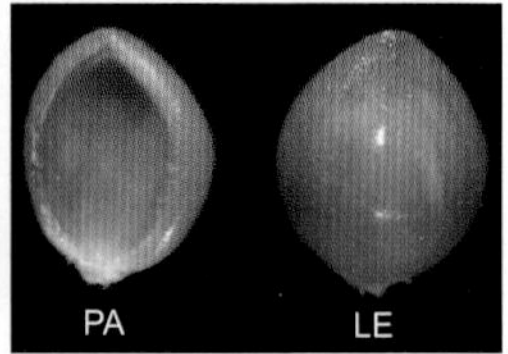

Upper floret

62. *PATIS*, BLACK-SEEDED RICE GRASS

Latin, *patus*, "pasture"

A conspicuous, clump-forming woodland grass with relatively broad, shining leaf blades, flowering in the summer; the inflorescence is a few-branched, few-flowered panicle with single-flowered spikelets; the glumes are as long as the floret body, while the floret is dorsally compressed, dark in color at maturity, and tipped by a slender straight awn. It differs from its relatives the mountain rice grasses (*Piptatheropsis*) in its longer lemmas and broader leaves, from rough-leaved rice grass (*Oryzopsis*) in its deciduous leaves, and from the needle grasses (*Hesperostipa* and *Nassella*) in its straight awns (Romaschenko et al. 2011). Three north temperate species, ours and two from eastern Asia.

Indian rice grass (*Eriocoma hymenoides* (Roem. and Schult.) Rydb.) is a common western US species that occasionally occurs as an introduction east to several Minnesota counties near the Wisconsin border; it may eventually be found in our state. It would key to *Patis* but has much narrower leaves less than 1 mm wide, larger panicles with dichotomously forking branches, and larger spikelets 5–10 mm long with strongly bearded florets.

Patis racemosa **(Sm.) Romasch., P. M. Peterson and R. J. Soreng, black-seeded rice grass**

racemosa, Latin, "having a raceme"

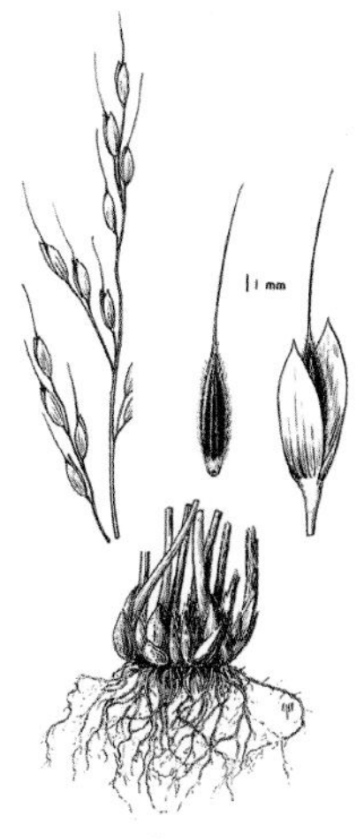

Illustration

Fairly common in the understory of mesic to dry to mesic, mostly hardwood forests; statewide except in the far north. The leaves are shiny and relatively broad. Fassett (1951) treats this as *Oryzopsis racemosa*, Barkworth et al. (2007) as *Piptatherum racemosum*.

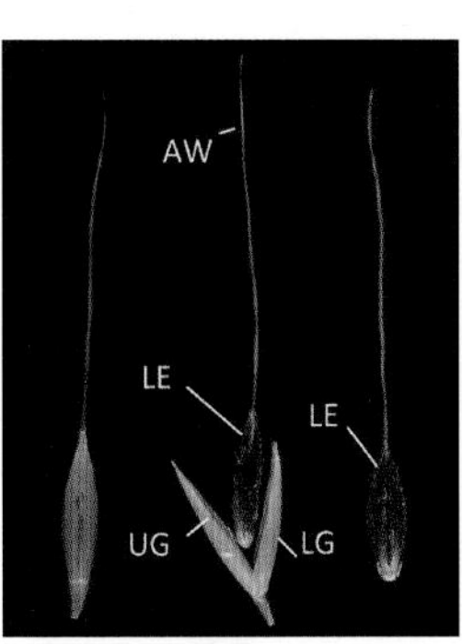

Spikelet

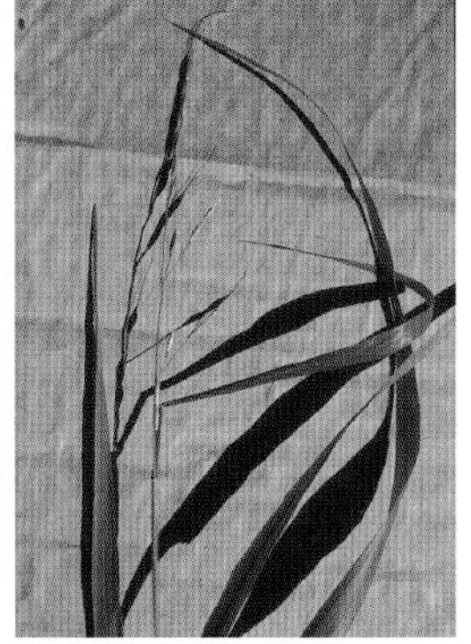
Upper leaves and inflorescence

Habit

63. *PHALARIS*, CANARY GRASS

Greek, "shining," referring to the shining fertile floret or possibly to the crest-like seed head

Perennials with panicles that expand at flowering time, then quickly contract and become spike-like as the grains mature; spikelets awnless, with one rather hard, smooth white floret and, attached to its base, two tiny, hairy rudimentary florets. Twenty-two species of temperate regions.

1. Panicle an open panicle 8–30 cm long, soon contracting and becoming elongate and spike-like; glumes 4–6 mm long; rhizomatous perennial; abundant wetland weed .... *P. arundinacea*
1. Panicle a small, compact, ovoid panicle 2–4 cm long; glumes 6–8 mm long; annual; rare weed .... *P. canariensis*

Phalaris arundinacea **L., reed canary grass**

arundinacea, "reed-like"

A Eurasian exotic, first collected in 1884, but present earlier (Lapham 1853). The most abundant, widespread, and pernicious exotic wetland grass in Wisconsin; often forming pure stands that displace sedge meadows and other wetland communities. Most commonly confused with Canada bluejoint (*Calamagrostis canadensis*), from which it differs in its broader leaves (10–20 mm vs. 4–8 mm wide) and slightly larger, awnless spikelets, the lemma lacking a callus beard.

Leaf ligular area

Portion of inflorescence

Open young inflorescence

Illustration

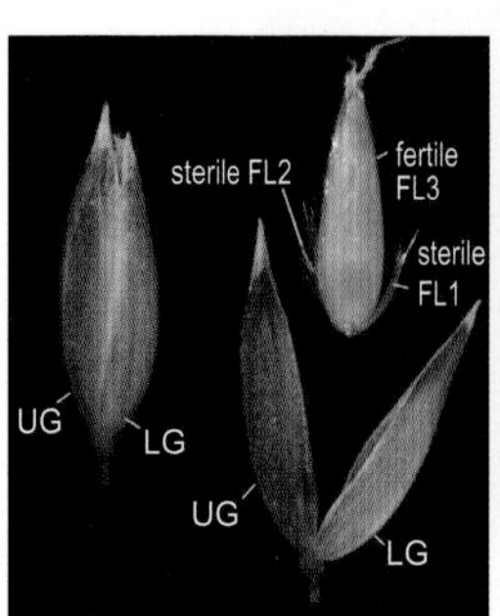

Spikelets

Contracted mature inflorescence

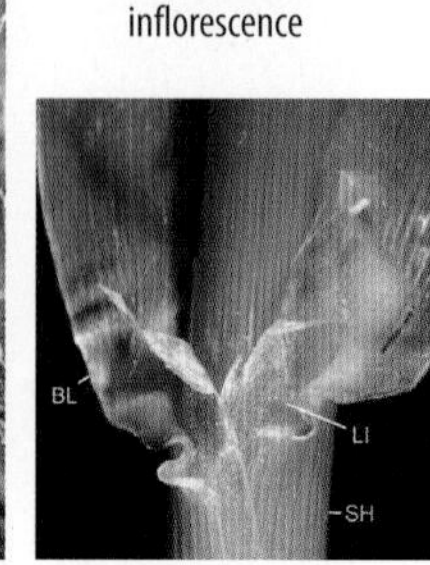

Leaf ligular area

Open young inflorescence

Reed canary grass dominating a wetland

Reed canary grass, dense network of rhizomes

Phalaris canariensis **L., common canary grass**
canariensis, "of the Canary Islands"

A rare European weedy exotic, first collected in 1894; much less common than its larger cousin; roadsides, railroads, and farmyards.

Illustration

Inflorescence

64. *PHLEUM*, TIMOTHY

phleos, Greek word for a kind of reed

A bulbous-based weedy perennial with pencil-like spikes. The spikelets are single-flowered with persistent glumes that are placed side by side as "mirror images" of each other; each glume has a hispid keel prolonged into a short awn, while the floret is awnless. Timothy could be confused with the meadow foxtails (*Alopecurus*), but in that genus the spikelets fall as a unit (including the glumes), the glumes are awnless, the lemma is awned, and the anthers are yellow. Fifteen species; Eurasia and northern North America.

Phleum pratense L., timothy

pratense, "of meadows"

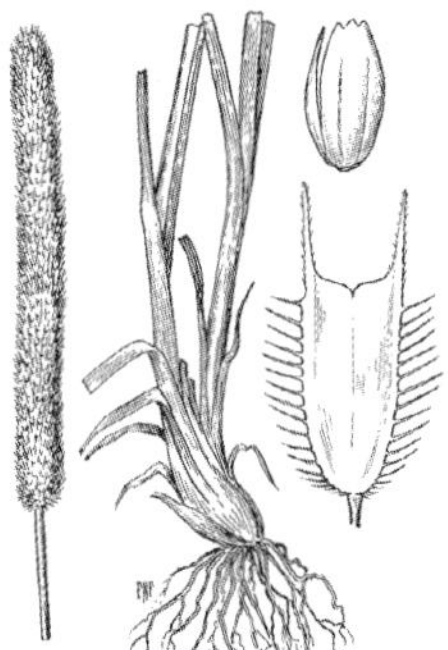

Illustration

A Eurasian exotic, first collected in 1880. A widespread and abundant hay and cover grass and a most characteristic weed of roadsides, meadows, pastures, and old fields. It is relatively benign and never becomes a long-term invasive dominant, in contrast to such species as smooth brome (*Bromus inermis*) and Canada bluegrass (*Poa compressa*).

Portion of inflorescence

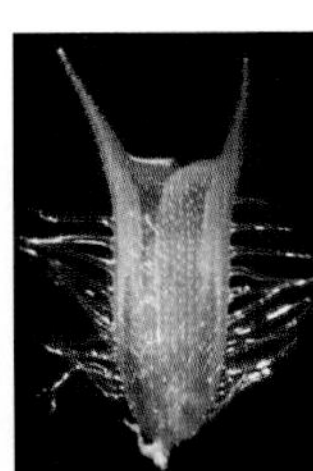

Spikelet

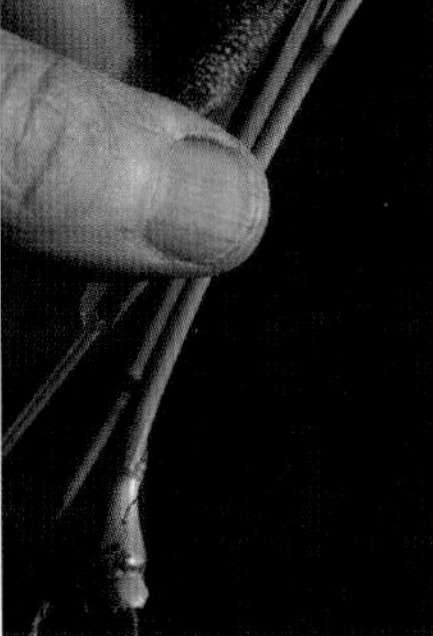

Bulbose culm base

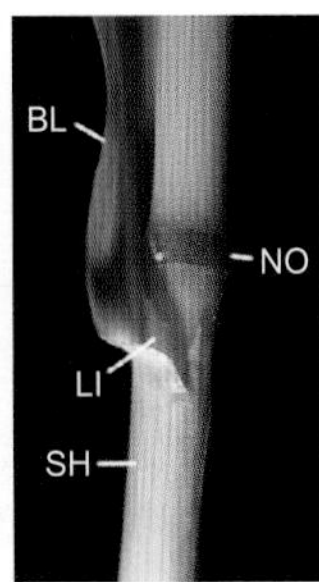

Leaf ligular area

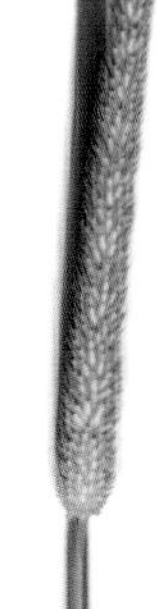

Inflorescence

65. *PHRAGMITES*, REED

Greek, *phragma*, "a fence or screen, hedge," hence growing in hedges

Robust reeds forming dense wetland clones via elongate rhizomes, the broad leaves often tending to "flag," that is, to all face the same direction. The inflorescence is a huge, pyramidal panicle that is smooth as it expands but becomes white and fluffy at maturity with hairs borne on the spikelet rachillas; spikelets many-flowered, with distant slender glumes and florets partly concealed by the abundant rachilla hairs. One to several species; cosmopolitan.

Phragmites australis **(Cav.) Trin., common reed**

australis, "southern"

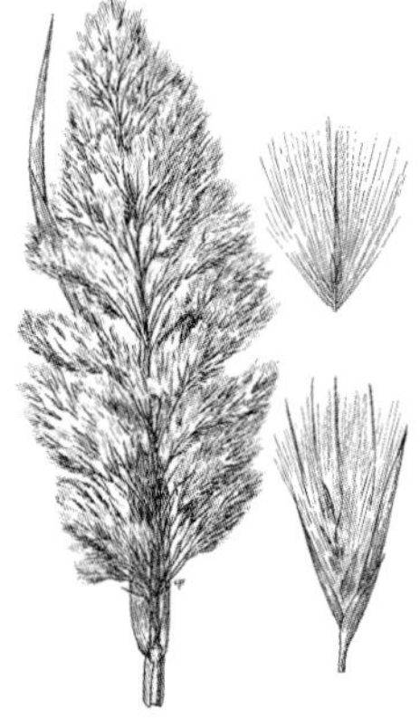

Illustration

Locally common large grass of wetlands statewide. The commonest form is an invasive ecotype introduced from Europe sometime during the twentieth century and now forming dominant, jungle-like stands in many places, such as in the wetlands on the western shore of Green Bay. The now-uncommon native strain of this species is *Phragmites australis* subsp. *americanus* Saltonstall, P. M. Peterson and Soreng (Saltonstall, Peterson, and Soreng 2004). It forms much less dense colonies, with leaf sheaths that fall early to reveal the shiny reddish culms surmounted by bare winter panicles (the fuzzy spikelets do not persist throughout the winter). The native also differs in its longer ligules (1–2 mm vs. 0.5–1 mm long) and larger glumes (the lower glumes are 3–6 mm vs. 2.5–5 mm long; the upper are 5–11 mm vs. 5–8 mm long) than the introduced strain. Fassett (1951) treats this as *P. communis.*

Inflorescence in early fall

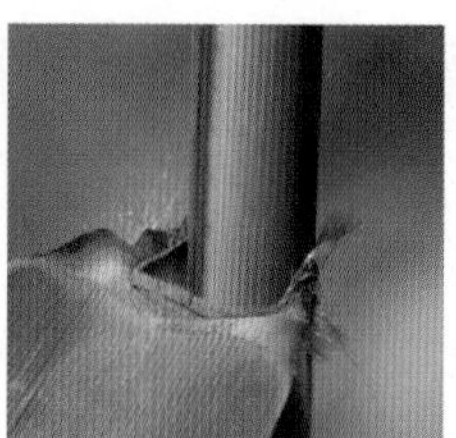

Leaf ligular area, Eurasian ecotype

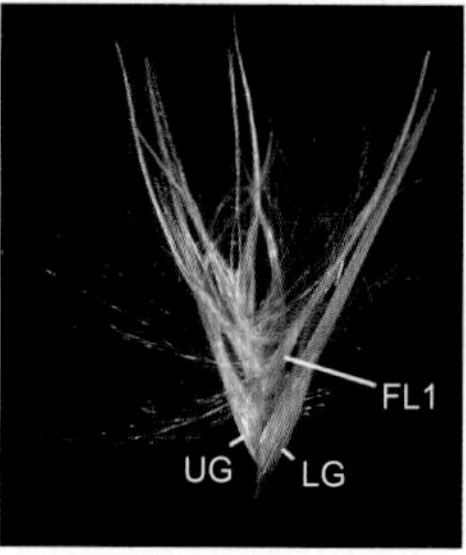

Spikelet showing long-hairy rachilla internodes

Inflorescence in midsummer

Common reed, dense stand of invasive Eurasian ecotype in midwinter

66. *PIPTATHEROPSIS*, MOUNTAIN RICE GRASS

Greek, *pipto*, "to fall," and *opsis*, "like" or "resembling"

Inconspicuous summer-flowering, narrow-leaved grasses with few-branched and few-flowered panicles with single-flowered spikelets; the glumes are as long as the whitish, membranous floret body, the lemma tipped by a slender awn. *Piptatheropsis* differs from black-fruited rice grass (*Patis racemosa*) and rough-leaved rice grass (*Oryzopsis asperifolia*) in its much narrower leaves and from the needle grasses (*Hesperostipa* and *Nassella*) in its strongly twisted awns (Romaschenko et al. 2011). Four North American species.

1. Awns 5–10 mm long, persistent; rare, dry open sandy areas, central and northern Wisconsin . . . . *P. canadensis*
1. Awns 1–2 mm long, quickly falling; occasional dry areas (often pine forests), except in southeastern Wisconsin . . . . *P. pungens*

Piptatheropsis canadensis (Poir.) Romasch., P. M. Peterson and R. J. Soreng, Canada mountain rice grass

canadensis, of or referring to Canada

SPECIAL CONCERN

Illustration

Rare in more or less open dry sand or moist peaty-sandy ground in northern and central Wisconsin; often associated with pines and sites that have had a history of fire. The first collection was made by Thomas Hartley in 1958 in Jackson County. The Wisconsin Department of Natural Resources (2014) treats this as *Oryzopsis canadensis*, Barkworth et al. (2007) as *Piptatherum canadense*.

Portion of inflorescence

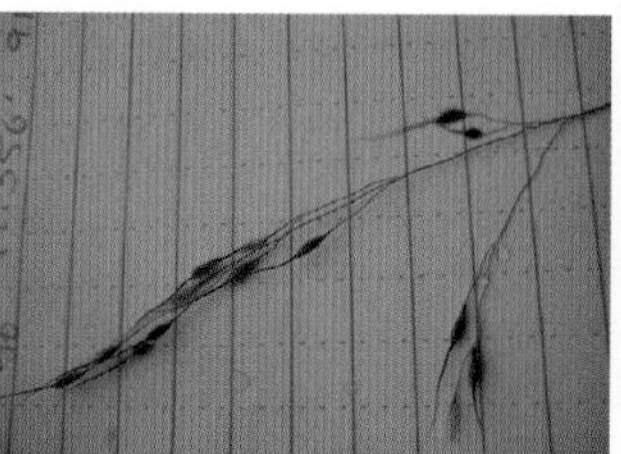

Inflorescence

Inflorescence

Piptatheropsis pungens (Torr. ex Spreng.) Romasch., P. M. Peterson and R. J. Soreng, mountain rice grass
pungens, "spiny, sharp-pointed"

Fairly common in dry to dry-to-mesic pine or less commonly oak forests or woodlands in all but southeastern Wisconsin; a small, spring-flowering, easily overlooked species that superficially resembles a species of *Dichanthelium* or *Panicum*. Fassett (1951) treats this as *Oryzopsis pungens*, Barkworth et al. (2007) as *Piptatherum pungens*.

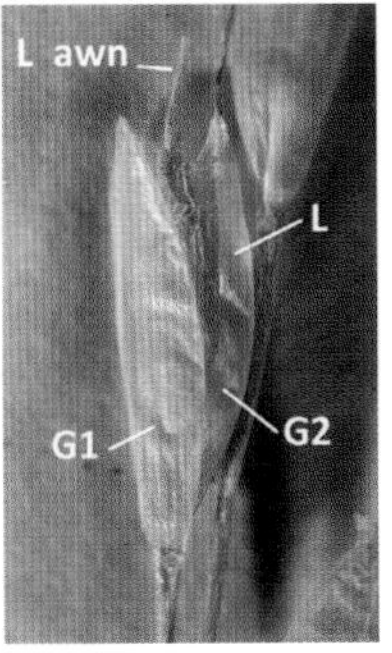

Illustration Inflorescence Spikelet

67. *POA*, BLUEGRASS

ancient Greek word for "grass" or "fodder"

A familiar cool-season grass genus, distinctive in its combination of these characters: panicle inflorescence, boat-shaped leaf blade apices, glumes shorter than the lowest floret, and lemmas usually with a tuft of cobwebby, crinkled hairs at the base, the lemma nerves not prominent and converging at the tip. To key out bluegrasses, one must pay particular attention to whether the lemmas have three or five nerves, whether these nerves are hairy, and, if so, whether the hairs are restricted to a particular part of the lemma. Yet, study of this difficult genus may be rewarded by the discovery of one of several rare native species known to occur in Wisconsin hardwood forests. Kentucky bluegrass (*Poa pratensis*) and Canada bluegrass (*P. compressa*) are the two commonest weedy species, while exotics bulblet bluegrass (*P. bulbosa*) and wood bluegrass (*P. nemoralis*) are becoming more frequent. Several other bluegrasses approach near to the Wisconsin border, including the introduced *P. chapmaniana* Scribn. in Illinois and the native *P. interior* Rydb. in the vicinity of Duluth, Minnesota. Five hundred species; cosmopolitan except in lowland tropical areas.

Boat-shaped tip of leaf blade in bluegrasses

1. Base of plant bulbous, the inflorescence producing bulblets; rare weed .... *P. bulbosa*
1. Base of plant not bulbous, the inflorescence never producing bulblets
   2. Base of floret lacking a tuft of cobwebby hairs
      3. Low, mat-forming, somewhat succulent annual weeds; lemmas five-nerved; lower glume one-nerved; common .... *P. annua*
      3. Erect, nonsucculent perennials; lemmas three-nerved; lower glume three-nerved
         4. Culms strongly flattened, with several exposed nodes per culm; upper glume 2–3 mm long; rachilla smooth; plants rhizomatous, the foliage green; common weed of dry open areas .... *P. compressa* (in part)
         4. Culms not flattened, with at most one exposed node per culm; upper glume 3–4.5 mm long; rachilla hairy; plants not rhizomatous, the foliage bluish whitened; clay bluffs and sandstone shores of Lake Superior .... *P. glauca* (in part)
   2. Base of floret with a tuft of cobwebby hairs
      5. Margins of lemma glabrous, even near the base
         6. Panicle branches one or two at lower nodes; lemma midribs smooth

7. Well-developed ligules of upper leaves 2.5–5 mm long; tips of lemmas obtuse; rare, eastern Wisconsin . . . . *P. languida*
7. Well-developed ligules of upper leaves 0.5–2.5 mm long; tips of lemmas acute; occasional, northern Wisconsin . . . . *P. saltuensis*
6. Panicle branches three to seven at lower nodes; lemma keels scabrous to hairy
8. Ligules 0.5–3 mm long; lemmas three-nerved; common . . . . *P. alsodes*
8. Ligules 3–10 mm long; lemmas five-nerved; uncommon . . . . *P. trivialis*
5. Margins of lemma hairy, at least near the base
9. Lemmas five-nerved; ligules 0.5–2.5 mm long
10. Lemma keels hairy almost to the end of the membranous part, the backs of the lemmas finely hairy throughout
11. Foliage a ghostly grayish-white color; spikelets 3.5–4.3 mm long; exotic, recently invading along salted interstate highway margins in southeastern Wisconsin . . . . *P. arida*
11. Foliage green; spikelets 2.5–3.5 mm long; native, rare, moist woods, southern Wisconsin . . . . *P. sylvestris*
10. Lemma keels hairy only about two-thirds of the way to the end of the membranous green part; backs of lemmas glabrous between the nerves
12. Plants rhizomatous; abundant weed and cultivar . . . . *P. pratensis*
12. Plants not rhizomatous; rare, moist woods, northern Wisconsin . . . . *P. wolfii*
9. Lemmas three-nerved; ligules various
13. Well-developed ligules 2.5–5 mm long; common native, wetlands . . . . *P. palustris*
13. Well-developed ligules 0.2–2.5 mm long
14. Rachilla smooth; panicle branches one to four (usually two) per node; upper glumes 2–3 mm long
15. Culms strongly flattened; panicle contracted, with erect branches; plants rhizomatous; well-developed ligules 1–2 mm long; common weed of dry open areas . . . . *P. compressa* (in part)
15. Culms not flattened; panicle open; plants not rhizomatous; well-developed ligules 0.5–1 (rarely 1.5) mm long; rare, forested seeps and fens . . . . *P. paludigena*
14. Rachilla hairy or scabrous; panicle branches two to five per node; upper glumes rarely 2.5 mm, usually 3–4.5 mm long
16. Well-developed ligules 1–2 mm long; at most one culm node exposed; foliage bluish whitened, the leaves ascending; native, clay bluffs and sandstone shores of Lake Superior . . . . *P. glauca* (in part)
16. Well-developed ligules 0.5–1 mm long; several culm nodes exposed; foliage green, the leaves often divergent; aggressive exotic in forests throughout the state . . . . *P. nemoralis*

Poa alsodes A. Gray, grove bluegrass

alsodes, "of woods"

Illustration

A common early spring–flowering species with relatively large spikelets and slightly curving glumes; in mesic hardwood or mixed forests; often along trails or old logging roads.

Habitat and habit

Poa annua L., annual bluegrass

annua, Latin, "yearly," usually referring to a plant that lives only one season

Illustration

A Eurasian exotic, first collected in 1878.
A low, cool-season weed in disturbed areas, including in lawns and sidewalk cracks and near buildings; abundant throughout. It can flower during mild spells in the winter but disappears during summer heat. The spikelets have only two or three rather distant florets, and the whole plant is somewhat succulent.

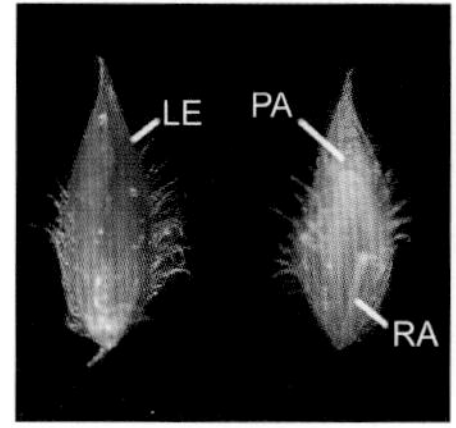

Florets

Spikelet

Habit

Poa arida Vasey, plains bluegrass

arida, "arid or dry"

Illustration

This salt-tolerant native of the Great Plains has been known for twenty years along freeways in the Chicago area (Swink and Wilhelm 1994) and was collected in 2008 by Gerould Wilhelm along Interstate 94 in Kenosha and Racine Counties (vPlants 2014). It is an early-flowering, soon-withering plant with ghostly grayish-white foliage.

Spikelet and floret

Habit

Poa bulbosa L., bulblet bluegrass

bulbosa, "bulbous, swollen"

A European exotic that is becoming more frequent, typically in lawns. It was introduced in the 1950s for turf, pasture, and erosion control but can be too aggressively weedy for such purposes (USDA NRCS 2014); it is seriously invasive in the western United States. First collected in 1984 by Gary Fewless and now locally common in the Green Bay area.

Illustration

Habit

Inflorescence

Habitat

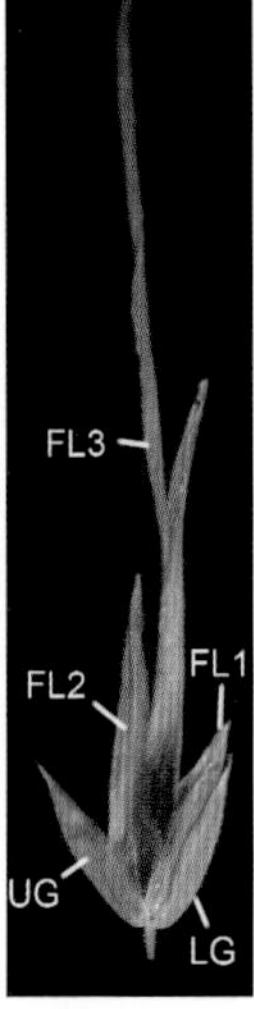

Spikelet, showing third floret modified into a bulblet

Poa compressa L., Canada bluegrass

compressa, "flattened, compressed"

Despite the common name, this is a European exotic, present early (Lapham 1853). It is abundant statewide in dry open areas: roadsides, fields, pastures, prairies, barrens, road outcrops, and dunes. The flattened culms and the unusually small, long-peduncled panicle are often diagnostic, but this species is highly variable.

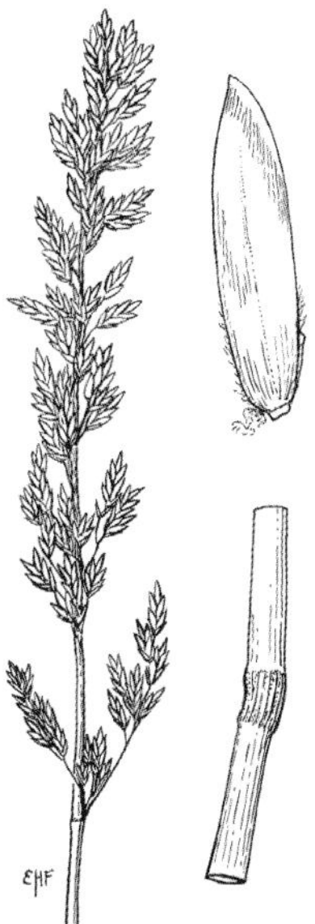

Illustration

Spikelet and floret

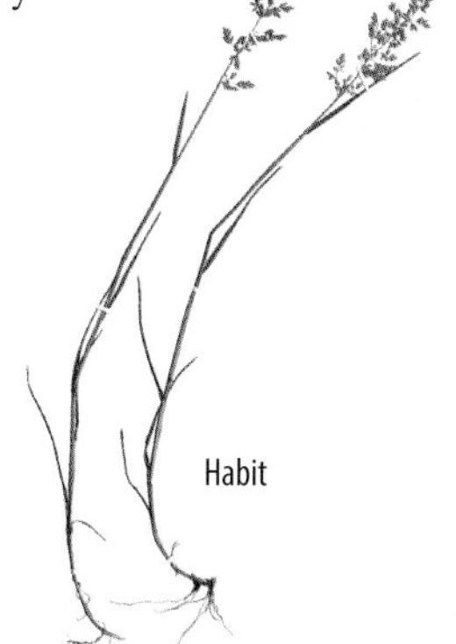

Habit

Poa glauca Vahl, glaucous bluegrass

glauca, Greek, "grayish or bluish green"

A boreal species that is locally common on wave-splashed sandstone ledges and clay bluffs of Lake Superior in the Apostle Islands and adjacent Bayfield Peninsula, but seemingly nowhere else in the state.

Illustration

Base of plant showing glaucous cast

Inflorescence

Poa languida **Hitchc., languid bluegrass**
languida, "weak"

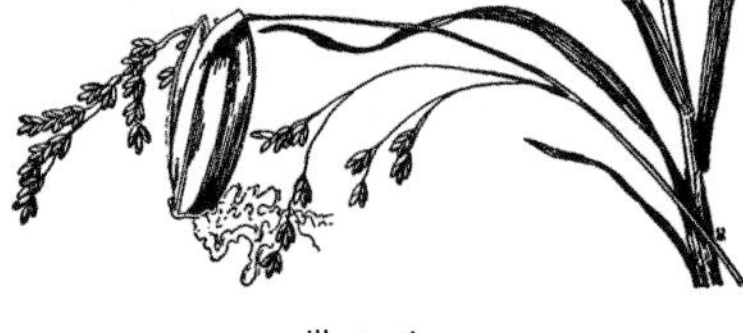
Illustration

Rare in rich hardwood forests in eastern Wisconsin. The long ligules and blunt lemmas are distinctive; it is closely related to *Poa saltuensis* and united by some with that species (Barkworth et al. 2007).

Poa nemoralis **L., wood bluegrass**
nemoralis, "growing in groves or woods"

Illustration

A European exotic, first collected in 1897. The narrow glumes and usually divaricate foliage are diagnostic. This species is becoming badly invasive in many northern hardwood and mixed conifer-hardwood forests in northern Wisconsin. For example, at Kemp Biological Station near Minocqua in Oneida County, it now dominates the understory of a wind-damaged hemlock–sugar maple stand. Gary Fewless (pers. comm.) reports it as increasing rapidly in forests on Washington Island, Door County.

Foliage with divaricate leaves

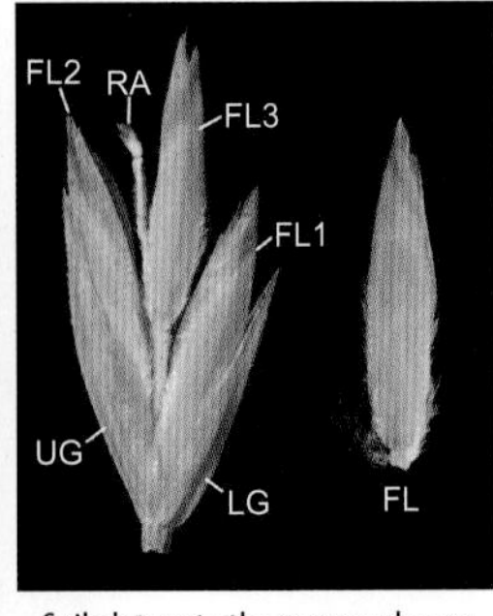

Spikelet; note the narrow glumes

Wood bluegrass habitat, large dense colony on shaded slope

Wood bluegrass, inflorescence

Poa paludigena **Fernald and Wiegand, bog bluegrass**

paludigena, "born of the marsh"

THREATENED

Rare in fens and cold, forested seeps; a small, elusive species that flowers and drops its fruits in about a one-week span in late June and early July.

Illustration

Inflorescence

Spikelets

Poa palustris **L., fowl meadow grass or marsh bluegrass**

palustris, "of marshes"

Common statewide in wetlands of all types. It is easily distinguished from other bluegrasses by its large stature; large, diffuse (untidy-looking) panicle; small spikelets; long ligules; and preference for wet habitats.

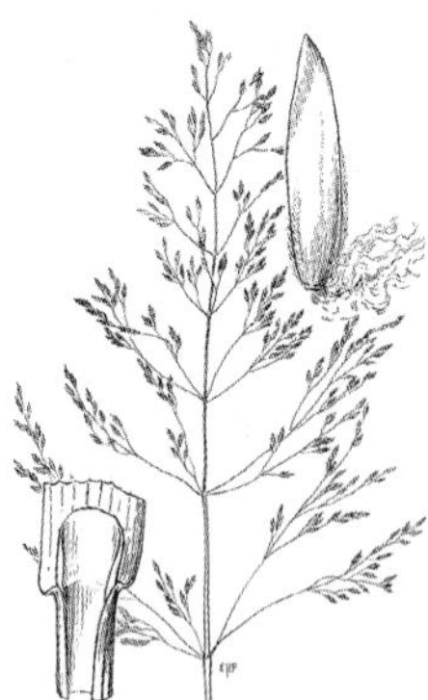

Illustration

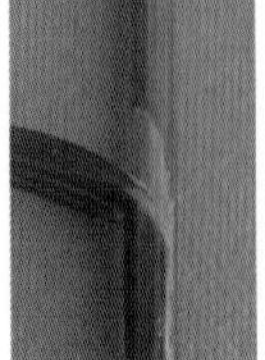

Leaf ligular area

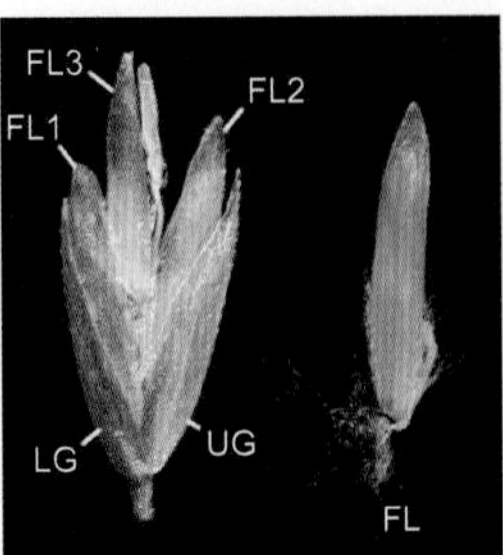

Spikelet and floret

Habit

Fowl meadow grass, inflorescence

Poa pratensis L., Kentucky bluegrass
pratensis, "of meadows"

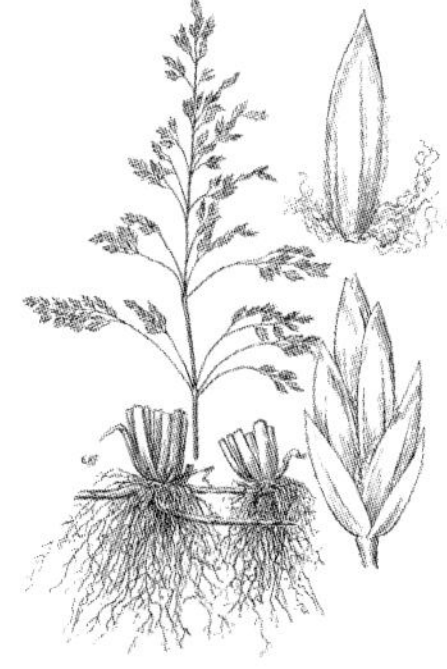
Illustration

Despite the common name, this is probably a European exotic, although present early (Lapham 1853). It is abundant statewide in moist to dry open areas: lawns, roadsides, fields, and pastures. Of all our bluegrasses, this species has the greatest density of cobwebby hairs on the floret, both at the base of the lemma and along its midvein and margins.

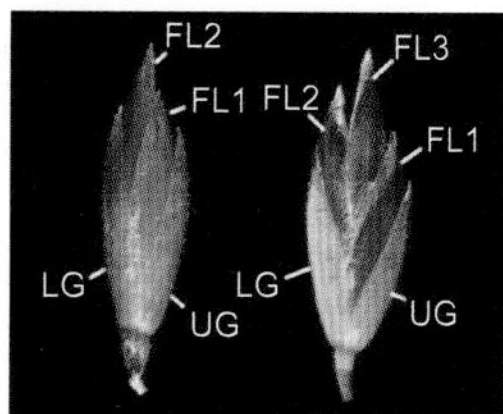

Spikelets (immature and mature)

Portion of inflorescence

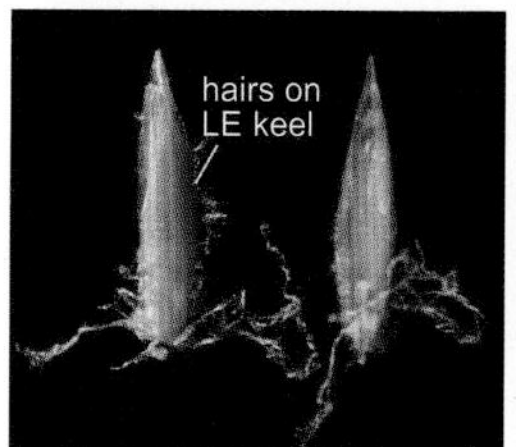

Florets

Habit

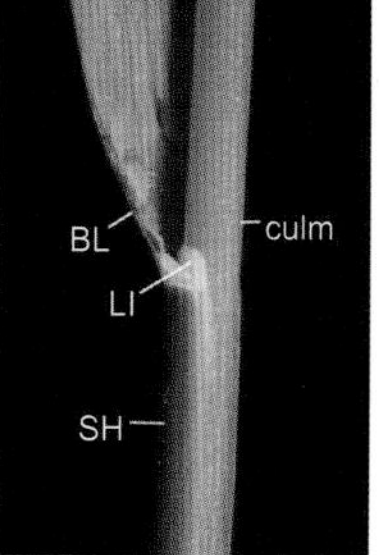

Leaf ligular area

Poa saltuensis Fernald and Wiegand, woodland bluegrass

saltuensis, "of bushy pastures"

Illustration

Occasional, rich hardwood and conifer-hardwood forests, mostly northern Wisconsin; occasionally in white cedar swamps and even in bogs. The relatively broad spikelets have some resemblance to those of fowl manna grass (*Glyceria striata*) and tend to cluster toward the ends of the drooping branches; closely related to *Poa languida*.

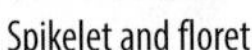

Spikelet and floret

Inflorescence

Poa sylvestris A. Gray, forest bluegrass

sylvestris, "growing in woods, forest-loving, wild"

SPECIAL CONCERN

Rare, rich, mesic hardwood forests in southern Wisconsin.

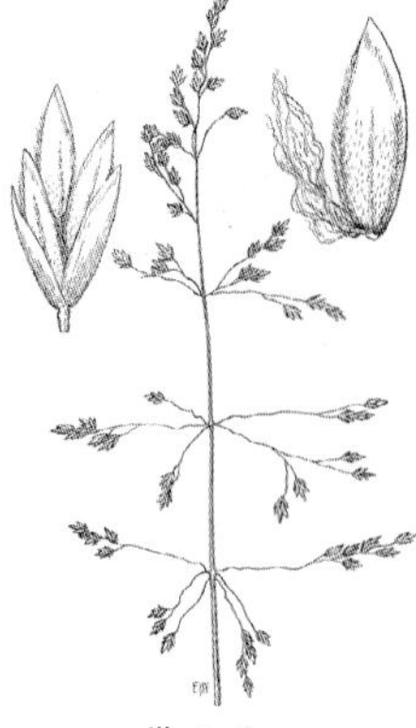

Illustration

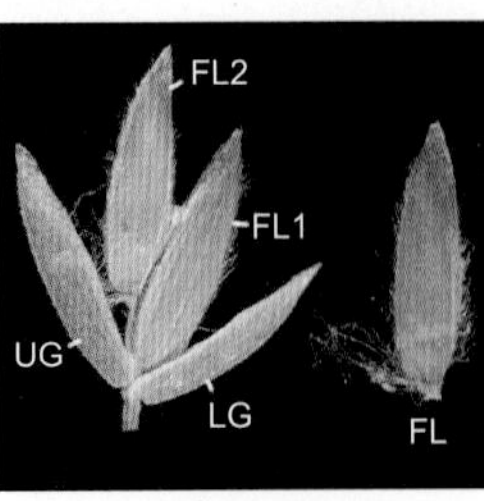

Spikelet and floret

Portion of inflorescence

Poa trivialis L., rough bluegrass

trivialis, "common, ordinary"

Illustration

A European exotic, present early (Lapham 1853). Uncommon, on roadsides and in lawns; also found in alder thickets, wet deciduous woods, sedge meadows, and streamsides.

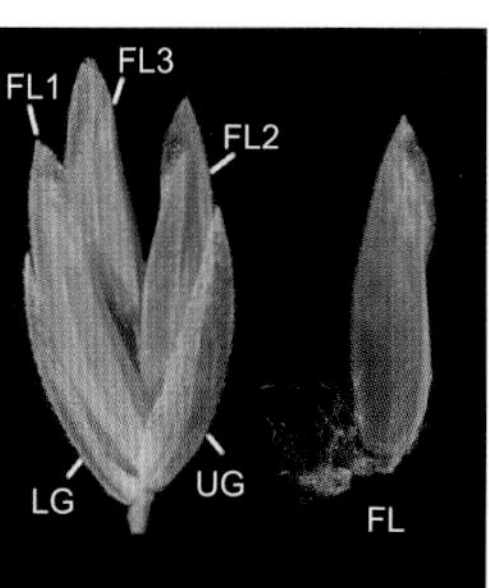

Spikelet and floret

Habit

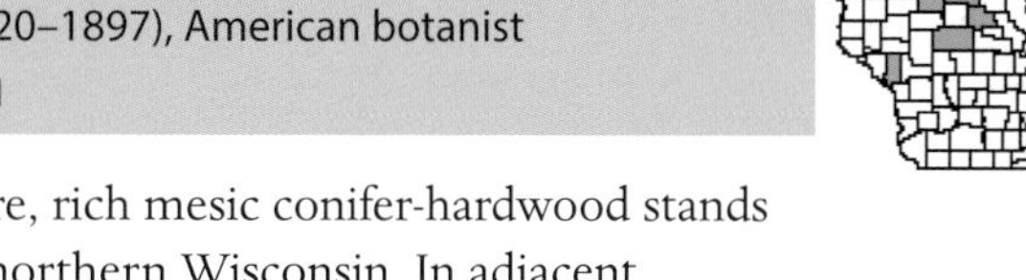

Poa wolfii Scribn., Wolf's bluegrass

wolfii, for John Wolf (1820–1897), American botanist

SPECIAL CONCERN

Habit

Rare, rich mesic conifer-hardwood stands in northern Wisconsin. In adjacent southeastern Minnesota, found in shaded dolomitic woods.

Spikelet

Spikelet and floret

68. *POLYPOGON*, POLYPOGON

polus, "much," and *pogon*, "beard"

Small weedy grasses with "awny" spike-like contracted panicles, the tiny single-flowered spikelets with awned glumes, the lemma and palea subequal in size and awnless. Could be mistaken at arm's length for a species of foxtail (*Setaria*), but differing most obviously in its branched inflorescence and awned glumes (in foxtails the awn-like bristles merely subtend the spikelets). Eighteen species, cosmopolitan in tropical and warm temperate areas.

1. Perennials; glume awns 2.5–4 mm long; apex of glumes acute to truncate, not at all lobed; ligules 1–3 mm long . . . . *P. interruptus*
1. Annuals; glume awns 4–10 mm long; apex of glumes rounded, slightly lobed; ligules 3–15 mm long . . . . *P. monspeliensis*

Polypogon interruptus **Kunth, ditch polypogon**

interruptus, "interrupted in some fashion"

Illustration

A western US species, here a rare introduction, first collected in 1903 in waste ground by Charles Goessl, then again in 1965 by Hugh Iltis along a path through a white cedar–ash–tamarack swamp.

Portion of inflorescence

Polypogon monspeliensis (L.) Desf., annual rabbit's-foot grass
monspeliensis, "of Montpellier," city in southern France, Latinized as *Mons Pessulanus*

Illustration

A rare southern European exotic, first collected in 2009 under an interstate highway bridge in Stevens Point by University of Wisconsin–Stevens Point student Leslie Day and still persisting there in August 2014.

Habit

69. *PUCCINELLIA*, ALKALI GRASS

for Benedetto Puccinelli (1808–1850), Italian botanist

Small, tufted salt-tolerant grasses with leaf sheaths fused near the base, merely overlapping above; panicle inflorescences of several-flowered spikelets, the florets falling from the glumes, the lemmas with three to five somewhat obscure parallel nerves and delicate, irregular, mostly blunt tips. Similar in morphology to the false manna grasses (*Torreyochloa*) and smaller species of "true" manna grasses (*Glyceria*), but the latter two genera have five to seven prominent nerves. *Puccinellia* differs from bluegrasses (*Poa*) in the pointed (not boat-shaped) leaf apices, smooth floret bases, and parallel (rather than convergent) lemma nerves. 120 species, north temperate areas; most are salt-tolerant halophytes.

1. Glumes obtuse at tip, 0.7–2.2 mm long; lemmas 1.7–2.5 mm long, obtuse to notched; occasional weed that is becoming common .... *P. distans*
1. Glumes acute at tip, 1–3 mm long; lemmas 2–3.5 mm long, narrowly obtuse; rare adventive .... *P. nuttalliana*

Puccinellia distans **(Jacq.) Parl., European alkali grass**

distans, "separated, apart, widely spaced," in reference to the long, exserted stamens, which are apart from each other

Illustration

A European exotic, first collected in 1882, and a common street weed in Milwaukee by 1939 (Shinners 1940); a locally common salt-tolerant weed of railroad rights-of-way, roadsides, shorelines, and waste ground generally that is spreading statewide.

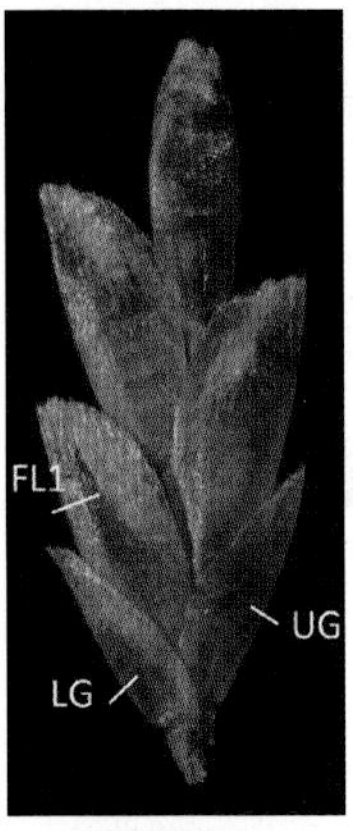

Spikelet

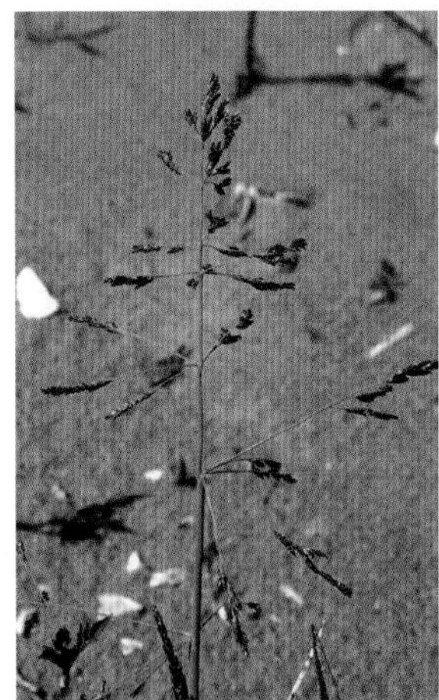

Inflorescence

Habitat along sidewalk, Stevens Point

Puccinellia nuttalliana (Schult.) Hitchc., Nuttall's alkali grass

nuttalliana, for Thomas Nuttall (1786–1859), English botanist who worked in the United States

Illustration

A common native of the Great Plains, extending east as a native to western Minnesota. In Wisconsin, cited by Shinners (1940) from a 1916 Goessl collection from Oconto County; collected again in 1941 by Shinners from a railroad yard and most recently in 2000 by Judziewicz in a saline wet clay ditch in Superior.

Habit

Inflorescence

70. *SCHIZACHNE*, FALSE MELIC GRASS

schizo, "to split," and *achne*, "chaff"

A spring-flowering woodland grass with fused leaf sheath margins; the inflorescence is a sparse open panicle of several-flowered spikelets, the florets awned from between two small apical teeth. False melic grass can be distinguished from Smith's melic grass (*Melica smithii*), which grows in the same habitat, as follows: it has longer (8–15 mm vs. 2–5 mm long), more divergent lemma awns, narrower leaf blades (2–5 mm vs. 5–13 mm wide), nonbulbous culm bases, and much less scabrous spikelet pedicels and leaf sheaths. False melic grass might also be mistaken for a species of brome (*Bromus*), but differs in that the callus of the floret is bearded rather than smooth. A single species ranging from northeastern Asia to North America.

Schizachne purpurascens **(Torr.) Swallen, false melic grass**

purpurascens, "becoming purple or purplish"

A common upland forest understory grass statewide, except in the far south; usually in mesic hardwoods with sugar maple dominant, but can also be found associated with pines, hemlock, and oaks. The purple glumes make a pleasing contrast with the green florets and help to identify this early spring–flowering species.

Illustration

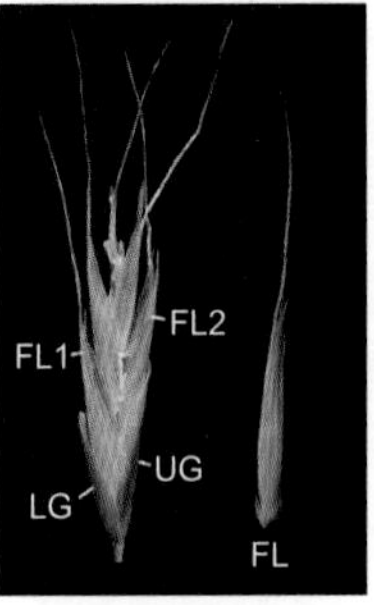

Spikelet and floret

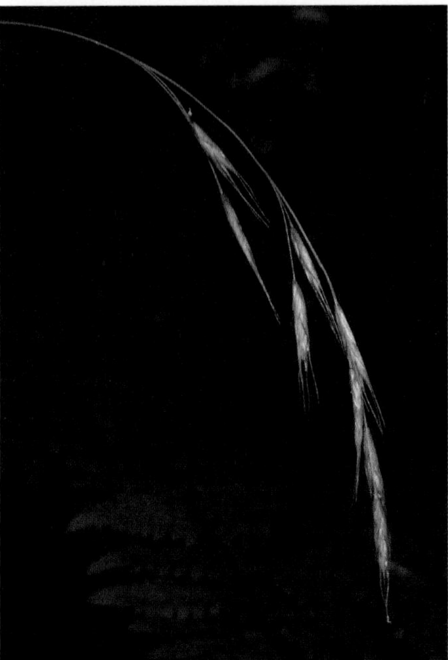
Portion of inflorescence showing spikelets with purple glumes and green florets

71. *SCHIZACHYRIUM*, LITTLE BLUESTEM

schizo, "to split" (in part), and *achyrion*, "chaff"

A slender, midsized, late summer–flowering perennial prairie grass with numerous solitary ascending lateral and terminal spike-like racemes of small awned spikelets. Sixty species; cosmopolitan, tropical to warm temperate regions.

Broomsedge bluestem (*Andropogon virginicus* L.) resembles this genus but has some of its inflorescences included in and concealed by the swollen uppermost leaf sheaths. It occurs in the Chicago area and should be looked for in southeastern Wisconsin.

Schizachyrium scoparium (Michx.) Nash, little bluestem

scoparium, Latin, "broom-like," from Latin, *scopa*, "broom"

Common in prairies, barrens, savannas, sandy old fields, and pastures, often spreading from plantations and appearing to move north along highways, such as on Judziewicz's property in Oconto County (Judziewicz 2004a), where over the course of several decades it has "taken over" a hayfield. This attractive grass has bluish-white emergent shoots in the early summer; in the fall, the foliage turns a bright orange as the rachilla hairs on these spikelets expand and give each raceme a white, puffy, feathery appearance; at this point it shatters and is dispersed by the wind. The orange foliage fades to pinkish and persists over the winter. Fassett (1951) treats this as *Andropogon scoparius*.

Illustration

Portion of inflorescence

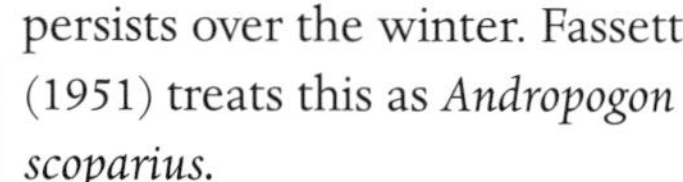

Spikelet with puffy rachilla hairs in fall

Winter habit, foliage turns orange

Little bluestem in summer, with glaucous culms

Little bluestem in fall, with puffy spikelets

72. *SCLEROCHLOA*, FAIRGROUNDS GRASS

Greek, *sclero*, "hard," and *chloa*, "grass"

Small, spring-flowering annuals with small, congested, one-sided racemes of several-flowered smooth spikelets with blunt, strongly seven- to nine-nerved lemmas. Two Eurasian species.

Sclerochloa dura (L.) P. Beauv., fairgrounds grass or hard grass

dura, Latin, "hard, tough, or durable"

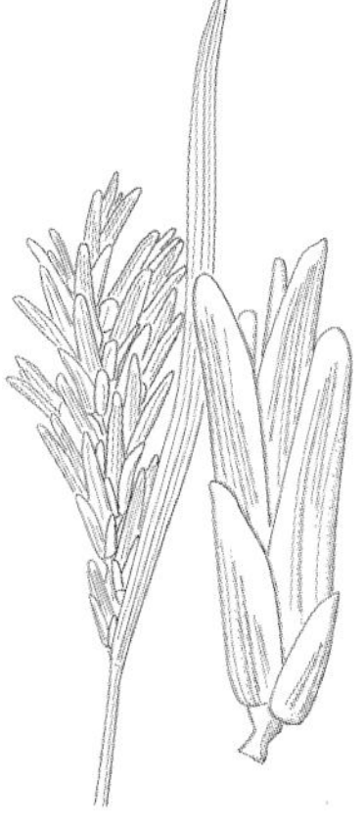

Spikelet illustration

An exotic, first collected in 2001 by Richard K. Rabeler (Cusick, Rabeler, and Oldham 2002); currently rare in eastern Wisconsin (north to Brown County), but expected to spread; in highly disturbed areas: fairgrounds, racetracks, lawns, and parking lot margins. In habit, habitat and flowering time reminiscent of annual bluegrass (*Poa annua*), but that species differs in its open panicles of spikelets with obscurely nerved florets.

Habit

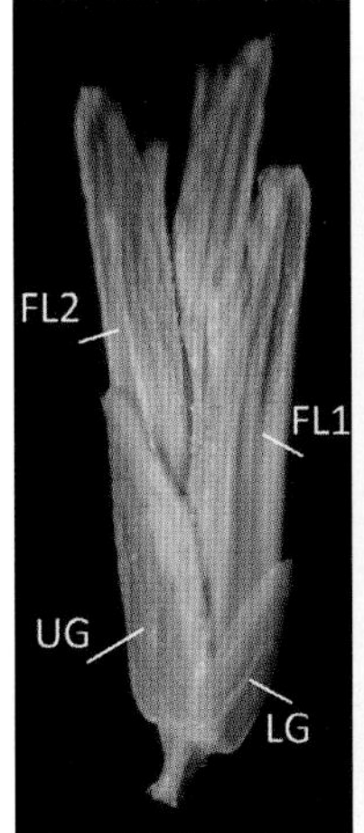

Spikelet

Habit

73. *SECALE*, RYE

Latin, "cereal grain"

Annuals or winter annuals; spike four-sided, nodding; one spikelet per node; two florets per spikelet; glumes and lemmas narrow, tapered, awned with prominent bristles all along the keels. Three Eurasian species.

Secale cereale L., rye

cereale, *cereus*, "waxen," and *alis*, "like"

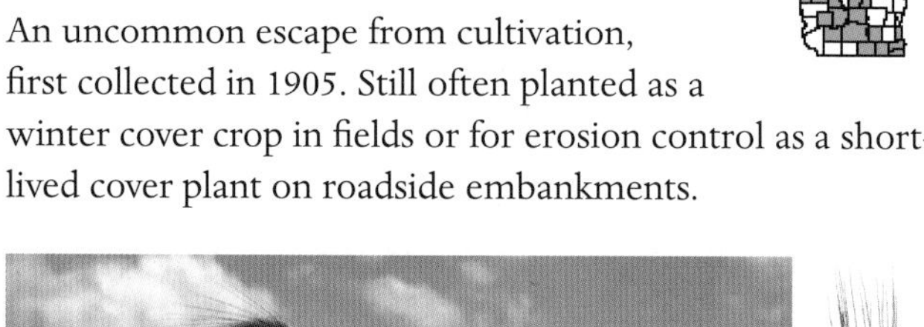

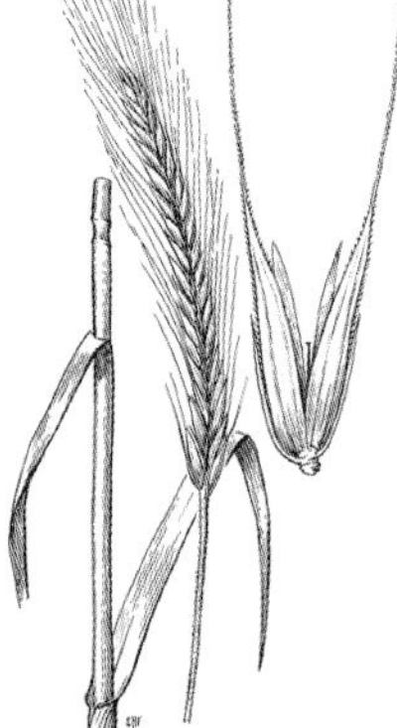

Illustration

An uncommon escape from cultivation, first collected in 1905. Still often planted as a winter cover crop in fields or for erosion control as a short-lived cover plant on roadside embankments.

Rye field

Inflorescence

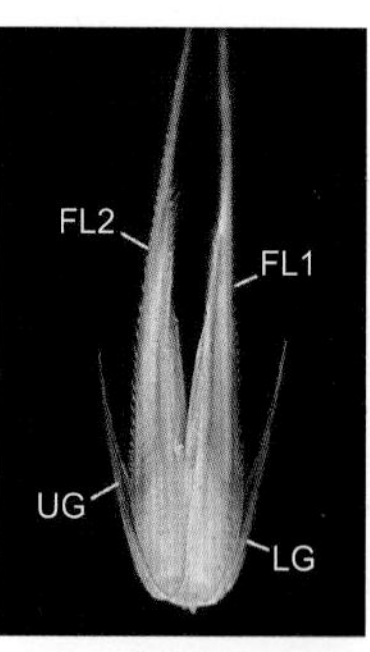

Spikelet

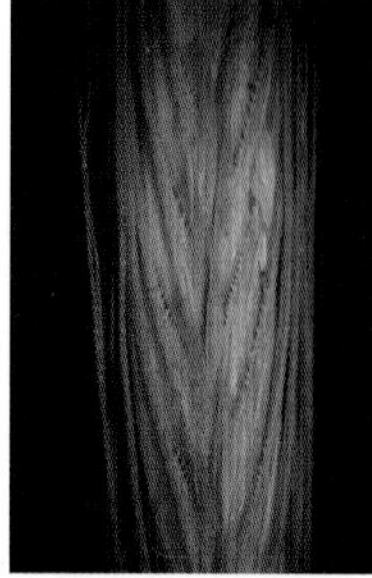

Portion of inflorescence

74. *SETARIA*, FOXTAIL

seta, "bristle"

Usually annuals; panicles usually densely cylindrical with fewer than five sterile branches forming clusters of bristles at base of spikelets, the latter falling free of the bristles, which persist on the axis of the inflorescence; upper floret plump, not strongly cross-wrinkled. Cultivated fountain grass (*Cenchrus compressus*) could be mistaken for foxtail, but it has spikelets that fall together with the bristles. All Wisconsin species are exotic. 140 species; cosmopolitan.

1. Spikelets subtended by four to twelve bristles
   2. Panicles arching and drooping, with bristles 8–12 mm long; leaf blades 10–20 mm wide; commoner in southern Wisconsin . . . . *S. faberi*
   2. Panicles erect, with bristles 3–8 mm long; leaf blades 5–10 mm wide; abundant statewide . . . . *S. pumila*
1. Spikelets subtended by one to three (rarely six) bristles
   3. Bristles retrorsely scabrous, making it uncomfortable to "stroke" them from bottom to top; occasional, southern Wisconsin . . . . *S. verticillata*
   3. Bristles antrorsely scabrous (or both antrorsely and retrorsely scabrous in the rare exotic *S. verticilliformis*), making it uncomfortable to "stroke" them from top to bottom
      4. Fertile lemma smooth and shiny, falling separately from glumes and lower lemma; uncommon . . . . *S. italica*
      4. Fertile lemma corrugated and wrinkled, falling as a unit with the glumes and lower lemma
         5. Panicles with whorled branches, the rachis visible; rare adventive . . . . *S. verticilliformis*
         5. Panicles densely spike-like, whorled branches not evident, the rachis obscured
            6. Panicles arching and drooping; leaf blades hairy above; spikelets 2.5–3 mm long; commoner in southern Wisconsin . . . . *S. faberi*
            6. Panicles erect; leaf blades scabrous but not hairy above; spikelets 1.7–2.3 mm long; statewide . . . . *S. viridis*

Setaria faberi R. A. W. Herrm., nodding foxtail

faberi, for Ernst Faber (1839–1899), German missionary who discovered it

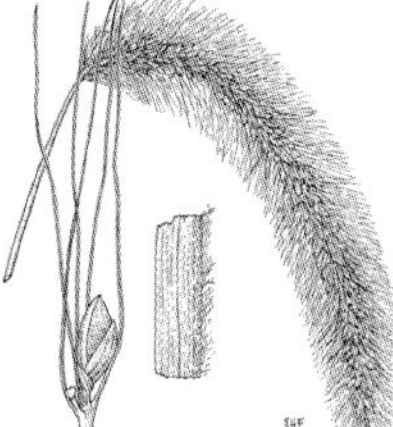

Illustration

East Asian, first collected in 1953 (first in United States: 1938). Since then it has become common statewide; in disturbed ground generally, but most characteristically a weed in the margins of agricultural fields.

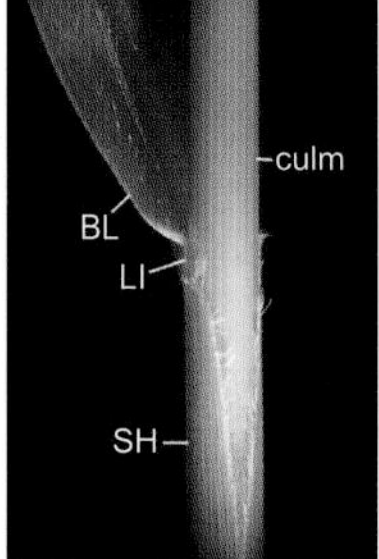

Leaf ligular area

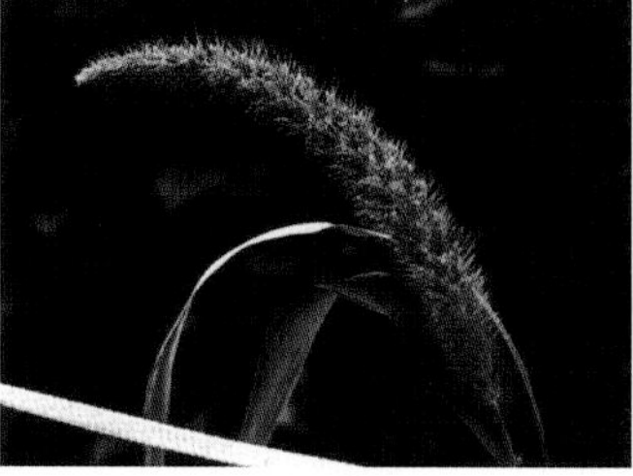

Inflorescence

Portion of inflorescence, showing bristles among the spikelets

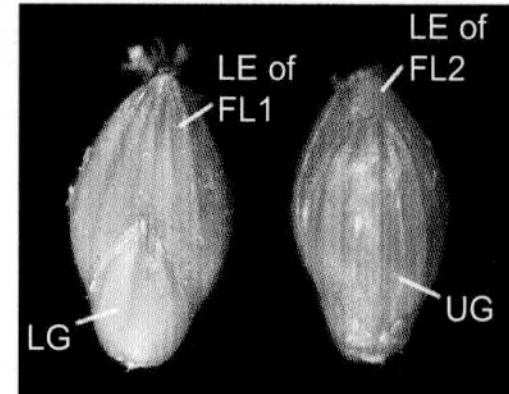

Spikelet

Habit

Setaria italica **(L.) P. Beauv., foxtail millet**

italica, "of Italy"

European, first collected in 1905 and not since 1988; an uncommon weed of roadsides and fields, perhaps no longer present in Wisconsin.

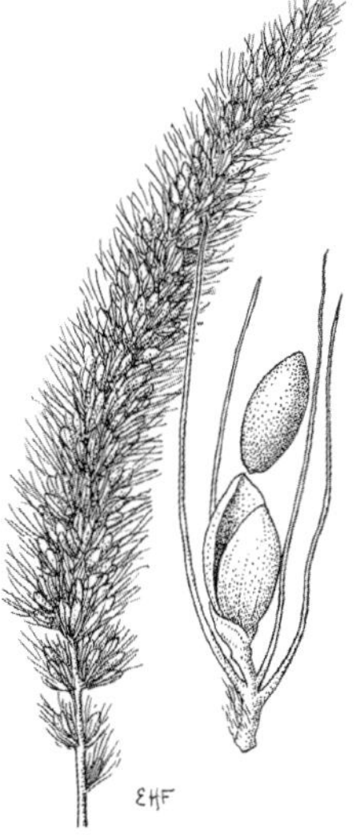

Illustration

Inflorescence

Habit

Setaria pumila **(Poir.) Roem. and Schult., yellow foxtail**
pumila, "dwarf"

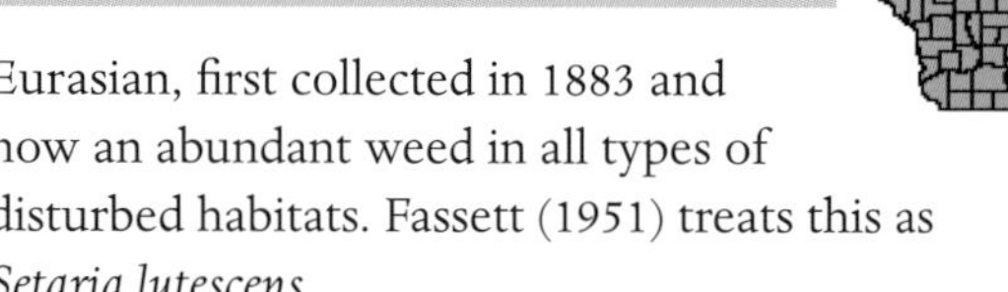

Eurasian, first collected in 1883 and now an abundant weed in all types of disturbed habitats. Fassett (1951) treats this as *Setaria lutescens.*

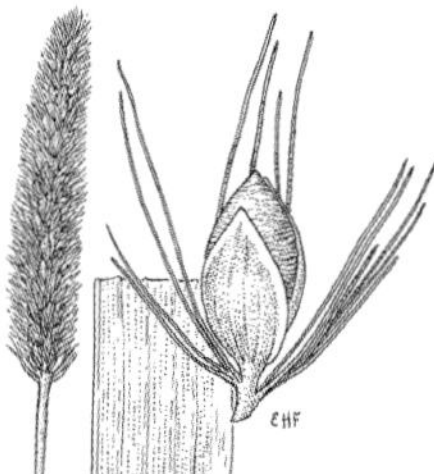

Illustration

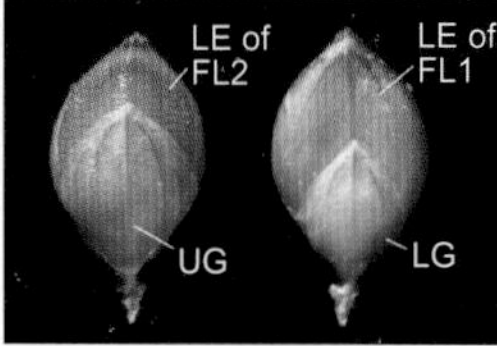

Spikelet

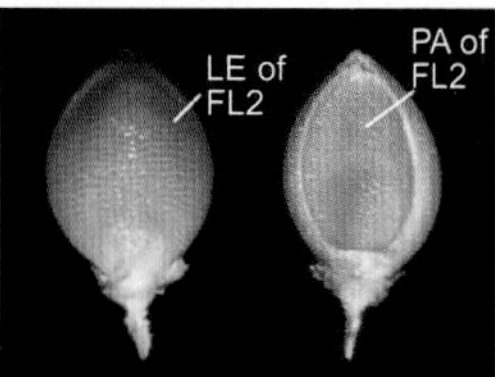

Upper floret

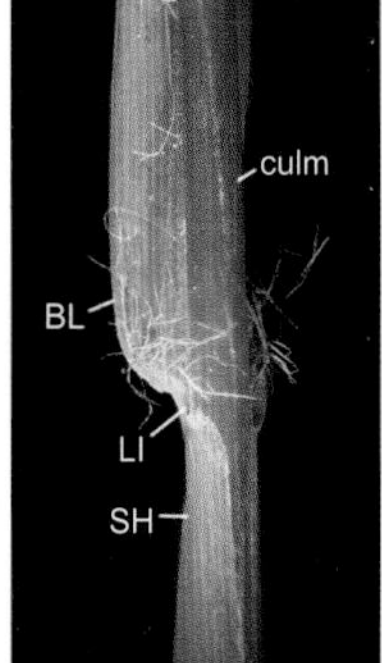

Leaf ligular area

Habit

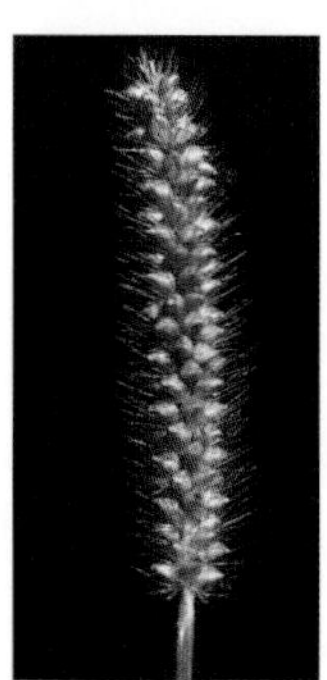

Inflorescence

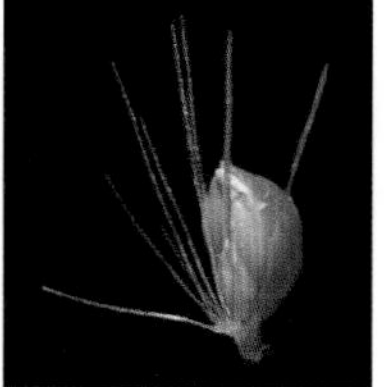

Spikelet, subtended by orange bristles

Setaria verticillata (L.) P. Beauv., bristly foxtail

verticillata, "whorled"

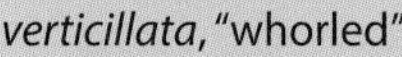

Illustration

European, first collected in 1908. An occasional weed of gardens, cornfields, roadsides, and railroad rights-of-way.

Spikelet, subtended by bristles

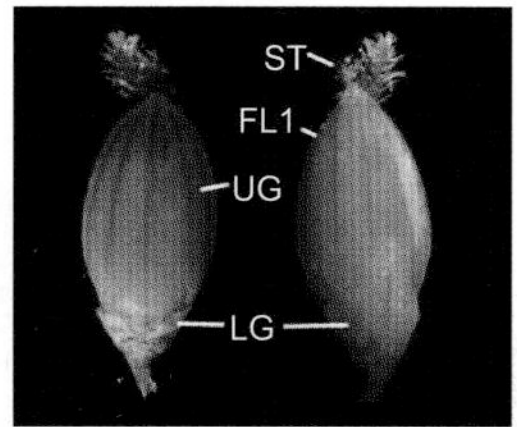

Spikelet

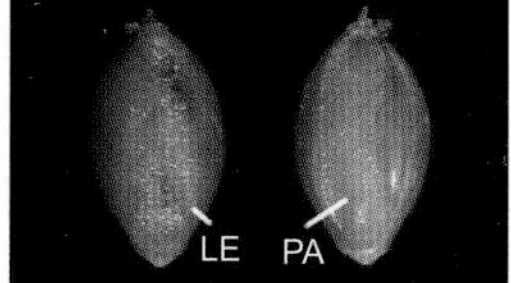

Upper floret

Portion of inflorescence

Setaria verticilliformis Dumort., barbed bristle grass

verticilliformis, from *verticillus* for the whirl of a spindle or whorled, and *forma*, "form, figure, shape"

European; known from gardens, roadsides, and cornfields; collected from 1964 to 1980 and perhaps no longer present in the state.

Setaria viridis (L.) P. Beauv., green foxtail

viridis, "green"

European, first collected in 1874; now an abundant weed of sidewalks, driveways, and gardens.

Illustration

Upper floret

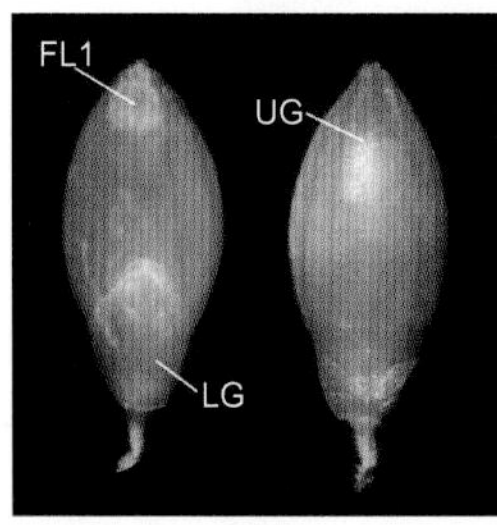

Spikelet

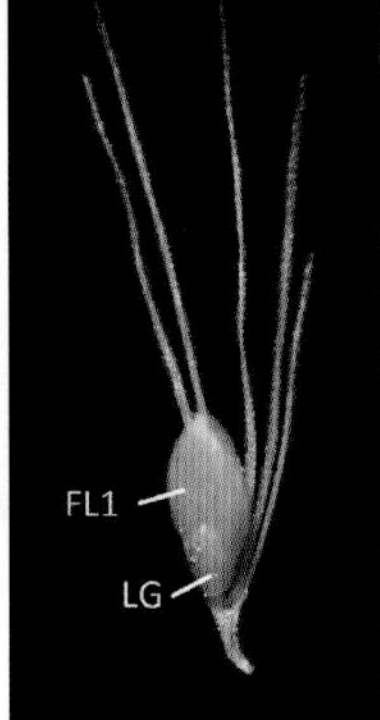

Spikelet, subtended by green bristles

Inflorescences

Habit

75. *SORGHASTRUM*, INDIAN GRASS

Greek, "a poor imitation of sorghum," for its resemblance to the *Sorghum* genus

Robust, late summer–flowering prairie perennials with ligule a membrane forming an erect auricle at the edge of the leaf; inflorescence a panicle of short, spike-like racemes, these hairy and often falling as a unit; spikelets awned, hairy, turning a rich golden-brown color at maturity. Eighteen species, most in North and South America, but a few in Africa.

Sorghastrum nutans (L.) Nash, Indian grass

nutans, "nodding or drooping," usually referring to the flowers

A common large prairie and savanna grass of southern and western Wisconsin, spreading from cultivation and along roadsides elsewhere.

Illustration

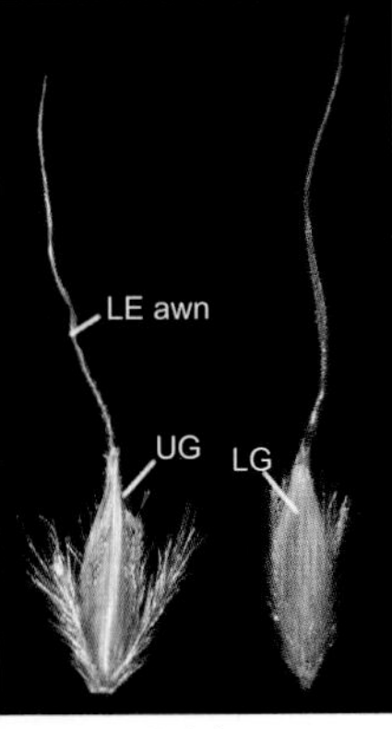

Spikelet

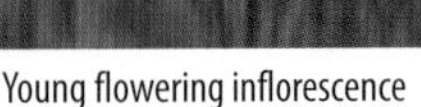

Young flowering inflorescence

Mature inflorescence

76. *SORGHUM*, SORGHUM OR BROOMCORN

Latinization of Italian *sorgo*, obscure

Robust, cultivated, late summer–flowering grasses, often with large, broad, corn-like leaves; inflorescence an open to contracted panicle of short, spike-like racemes, the spikelets heavy, turgid, and sparsely hairy. Twenty-five species, mostly in the Old World.

1. Plants rhizomatous perennials; spikelets deciduous, 1.5–2.3 mm wide; mature grain concealed .... *S. halepense*
1. Plants annuals; spikelets persistent or tardily deciduous, 2–4 mm wide; mature grain exposed
   2. Sessile spikelets 2–3 mm wide .... *S.* × *almum*
   2. Sessile spikelets 3–4 mm wide .... *S. bicolor*

Sorghum × *almum* Parodi (pro sp.), Columbus grass

A European exotic, first collected in 1981; this recent cornfield weed is still uncommon. It is a hybrid of *S. bicolor* and *S. halepense*.

Sorghum bicolor (L.) Moench, broomcorn, sorghum

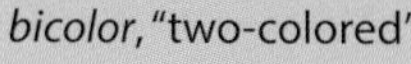

bicolor, "two-colored"

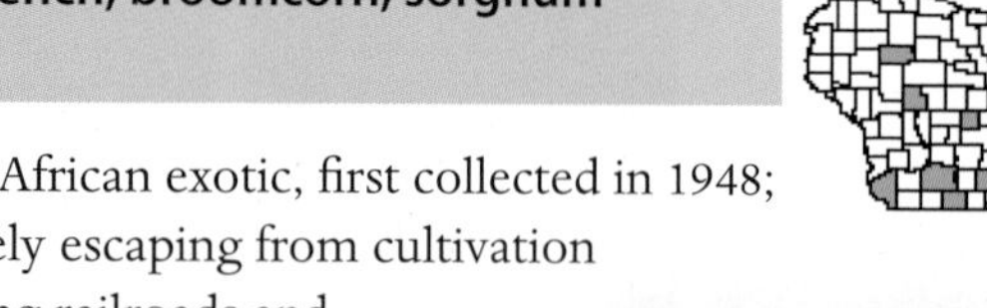

An African exotic, first collected in 1948; rarely escaping from cultivation along railroads and highways. Fassett (1951) treats this as *Sorghum vulgare*.

Illustration

Inflorescences

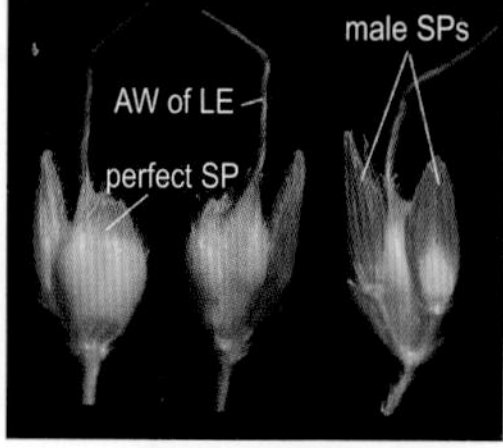

Spikelet arrangements: *left two*, typical spikelet pairs; *far right*, triple spikelet at end of branch

Broomcorn, cultivated plants

Broomcorn,
fruiting inflorescence

Sorghum halepense **(L.) Pers., Johnson grass**

halepense, "of Aleppo"

A Eurasian exotic, first collected in 1939; rarely escaping along railroads and highways.

Illustration

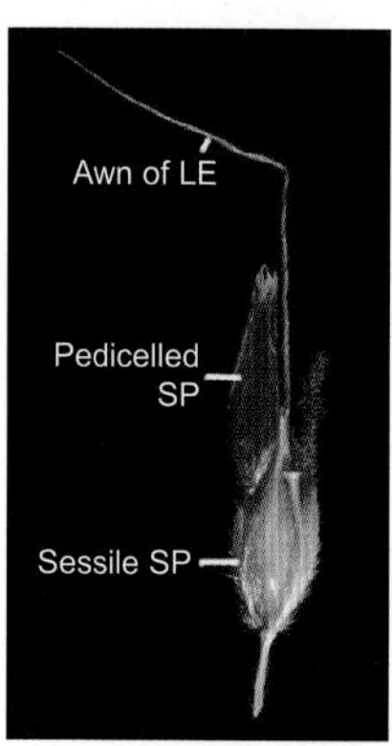

Spikelets

Habit

77. *SPARTINA*, CORDGRASS

Greek, *spartine*, "cord"

Vigorous rhizomatous wetland perennials, blades usually inrolled toward the tip; spikes several to numerous, ascending, pinnately arranged; spikelets smooth, pale, strongly laterally compressed, the upper glume elongate and awned, the single floret awnless. About fifteen species; cosmopolitan.

Spartina pectinata Link, prairie cordgrass

pectinata, Latin, "comb-like"

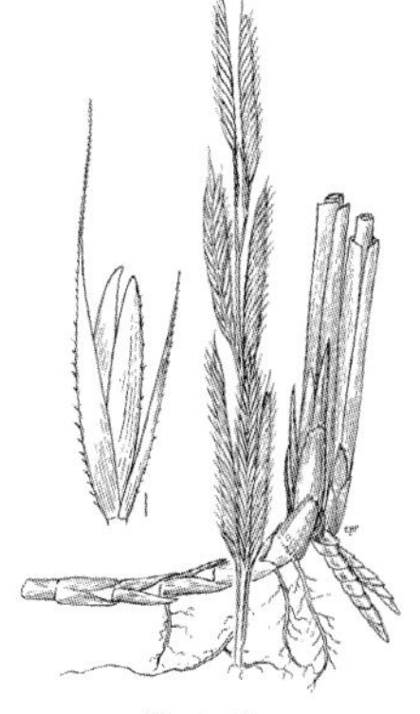

Illustration

Common in southern Wisconsin in wet prairies, sedge meadows, and fens; marshes along the west shore of Green Bay; sandy shores of inland lakes. A robust, coarse-leaved wetland grass that often forms pure stands, these sometimes nearly all sterile.

Floating leaves in flowing water

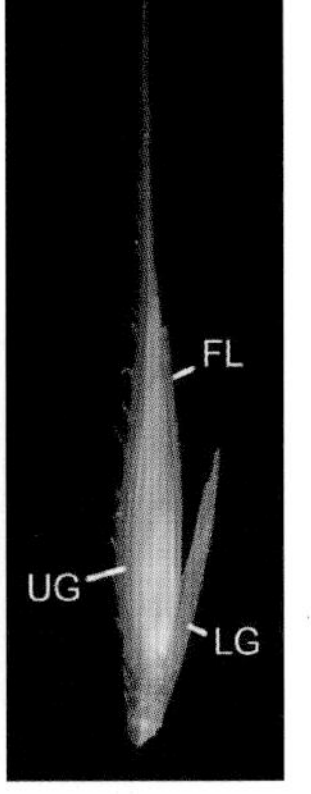

Spikelet

Habit

Portion of inflorescence

78. *SPHENOPHOLIS*, WEDGE OR SATIN GRASS

sphen, "wedge," and *pholis*, "scale," referring to the broadly ovate or wedge-shaped second glume

Small perennials; inflorescence a loose, contracted panicle of small, shining, two-flowered spikelets that fall as a unit from the tops of their stalks. The glumes of this genus are subtly distinctive; they are of roughly equal length but have completely different shapes. The lower glume is narrowly lanceolate and broadest near the middle or base, while the upper glume is much broader and looks top-heavy and wedge-shaped (hence the common name). June grass (*Koeleria*) is a related genus differing in its monomorphic glumes, minutely hairy panicle axis, tendency for the florets to fall from the persistent glumes, and dryland habitat. Six North American species.

Shiny wedge grass (*Sphenopholis nitida* (Biehl.) Scribn.) is not known from Wisconsin but may eventually be found here, since it occurs in dry upland oak forests in two adjacent Illinois border counties; there is an old, unverified report from Whitefish Bay, Milwaukee County (Brues and Brues 1911). It differs from the two species below in its distinctly wider lower glume (ca. 0.5 mm on a side vs. less than 0.3 mm wide) and larger anthers frequently at least 1 mm long.

1. Upper glume slightly obovate, obtuse to acute at the tip; panicle more or less open; common in wetlands statewide . . . . *S. intermedia*
1. Upper glume strongly obovate, truncate to slightly lobed at the tip; panicle contracted; occasional in dry forests and prairies, southern Wisconsin . . . . *S. obtusata*

Sphenopholis intermedia **(Rydb.) Rydb., satin wedge grass**
intermedia, "intermediate," indicating an observation that a species was probably considered as being halfway or partway between two others with regard to some particular characteristic

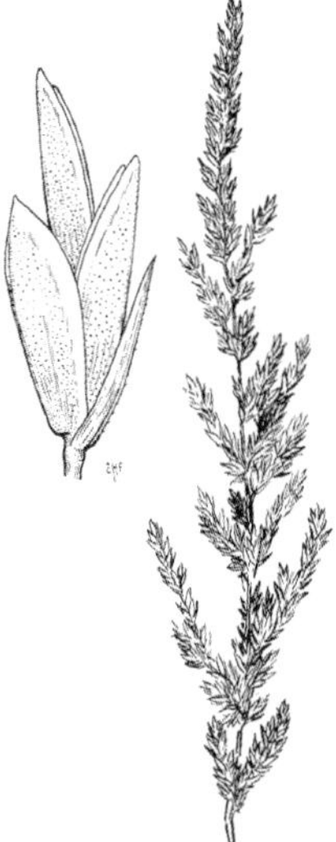

Illustration

Common on wetland margins, especially calcareous ones; fens, sedge meadows, streambanks, and wave-splashed sandstone and dolomite bluffs along Lakes Superior and Michigan.

Spikelet

Inflorescence, showing open, shiny young inflorescence

Mature contracted inflorescence

Sphenopholis obtusata (Michx.) Scribn., prairie wedge grass
obtusata, Latin, *obtus*, "dull or blunt"

Illustration

Occasional in southern Wisconsin in dry upland woods and prairies.

Portion of inflorescence

Spikelet

79. *SPOROBOLUS*, DROPSEED

Greek, *spora*, "seed," and *ballein*, "to cast forth," referring to dropping seeds at maturity

Tufted perennials with inrolled leaf blades, the ligule minute but with tufts of stiff, spreading hairs at the sheath summits; and panicles often contracted and partially enclosed in the uppermost leaf sheath. The spikelets are single-flowered, thinly membranous, and awnless, with subequal one-nerved glumes and lemma. The seed drops free from the grain, which often has a sticky coating. With its "minimalist" single-flowered spikelets (no awns, smooth, just one floret, one-nerved bracts), this genus is not easily confused with any other (Shinners 1941). 160 species; cosmopolitan.

1. Base of leaf blade conspicuously white-bearded; statewide, but commoner in open sandy areas in central and southern Wisconsin .... *S. cryptandrus*
1. Base of leaf blade not bearded or with a wispy tuft of a few hairs
    2. Spikelets 1.5–3 mm long; glumes about equal in length; plants annual; dry open areas, mostly southern Wisconsin .... *S. neglectus*
    2. Spikelets 3–7 mm long; glumes unequal; plants annual or in most species perennial
        3. Florets glabrous
            4. Panicle contracted, partly or wholly concealed within the uppermost leaf sheath; glumes shorter than the lemma; southern Wisconsin .... *S. compositus*

4. Panicle open and exserted from the uppermost leaf sheath; one glume equaling or longer than the lemma; southern and western Wisconsin .... *S. heterolepis*

3. Florets hairy

5. Plants 50–100 cm tall, perennial; rare, sand prairie, southwestern Wisconsin .... *S. clandestinus*

5. Plants 10–40 cm tall, annual; common statewide in dry disturbed areas such as gravel road shoulders and pavement cracks .... *S. vaginiflorus*

Sporobolus clandestinus **(Biehler) Hitchc., hidden dropseed**

clandestinus, "hidden"

A southern US grass at its far northern range limit here, known from just one collection made in 1940 in a sandy prairie two miles south of Bell Center, Crawford County, by Norman Fassett, John Catenhausen, and Lloyd Shinners. Shinners (1941) notes that at this site it is "abundant, but confined to an area no larger than a city block"; in 2012 Judziewicz drove past this site, which is on private property, and observed a small remnant prairie still extant.

Illustration

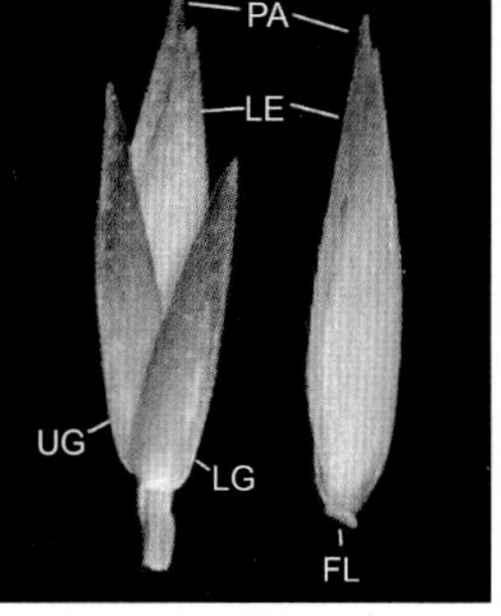

Spikelet and floret

Habit

Sporobolus compositus **(Poir.) Merr., meadow dropseed**

compositus, "composite, of many parts"

Occasional, mesic to more frequently dry prairies in southwestern Wisconsin, spreading along roadsides and railroads to southeastern Wisconsin by the 1930s (Shinners 1940). Fassett (1951) treats this as *Sporobolus asper.*

Illustration

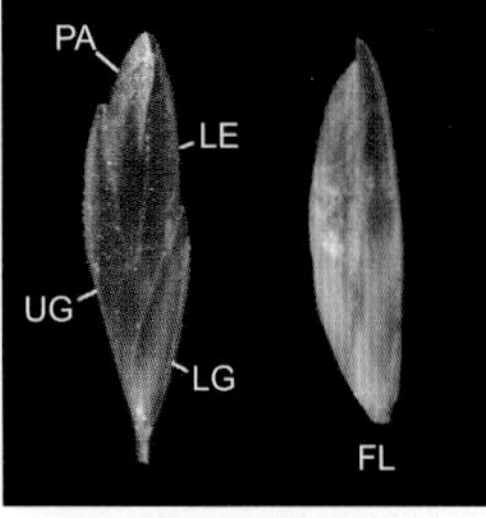

Spikelet and floret

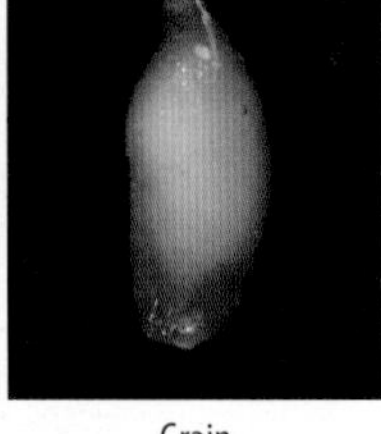
Grain

Habit

Sporobolus cryptandrus (Torr.) A. Gray, sand dropseed

cryptandrus, *cryptos*, "hidden," and *anthos*, "flower"

Illustration

Common in barrens, on dunes, and in dry prairies in southwestern and central Wisconsin; as a native plant most characteristic of sand terraces and prairies along major rivers; by 1941 spreading nearly statewide via railroads and highways (Shinners 1941). The large panicle emerging from the swollen, often arching leaf sheath is characteristic.

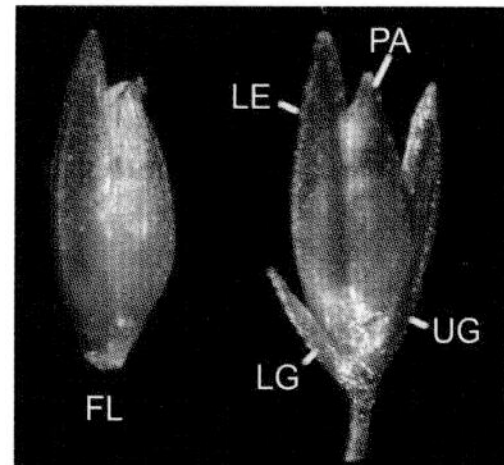

Floret and spikelet

Leaf ligular area showing "beards" of hairs at the sheath summit

Habitat and inflorescence

***Sporobolus heterolepis* (A. Gray) A. Gray, prairie dropseed**

heterolepis, *heteros*, "varied," and *lepis*, "scale"

Illustration

Occasional to fairly common in undisturbed dry to dry to mesic prairies, often on the lower parts of steep slopes and on river sand terraces; the most conservative of our large prairie grasses. Densely tussock-forming and with abundant fine leaf blades, clumps of this grass have been compared to Moe Howard's (of Three Stooges fame) "sugar bowl" haircut. The diffuse, wispy inflorescence is sometimes difficult to see even at short distances, and when it is fresh it has a distinctive smell that has been compared to buttered popcorn, melted crayons, and cilantro.

Spikelet and floret

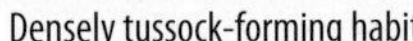
Densely tussock-forming habit

Portion of inflorescence

Tussock after burn

Sporobolus neglectus Nash, small dropseed

neglectus, "neglected or overlooked"

Uncommon in southern Wisconsin; purportedly native, but found mainly on gravelly or sandy roadsides, quarries, and railroad rights-of-way. A few collections, however, have been made in dry prairies.

Illustration

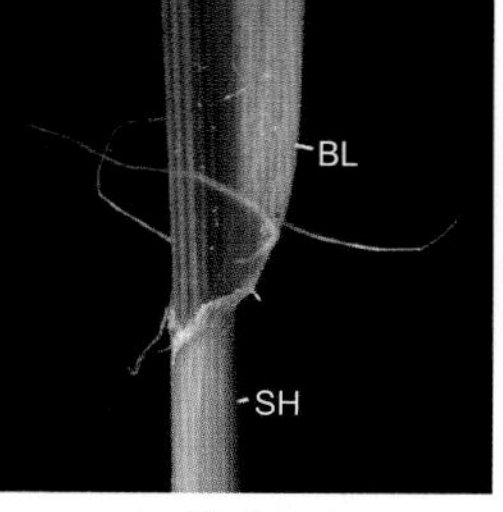

Leaf ligular area

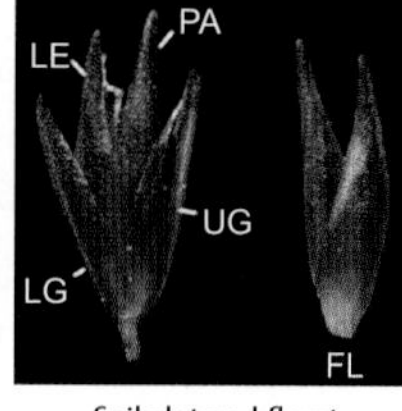

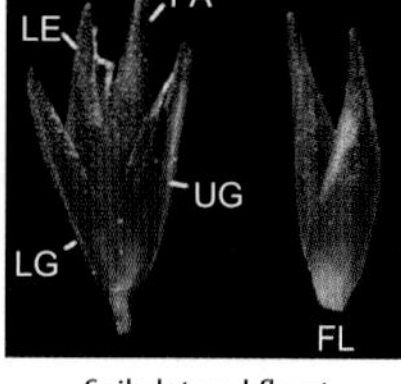

Spikelet and floret

Portion of inflorescence

Sporobolus vaginiflorus (Torr. ex A. Gray) A. W. Wood, poverty dropseed

vaginiflorus, "with a sheath and flower"

Like *Sporobolus neglectus*, a supposedly native species, but a majority of collections have been made from disturbed sites such as railroad rights-of-way. It is most characteristic of dry sandy or gravelly highway shoulders and asphalt or sidewalk cracks, and in such habitats it has spread over the entire state.

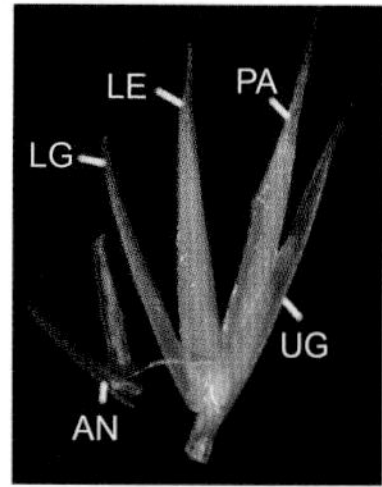

Spikelet and floret

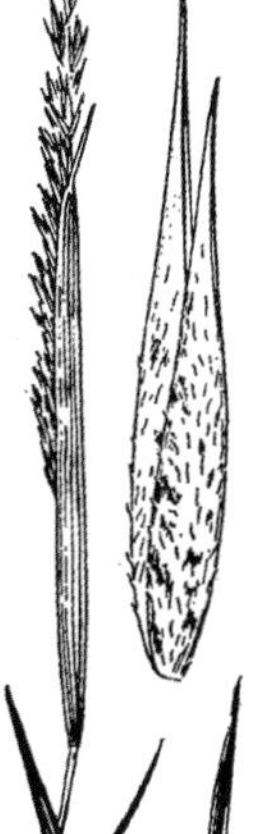

Illustration

Portion of inflorescence

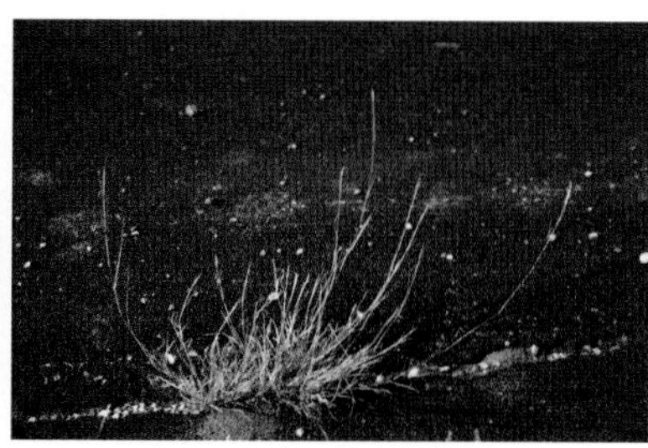

Habitat and habit in pavement crack

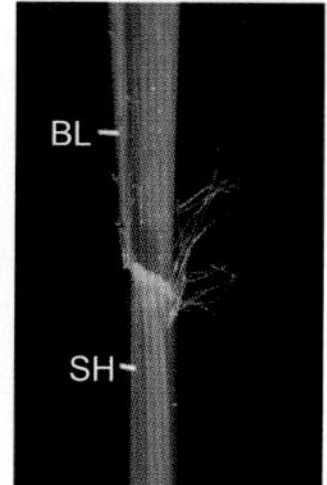

Leaf ligular area

80. *TORREYOCHLOA*, FALSE MANNA GRASS

for John Torrey (1796–1873), American chemist and botanist who coauthored with Asa Gray *A Flora of North America*, and *chloa*, "grass"

Native small tufted grass, sometimes producing rafts of floating leaves in still or quietly flowing water; inflorescence a panicle of several-flowered spikelets, the florets falling from the glumes, the lemmas with five to seven prominent raised parallel nerves and acute tips. Very similar in morphology to the exotic alkali grasses (*Puccinellia*), but differing in the acute lemmas with more numerous, prominently raised nerves. From the smaller species of "true" manna grasses (*Glyceria*), differing in that the leaf sheaths are fused only near the base and merely overlapping above; and differing from the bluegrasses (*Poa*) in the pointed (not boat-shaped) leaf apices, smooth floret bases, and parallel raised (rather than convergent and obscure) lemma nerves. Four to five species of northeastern Asia and North America.

1. Leaf blades 2–4 mm wide; ligules 3–6 mm long; larger lemmas 2–2.5 mm long . . . . . . *T. fernaldii*
1. Leaf blades 5–10 mm wide; ligules 5–10 mm long; larger lemmas 2.7–3.5 mm long . . . *T. pallida*

Torreyochloa fernaldii (Hitchc.) G. L. Church, Fernald's false manna grass

fernaldii, for Merritt Lyndon Fernald (1873–1950), American botanist

Illustration

Fairly common (although inconspicuous) in spring seeps, cold streamsides, ditches, and lakeshores; mostly in the northern third of Wisconsin, but also in small cold streams in the bed of Glacial Lake Wisconsin in the central part of the state. Fassett (1951) treats this as *Glyceria pallida* var. *fernaldii*, Barkworth et al. (2007) as *Torreyochloa pallida* var. *fernaldii*.

Fernald's false manna grass, habit

Fernald's false manna grass, floating leaves

Torreyochloa pallida (Torr.) G. L. Church, pale false manna grass
pallida, "pale"

Illustration

Uncommon in about the same habitats as *Torreyochloa fernaldii*, but more characteristic of marshes, vernal ponds, and sedge meadows rather than streamsides; northern Wisconsin, not present in the bed of Glacial Lake Wisconsin. Fassett (1951) treats this as *Glyceria pallida* var. *pallida*.

Floating leaves

81. *TRIDENS*, TRIDENS

tres, "three," and *dens*, "tooth"

Small tufted perennials with panicles of plump, purple, several-flowered spikelets; the lemmas are distinctive: they have three raised parallel hairy nerves that project from the lemma apex as three short teeth or mucros. Seven species, North and South America.

Tridens flavus (L.) Hitchc., purple-top

flavus, "yellow"

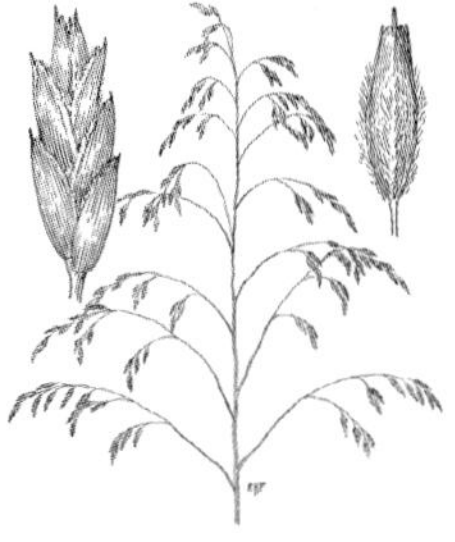
Illustration

Native to the southern United States as far north as Illinois; first collected in 1961 by Hugh Iltis. An uncommon weed in far southwestern Wisconsin, on sandy or gravelly roadsides.

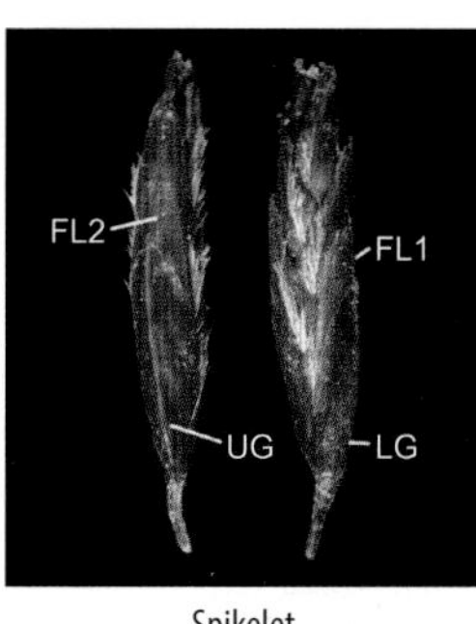

Spikelet

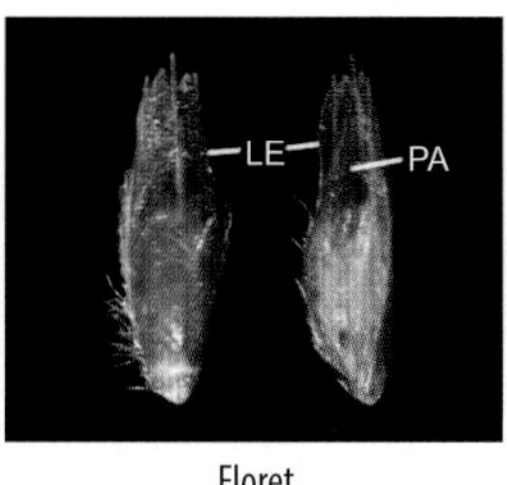

Floret

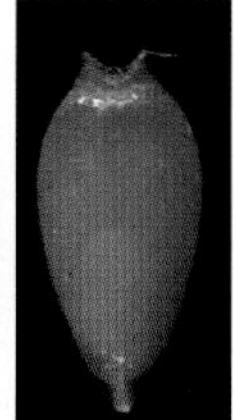
Grain

82. *TRIPLASIS*, SAND GRASS

triplasios, "thrice as many"

Slender annuals with sparse, few-branched panicles usually partly included in the upper leaf sheaths; spikelets slender, purple, several-flowered, the lemma hairy, tipped by a minute awn or mucro, the paleas bowed out (curving) and densely villous on their keels. Two North American species.

Triplasis purpurea **(Walter) Chapm., purple sand grass**

purpurea, Greek, "purple"

An occasional annual or short-lived perennial of open sand of prairies, river terraces, roadside blowouts, and railroad margins; first collected in 1916. It is not certain whether this southern US species was originally native to Wisconsin; however, it appears to be moving northward.

Illustration

Inflorescence

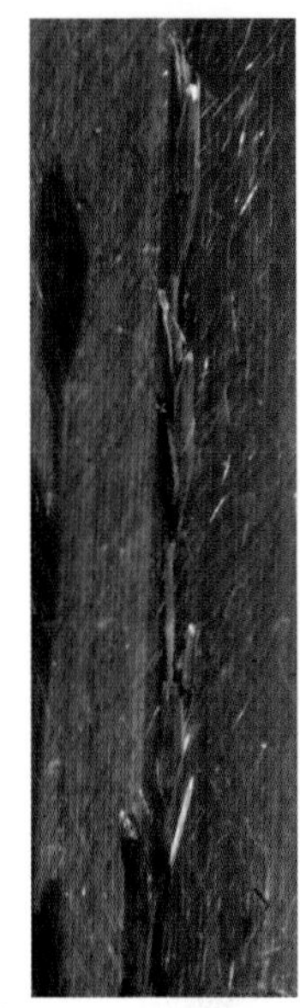

Hairy leaf blade and purple spikelets

Habit

83. *TRISETUM*, TRISETUM

tres, "three," and *seta*, "bristle"

Small, tufted, softly hairy perennials; panicles spike-like, two-flowered, the glumes thin and about as long as the spikelet; the rachilla internodes hairy, the lemmas with short, wavy awns from the middle of their backs. Superficially resembling meadow foxtail (*Alopecurus*) or timothy (*Phleum*), but with two instead of just one floret. Melic-oats (*Graphephorum melicoides*) was previously placed in this genus. Seventy-five species; cosmopolitan except in tropical areas.

Trisetum spicatum (L.) K. Richt., spike trisetum

spicatum, "spiked"

THREATENED

Illustration

Rare, wave-splashed, north-facing sandstone ledges and cliffs of the Apostle Islands and the nearby Bayfield Peninsula (Judziewicz and Nekola 2000 [as 1997]). Janet Marr's 2012 survey of Apostle Islands populations indicated that populations of this species were stable and not in decline (pers. comm.).

Young inflorescence

Portion of inflorescence

Habitat and habit

Mature inflorescence

84. *TRITICUM*, WHEAT

Latin, "wheat"

An annual with a flattened spike inflorescence with one spikelet per node, each spikelet with two to five florets, the glumes and lemmas truncate at top, awned or not. Twenty-five Eurasian species.

Triticum aestivum L., wheat

aestivum, "of summer"

Occasional weed or escape from cultivation along roads, railroads, or edges of agricultural fields; first collected in 1886.

Illustration

AW of FL1
UG
LG

Spikelet

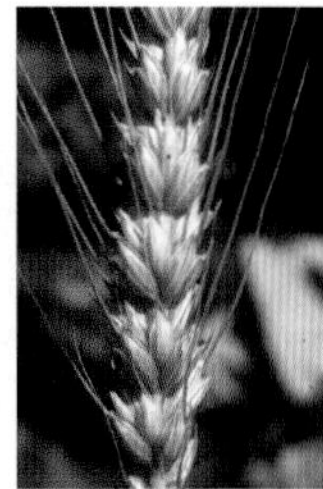

Inflorescence showing long- and short-awned cultivars

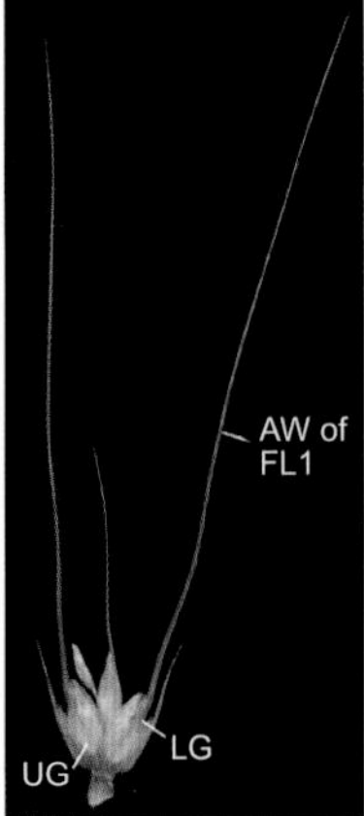

Spikelet

Inflorescence showing long- and short-awned cultivars

Wheatfield

85. *VENTENATA*, VENTENATA GRASS

for Étienne Pierre Ventenat (1757–1808), French botanist

Annuals with panicle inflorescences; spikelets with two to several florets, the lowest floret remaining attached to the glumes. Five species, Mediterranean region.

Ventenata dubia (Leers) Coss. and Durieu, ventenata grass

dubia, "doubtful"

A rare roadside European adventive, collected in waste ground in 1981 by Judziewicz in Oconto County with *Puccinellia distans* and *Apera interrupta* (Solheim and Judziewicz 1984).

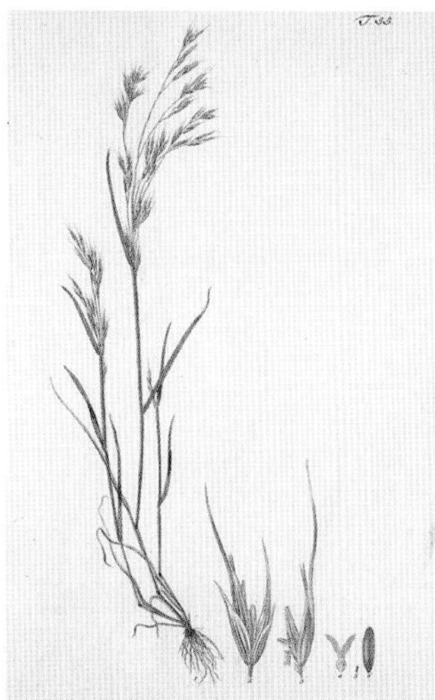

Inflorescence

Portion of Inflorescence

86. *ZEA*, CORN OR MAIZE

Greek for another cereal name

Annual cereals with large, broad leaves and sexes segregated in different parts of the same plant (monoecious): the male spikelets in terminal branched "tassels," and the female spikelets in lateral unbranched "ears." Five species of Mexico and Central America.

***Zea mays* L., corn, maize**

mays, from the Spanish word for Indian corn, *maíz*

Sporadic escape from cultivated fields; first collected in 1937, but cultivated by Native Americans for hundreds of years in Wisconsin before European settlement.

Illustration

Male inflorescence (terminal tassel)

Female inflorescence (lateral ear)

87. *ZIZANIA*, WILD RICE

from the adaptation of Greek *zizanion*, an ancient name of some wild grains

Pulpy aquatic annuals, often with the lower blades floating; sexes segregated in different parts of the same plant (monoecious), the long-awned female spikelets on erect branches in the upper half of the panicle, the awnless male spikelets pendant on spreading lower branches. Four species of North America and northeast Asia.

1. Female lemma delicate and membranous, scabrous between the nerves as well as on them (use a hand lens), 0.5–1 mm wide; widest leaves 1.5–5 cm wide; plants usually 2–4 m tall .... *Z. aquatica*
1. Female lemma firm, smooth between the nerves, 1–2 mm wide; widest leaves 0.5–2 cm wide; plants usually 0.5–1.5 m tall .... *Z. palustris*

Zizania aquatica **L., southern wild rice**

aquatica, "growing in or near water"

Fairly common in marshes or shallow water of marshes and sloughs. This species is Wisconsin's largest native grass; our plants are often 2–3.5 m tall and may attain heights of 4.3 m, as measured in 1981 from specimens collected in Marquette County. The caryopsis does not thresh free from the florets in this species, so it is not used as a human food source. Hybrids (or at least introgressions) with *Zizania palustris* occur, such as, for example, the large stands just inland from the levee on the west bank of the Wisconsin River in Stevens Point.

Male spikelets

Judziewicz with 4.3 m tall Marquette County specimen collected by Hugh H. Iltis and agrostology class in 1981

Inflorescence, yellow stamens releasing pollen

Male (*left*) and female (*right*) spikelets

Southern wild rice, habit

Zizania palustris **L., northern wild rice**

palustris, "of marshes"

Illustration

Fairly common in shallow water of lakes and streams, especially characteristic of slow-moving waters such as the "thoroughfares" connecting northern lakes; a culturally important plant to many Native Americans, especially in the lower reaches of the Bad River and Kakagon Sloughs near Lake Superior in Ashland County (Meeker 1993). It is often planted for both wildfowl and human food in many lakes where it did not naturally occur. The caryopsis does thresh free from the florets in this species, so it is used as a human food source. Fassett (1951) treats this as *Zizania aquatica* var. *angustifolia* and *Z. aquatica* var. *interior*.

Female spikelets

Kakagon Sloughs rice bed, Ashland County

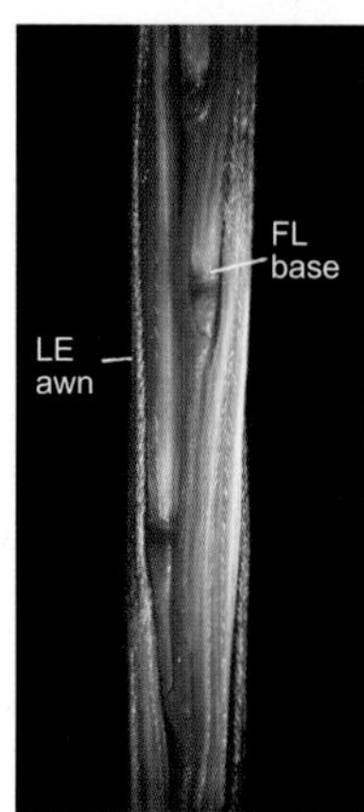

Female spikelets

Plants in flower

Glossary

achene: The fruit of a sedge (and other families), in which the seed is not fused to the fruit coat, leaving a small air space in between.

adventive: Exotic grasses that have been accidentally introduced to Wisconsin as small, short-lived populations that have not persisted and have not become naturalized (permanently established). See also *exotic* and *native.*

anther: The tubular, pollen-producing sacks of the stamens.

anthropogenic: Caused by human activities.

antrorsely barbed or *antrorsely scabrous:* Said of a plant surface covered with minute pointed structures directed upward, making it hard to stroke the surface downward.

apex (pl. *apices*): The tip of an organ.

auricles: A pair of narrow ear- or horn-like extensions of the leaf from the base of the blade or top of the sheath margin; these tend to wrap around the culm above the "V neckline" formed by the edges of the sheath.

awn: A bristle-like or needle-like extension of the midrib of either the lemma or the glume; the awn lacks the softer, thinner, internerve tissue on both sides. Awns are found in many grasses. Lemmas can be awned from near the base, from the middle of the back, from between two apical teeth, or directly from the apex. See also *bristles.*

axillary: In a lateral position; in grasses, most commonly refers to inflorescences that are produced along a leafy culm rather than terminally.

axis: The central stalk of the inflorescence on which are produced spikelets or branches that ultimately bear spikelets.

beard: A cluster of long, straight, upward-pointing hairs at the base or callus of a floret.

bisexual (said of a floret or spikelet): Having the stamens and pistil produced in the same floret. Also known as "perfect." The great majority of Wisconsin grasses have bisexual spikelets. See also *male* and *female.*

blade: The expanded portion of the leaf that extends away from the sheath and ligule; flat in the majority of Wisconsin grasses.

bristles: A group of pedicels that fail to bear spikelets and that have become modified to resemble large, stiff hairs, bearing tiny forward- or backward-projecting barbs. Bristles are not part of their associated spikelets; foxtails (*Setaria*) and cultivated fountain grasses (*Cenchrus*) are the only Wisconsin grasses with bristles. See also *awn.*

bulb: The swollen culm base of certain grasses such as timothy (*Phleum pratense*).

bulbose: With a swollen, bulblike culm base.

callus: The sharply pointed base of the florets of many grasses; often bearded with a dense tuft of upward-pointing hairs.

caryopsis: See *grain.*

cilia: Long straight hairs, often found along the margins (like eyelashes) of various plant organs.

cleistogamous: In grasses, inflorescences that flower while they are enclosed or hidden within leaf sheaths.

convergent: In grass spikelet terminology, often refers to nerves that gradually converge from the base to the apex of the lemma.

cordate: Heart-shaped.

culm: The above-ground stem of a grass.

cultivar: A variety cultivated for either agricultural or horticultural purposes.

decumbent: Reclining at the base, but with the tip turned up.

digitate: Spreading like the fingers from the palm of the hand; used to describe certain inflorescences such as crabgrasses (*Digitaria ischaemum* and *D. sanguinalis*). Also known as "radiate."

dioecious (said of a plant): Having the stamens and pistil produced in differently shaped spikelets on different plants. A plant is thus either functionally "male" or "female." See also *monoecious* and *bisexual.*

divaricate: With branches or spikelets widely spreading or diverging.

dorsally compressed: Usually referring to spikelet compression; with the more flattened sides centered on the midrib of the glumes and lemmas, like the human body. See also *laterally compressed* and *terete.*

elliptical: Said of an organ such as a spikelet that is widest near the middle and equally rounded at both ends; the three-dimensional analog is "ellipsoid."

exotic: Grasses believed to have been inadvertently or purposely introduced to Wisconsin from the time of European settlement (ca. 1830) to the present day and are naturalized in the state; that is, they have established self-perpetuating populations as if they are native. See also *adventive* and *native.*

exserted: With inflorescences, spikelets, or stamens protruding and easily visible.

fascicle: A dense cluster of branches, leaves, or spikelets.

female: Florets or spikelets that bear flowers with pistils only. See also *bisexual* and *unisexual.*

floret: Portion of the grass spikelet composed of the lemma, the palea (usually present), and the enclosed flower or fruit. The lemma is the most visible part of a floret.

glabrous: Lacking hairs.

glands: Tiny, sometimes almost microscopic swellings (use a hand lens).

glaucous: Covered with a white, waxy coating.

glomerule: A crowded, congested group of spikelets.

glume: One of two empty scales at the base of a spikelet, often resembling a lemma and with an odd number of nerves (a midrib plus zero to several pairs of lateral nerves).

grain: The fruit of a grass, consisting of a seed fused to the fruit coat. Also known as a "caryopsis."

hirsute: With stiff, straight hairs.

hispid: With stout, stiff, almost prickly hairs.

inflorescence: In grasses, the collective term for all spikelets along with the axis and branches that bear them.

internode: The portion of the grass culm (stem) between successive leaf nodes, or portions of the inflorescence between successive spikelets, whorls of spikelets, or branches.

involucre: A bract or cluster of bracts partially or wholly surrounding an inflorescence.

keel: A ridge formed by the folded midrib of a glume, lemma, or leaf blade. See also *midrib*.

lanceolate: Lance-shaped, referring to the shape of an organ such as a leaf, glume, or lemma that is between about three to ten times as long as wide and is widest below the middle. See also *linear* and *ovate*.

laterally compressed: Usually referring to spikelet compression; with the spikelet parts folded at the midrib, comparable to a fish such as a perch or a bluegill. See also *dorsally compressed* and *terete*.

lemma: The lower, larger, dominant scale in a floret, enclosing the palea; always with an odd number of nerves (a midrib plus zero to several pairs of lateral nerves).

ligule: A semicircular membrane or line of hairs (or a membrane tipped by hairs) at the summit of the sheath and base of the blade (on its upper side) that extends upward a short distance along the culm.

linear: Line-shaped, referring to the shape of an organ such as a leaf, glume, or lemma that is at least ten times as long as wide, with parallel sides; often much narrower. See also *lanceolate* and *ovate*.

lodicules: Two or three tiny, delicate, transparent, scale-like, short-lived, modified petals found at the base of the pistil of the grass flower; they swell with water at pollination time, forcing the lemma and palea apart and facilitating pollination.

male: Florets or spikelets that bear flowers with stamens only. See also *bisexual* and *unisexual*.

meristematic: The area of actively dividing cells near the growing tip of the plant.

mesic: Moist, well-drained sites with soils that are neither too dry nor too wet.

midrib: The central nerve of a glume, lemma, or leaf blade; not strongly folded. Also known as the "midnerve" or "midvein." See also *keel*.

monoecious (said of a plant): Having the stamens and pistil produced in differently shaped spikelets on the same plant. See also *dioecious* and *bisexual*.

monomorphic: Having just one type of growth form.

mucro: A very short, needle-like outgrowth of a nerve; in grasses, most commonly occurring on the lemmas.

native: Grasses believed to have been present in Wisconsin prior to the time of European settlement (ca. 1830). See also *adventive* and *exotic*.

nerves: The major vascular bundles in a leaf blade, glume, lemma, or palea; nerves are often slightly raised and usually converge at the tip of the glumes or lemma. Also known as "veins."

node: Level or region of a culm (stem) at which the leaf sheath is attached (often with a slightly swollen area), or level of the inflorescence axis at which one or more spikelets are attached.

obovate: Egg-shaped (see *ovate*), but with the widest portion occurring near the apex instead of near the base.

obscure: Not easily seen; in grasses, often refers to poorly developed, barely visible nerves of the lemmas of some species.

ovate: Egg-shaped, referring to the outline shape of an organ such as a leaf, glume, or lemma that is less than three times as long as wide and usually broader than that; "ovoid" is the term in three dimensions. See also *lanceolate* and *linear*.

palea: The upper, narrower, more hidden scale in a floret, partially enclosed by the lemma; always with an even number of nerves, usually just two submarginal keels. An exception is *Leersia*, with a three-nerved palea.

panicle: An inflorescence bearing several major branches along a central axis, with some of the branches rebranching at least once; the inflorescence is often egg-shaped (ovoid). See also *spike* and *raceme*; many "spikes" are actually strongly contracted and spike-like panicles.

panicoid: Having a spikelet that is either *dorsally compressed* or *terete*, with two florets, the lower floret membranous and usually either sterile or male, the upper floret hardened and fertile.

papillose: Minute wart-shaped projections, often at the base of pilose hairs.

pedicel: The stalk of an individual grass spikelet, usually thin and thread-like. If no pedicel is present, the spikelet is said to be sessile.

peduncle: The stalk of the inflorescence.

perigynium: The inflated flask-shaped bract surrounding the fruit of the sedge (Cyperaceae) genus *Carex*.

pilose: With soft, fairly straight long hairs.

pistil: The female part of the grass flower, exserted from the floret during the very short period of flowering; most all Wisconsin grasses have pistils with two feathery stigmas.

pseudoligule: Hairs at the base of the blade, adjacent to the ligule.

puberulent: Minutely hairy, such that a hand lens is usually needed to see the hairs.

pulvinus: A swelling of the inflorescence axis where each branch is attached.

raceme: Inflorescence composed of an elongate main axis, bearing spikelets on pedicels that arise at intervals along the main axis directly, without intervening rebranching. See also *spike* and *panicle*.

rachilla: The tiny, unbranched stalk or axis running through a spikelet from the lower glume to the upper glume to successive floret(s); sometimes prolonged as a minute sterile bristle past the uppermost floret.

rachis: The axis of the inflorescence.

radiate: See *digitate*.

retrorsely barbed or *retrorsely scabrous:* Said of a plant surface covered with minute pointed structures directed downward, making it hard to stroke the surface upward. See also *antrorsely barbed*.

rhizome: A laterally spreading underground stem with leaves reduced to semiencircling sheaths.

rosette: A dense circular cluster of basal leaves.

rudimentary: A reduced, sterile, irregularly shaped structure appearing in the position of a spikelet or floret. Rudimentary florets can occur at either the base or (more frequently) the tip of a spikelet. Also known as "sterile" or "vestigial."

scabrid: Minutely or only slightly *scabrous*.

scabrous: Said of a surface that is rough to the touch, like sandpaper, but bearing no obvious hairs, even under a hand lens.

sessile: Said of an organ (such as a spikelet) with no stalk or pedicel.

sheath: Portion of a grass leaf that wraps around the stem (culm) from node to blade. A majority of grasses have an open sheath that wraps around the stem like a coat or robe, one edge lying over the other (overlapping), forming a long "V neckline," with separable edges. A minority of grasses have a closed or fused sheath that forms a tube below a short V neckline like a pullover sweater, the edges fused together and not separable (as in the bromes, *Bromus*).

smooth: A surface that is hairless and, in addition, not scabrous.

spike: An inflorescence in which all spikelets are borne directly on the inflorescence axis; they have no stalks (pedicels). See also *panicle* and *raceme*; some racemes and panicles can be strongly contracted and spike-like, but on close examination it will be found that the spikelets are stalked.

spikelet: The basic unit of the grass inflorescence; unbranched and consisting of a pair of glumes (rarely glumeless) plus one or more florets above the glumes, all on an axis called the "rachilla."

stamens: The male part of the grass flower, exserted from the floret during the very short period of flowering; most all Wisconsin grasses have three stamens.

stigmas: In most grasses, two feathery pollen-receptive parts of the pistil.

stolon: A laterally spreading stem that runs horizontally along the surface of the ground. Also called a "runner."

subtend: To lie directly below another structure, often partially surrounding or embracing that structure.

terete: Usually referring to a spikelet that is plump, uncompressed, and round in cross section, like the body of a snake. See also *dorsally compressed* and *laterally compressed.*

truncate: With a horizontal, truncated apex.

unisexual: Florets or spikelets that bear flowers that have only stamens or a pistil.

villous: Covered with long, soft, wavy hairs.

whorled: With three or more parts converging at a common point; in grasses, most commonly refers to branches of the inflorescence.

References

Barkworth, M. E., L. K. Anderton, K. M. Capels, S. Long, and M. B. Piep. 2007. *Manual of the Grasses for North America*. Logan, UT: Intermountain Herbarium and Utah State University Press.

Black, M. R., and E. J. Judziewicz. 2009. *Wildflowers of Wisconsin and the Great Lakes Region*. 2nd ed. Madison: University of Wisconsin Press.

Brues, C. T., and B. B. Brues. 1911. "The Grasses of Milwaukee County, Wisconsin." *Transactions of the Wisconsin Academy of Sciences, Arts and Letters* 17:57–76.

Chemisquy, M. A., L. M. Giussani, M. A. Scataglini, E. A. Kellogg, and O. Morrone. 2010. "Phylogenetic Studies Favour the Unification of *Pennisetum*, *Cenchrus* and *Odontelytrum* (Poaceae): A Combined Nuclear, Plastid and Morphological Analysis, and Nomenclatural Combinations in *Cenchrus*." *Annals of Botany* (Oxford), n.s., 106:107–30.

Cholewa, A. 2011. "Comprehensively Annotated Checklist of the Flora of Minnesota, Version 2011.2." Unpublished list, University of Minnesota.

Clark, L. G., and R. W. Pohl. 1996. *Agnes Chase's First Book of Grasses*. Washington, DC: Smithsonian Institution Press.

Cochrane, T. S., K. Elliot, and C. Lipke. 2006. *Prairie Plants of the University of Wisconsin–Madison Arboretum*. Madison: University of Wisconsin–Madison Arboretum.

Cochrane, T. S., and H. H. Iltis. 2000. "Atlas of the Wisconsin Prairie and Savanna Flora." *Wisconsin Department of Natural Resources Technical Bulletin* 191:1–226.

Columbus, J. T. 1999. "An Expanded Circumscription of *Bouteloua* (Gramineae: Chloridoideae): New Combinations and Names." *Aliso* 18:61–65.

Curtis, J. T. 1971. *The Vegetation of Wisconsin: An Ordination of Plant Communities*. Madison: University of Wisconsin Press.

Cusick, A. W., R. K. Rabeler, and M. J. Oldham. 2002. "Hard or Fairgrounds Grass (*Sclerochloa dura*, Poaceae) in the Great Lakes Region." *Michigan Botanist* 41:125–35.

Darbyshire, S. J. 1993. "Realignment of *Festuca* Subgenus *Schedonorus* with the Genus *Lolium* (Poaceae)." *Novon* 3:239–43.

Darke, R. 1999. *The Color Encyclopedia of Ornamental Grasses: Sedges, Rushes, Restios, Cat-tails, and Selected Bamboos*. Portland, OR: Timber Press.

Fassett, N. C. 1951. *Grasses of Wisconsin*. Madison: University of Wisconsin Press.

Finot, V., P. M. Peterson, R. J. Soreng, and F. O. Zuloaga. 2005. "A Revision of *Trisetum* and *Graphephorum* (Poaceae: Pooideae: Aveninae) in North America North of Mexico." *Sida* 21:1419–53.

Flora of North America Editorial Committee, ed. 2003. *Flora of North America North of Mexico.* Vol. 25. *Magnoliophyta: Commelinidae (in Part): Poaceae, Part 2.* New York: Oxford University Press.

———, ed. 2007. *Flora of North America North of Mexico.* Vol. 24. *Magnoliophyta: Commelinidae (in Part): Poaceae, Part 1.* New York: Oxford University Press.

Freckmann, R. W. 1972. "Grasses of Central Wisconsin." *Reports on Fauna and Flora of Wisconsin* 6:1–81. Museum of Natural History, University of Wisconsin–Stevens Point.

———. 1977. "An Introduction to Grass Identification." *Newsletter of the Botanical Club of Wisconsin* 9(4): 1–35.

———. 1978. "A Checklist of the Vascular Plants of the 'Whiting Triangle,' in Central Wisconsin." *Proceedings of the Iowa Academy of Science* 86:44–49.

Freckmann, R. W., F. D. Bowers, and L. Echola. 1989. "*Molinia caerulea*, Moorgrass, New to Wisconsin." *Michigan Botanist* 27:88–89.

Freckmann, R. W., and D. M. Reed. 1979. "*Glyceria maxima*, a New, Potentially Troublesome Wetland Weed." *Newsletter of the Botanical Club of Wisconsin* 11:30–35.

Ginsburg, R. 2002. *Lloyd Herbert Shinners: By Himself.* Fort Worth: Botanical Research Institute of Texas.

Gleason, H. A., and A. Cronquist. 1991. *Manual of Vascular Plants of the Northeastern United States and Adjacent Canada.* 2nd ed. New York: New York Botanical Garden Press.

———. 1998. *Illustrated Companion to the Manual of Vascular Plants of the Northeastern United States and Adjacent Canada.* New York: New York Botanical Garden Press.

Grasses of Iowa. 2014. Ecology, Evolution, and Organismal Biology Department, Iowa State University, Ames. http://www.eeob.iastate.edu/research/iowagrasses.

Greenlee, J. 1992. *The Encyclopedia of Ornamental Grasses: How to Grow and Use Over 250 Beautiful and Versatile Plants.* New York: Michael Friedman Publishing Group.

Hipp, A. 2008. *Field Guide to Wisconsin Sedges.* Madison: University of Wisconsin Press.

Hitchcock, A. S., and A. Chase. 1951. *Manual of the Grasses of the United States.* Washington, DC: U.S. Department of Agriculture.

Hoffman, R. 2002. *Wisconsin's Natural Communities: How to Recognize Them, Where to Find Them.* Madison: University of Wisconsin Press.

Janik, E. 2007. "Citizen Scientist: Wisconsin's First Renaissance Man, Increase A. Lapham Merits Renown as Surveyor, Engineer, Mapmaker, Historian, Self-Taught Botanist, Geologist, Forester, Meteorologist and a Prescient Conservation Voice." *Wisconsin Natural Resources Magazine*, February, 13–16.

Judziewicz, E. J. 2002. "Flora and Vegetation of the Grand Traverse Islands (Lake Michigan), Door and Marinette Counties, Wisconsin and Delta County, Michigan." *Michigan Botanist* 40:81–208.

———. 2004a. "Flora and Vegetation of Butternut Pines, Oconto County, Wisconsin." *Michigan Botanist* 43:35–58.

———. 2004b. "Vascular Plants of Kaukamo Spruces, a 20-Acre Tract in the Lake Superior Lowlands of Bayfield County, Wisconsin." *Michigan Botanist* 43:401–19.

Judziewicz, E. J., and R. G. Koch. 1993. "Flora and Vegetation of the Apostle Islands National Lakeshore and Madeline Island, Ashland and Bayfield Counties, Wisconsin." *Michigan Botanist* 32:43–189.

Judziewicz, E. J., and J. C. Nekola. 2000 [as 1997]. "Recent Wisconsin Records for Some Interesting Vascular Plants in the Western Great Lakes Region." *Michigan Botanist* 36:91–118.

King, M., and P. Oudolf. 1998. *Gardening with Grasses*. Portland, OR: Timber Press.

Lapham, I. A. 1853. "The Grasses of Wisconsin." *Transactions of the Wisconsin State Agricultural Society* 3:397–488.

Manual of North American Grasses. 2014. http://herbarium.usu.edu/grassmanual.

Meeker, J. E. 1993. "The Ecology of Wild Rice (*Zizania palustris* var. *palustris*) in the Kakagon Sloughs, a Riverine Wetland on Lake Superior." PhD diss., University of Wisconsin–Madison.

Oaks, A. J. 1990. *Ornamental Grasses and Grasslike Plants.* New York: Van Nostrand Reinhold.

Peterson, P. M., K. Romaschenko, N. Snow, and G. Johnson. 2012. "A Molecular Phylogeny and Classification of *Leptochloa* (Poaceae: Chloridoideae: Chlorideae) *Sensu lato* and Related Genera." *Annals of Botany* 109(7): 1317–30.

Plants of Wisconsin. 2014. Robert W. Freckmann Herbarium, University of Wisconsin–Stevens Point. http://wisplants.uwsp.edu/Wisplants.html.

Pohl, R. W. 1966. "The Grasses of Iowa." *Iowa State Journal of Science* 40(4): 342–573.

Romaschenko, K., P. M. Peterson, R. J. Soreng, O. Futorna, and A. Susanna. 2011. "Phylogenetics of *Piptatherum* s.l. (Poaceae: Stipeae): Evidence for a New Genus, *Piptatheropsis*, and Resurrection of *Patis*." *Taxon* 60:1703–16.

Saltonstall, K., P. M. Peterson, and R. J. Soreng. 2004. "Recognition of *Phragmites australis* subsp. *americanus* (Poaceae: Arundinoideae) in North America: Evidence from Morphological and Genetic Analysis." *Sida* 21:683–92.

Shinners, L. H. 1940. "Notes on Wisconsin Grasses I: Additions to the Grass Flora." *American Midland Naturalist* 24:757–60.

———. 1941. "Notes on Wisconsin Grasses II: *Muhlenbergia* and *Sporobolus*." *American Midland Naturalist* 26:69–73.

———. 1943. "Notes on Wisconsin Grasses III: *Agrostis*, *Calamagrostis*, *Calamovilfa*." *American Midland Naturalist* 29:779–82.

———. 1944. "Notes on Wisconsin Grasses IV: *Leptoloma* and *Panicum*." *American Midland Naturalist* 32:164–80.

Skawinski, P. M. 2011. *Aquatic Plants of the Upper Midwest: A Photographic Field Guide to Our Underwater Forests*. Privately published.

Solheim, S. L., and E. J. Judziewicz. 1984. "Four Noteworthy Wisconsin Plants." *Phytologia* 54:490–92.

Soreng, R. G., G. Davidse, P. M. Peterson, F. O. Zuloaga, E. J. Judziewicz, T. S. Filgueiras, O. Morrone, and K. Romaschenko. 2014. "A Worldwide Phylogenetic Classification of the Poaceae (Gramineae)." In *Catalogue of New World Grasses*. http://www.tropicos.org/Project/CNWG.

Stevens, P. F. 2001 onward. Angiosperm Phylogeny Website. Version 12, July 2012 [and more or less continuously updated since]. http://mobot.org/MOBOT/research/APweb.

Swink, F., and G. Wilhelm. 1994. *Plants of the Chicago Region*. Lisle, IL: Morton Arboretum.

Thiers, B. M. 2014. Index Herbariorum: A Global Directory of Public Herbaria and Associated Staff. New York Botanical Garden's Virtual Herbarium, http://sweetgum.nybg.org/ih.

Tropicos Plant Database. 2014. Missouri Botanical Garden. http://www.tropicos.org.

USDA NRCS (Natural Resources Conservation Service). 2014. Plant Guides. http://plant-materials.nrcs.usda.gov.

USDA Plants Database. 2014. http://plants.usda.gov.

Valdes, B., and H. Scholz. 2006. "The Euro+Med Treatment of Gramineae—a Generic Synopsis and Some New Names." *Willdenowia* 36:657–69.

Voss, E. G., and A. A. Reznicek. 2012. *Field Manual of the Michigan Flora*. Bloomington Hills, MI: Cranbrook Institute of Science.

vPlants. 2014. A Virtual Herbarium of the Chicago Region. Morton Arboretum, Field Museum, Chicago Botanical Garden. http://www.vplants.org.

Waller, D. M., and T. P. Rooney, eds. 2008. *The Vanishing Present: Wisconsin's Changing Lands, Waters, and Wildlife*. Chicago: University of Chicago Press.

Weakley, A. S., R. J. LeBlond, B. A. Sorrie, C. T. Witsell, L. D. Estes, K. Gandhi, K. G. Mathews, and A. Ebihara. 2011. "New Combinations, Rank Changes, and Nomenclatural and Taxonomic Comments in the Vascular Flora of the Southeastern United States." *Journal of the Botanical Research Institute of Texas* 5:437–55.

Wisconsin Department of Natural Resources. 2014. Wisconsin's Rare Plants. http://dnr.wi.gov/topic/endangeredresources/plants.asp.

Wisflora. 2014. Wisconsin Vascular Plant Species. Wisconsin State Herbarium, Madison. http://www.botany.wisc.edu/wisflora.

Illustration Credits

There are over 1,100 illustrations in this book, and here we credit these images to the many contributors who made the book possible. We are very grateful for their contributions. In the credits listed below, each illustration in the book is assigned a code to indicate its location. The first number in this code represents the page number on which the illustration appears. The letter represents the location on that page. For illustrations that appear in the field guide section, *a* represents the first entry on the page, *b* the second, and so on; for illustrations that appear in the morphology, agrostology, and Wisconsin plant communities chapters, as well as those illustrations that appear in the keys, letters are assigned from left to right or from top to bottom. For illustrations that appear in the field guide section (most images in the book), a final code number is assigned. This final number represents an illustration's position relative to other images in the entry and is assigned from left to right and then top to bottom.

William S. Alverson: 187a1, 266a4

Derek Anderson: 167a5, 220a4

Merel R. Black: ii, 21c, 22a, 191a1, 223a4, 319a1

Matt Bushman, U.S. Forest Service: 266a3

Eric J. Epstein, Wisconsin Natural Heritage Inventory: 36a, 37a, 45a, 46a, 49a, 50a, 51a, 320a3

Norman C. Fassett, *Grasses of Wisconsin*: 71f, 76a, 76d, 77a, 85f, 99a1, 111a1, 136a2, 139b1, 204b1, 211a1, 216a1, 247b1, 250a1, 270a1, 274a1, 283a1, 310a1

Gary Fewless: 29b, 287a4

Robert W. Freckmann: 4a, 4b, 5a, 5b, 5c, 6a, 6b, 6c, 7a, 8a, 8b, 9a, 9b, 11b, 15a, 17a, 21a, 21b, 35a, 37b, 39a, 42a, 42b, 45b, 52a, 52b, 54b, 60b, 86c, 91a2, 93a3, 94b3, 95b4, 98a3, 99a3, 99a4, 101a5, 101a6, 105a2, 110b2, 110b4, 116a5, 117a1, 118b2, 127a3, 127a5, 127a6, 128a2, 128a3, 129a1, 132a4, 134a4, 135a1, 135a2, 135a3, 139a1, 141a2, 141a3, 141a4, 144a3, 147a3, 154a2, 160a4, 161a1, 161b3, 162a3, 162b4, 164a4, 165a5, 166a5, 166a6, 169a1, 169a3, 170a2, 173a4, 173a5, 178a4,

178a5, 185b2, 186a1, 187a2, 188b4, 188b5, 190a1, 192a7, 200b1, 209a6, 210a1, 210a2, 210a3, 211a3, 211b3, 211b6, 212a1, 213a1, 215a1, 221a1, 222a2, 222a3, 230a1, 234a5, 235a4, 242a3, 242a4, 243a6, 250a3, 252a4, 252a5, 254a4, 254a5, 254a6, 255b5, 256a1, 260a2, 260a4, 260a6, 260a8, 261a1, 261a2, 263a4, 268a1, 270a2, 271a4, 272a3, 277b3, 277b5, 288a3, 291a2, 291a3, 292a5, 292a6, 294a5, 294a6, 295a3, 295a4, 296a2, 297a1, 297a2, 298a3, 299a2, 299a4, 301a4, 305a4, 309a1, 309a2, 314a3, 314a5, 314a6, 316a2, 316a3, 318a1, 318a3, 318a4, 320a2, 320a5

Elsie H. Froeschner: 68a, 68b, 68c, 68d, 68e, 68f, 69a, 69b, 69c, 69d, 69e, 69f, 69g, 70a, 70b, 70c, 70d, 70e, 70f, 71a, 71b, 71c, 71d, 71e, 71g, 72a, 72b, 72d, 72e, 73a, 73b, 73c, 73d, 73e, 73f, 74a, 74b, 74c, 74d, 74e, 74f, 75a, 75b, 75c, 75e, 75f, 75g, 76b, 76c, 76f, 77c, 77d, 78a, 78c, 78d, 78e, 78f, 79a, 79b, 79c, 79d, 79e, 80b, 80c, 80d, 80e, 80g, 81a, 81b, 81e, 81g, 82a, 82b, 82c, 82d, 82e, 83a, 83b, 83c, 83e, 83f, 84a, 84b, 84c, 84d, 84e, 84f, 84g, 85a, 85b, 85c, 85d, 86a, 86b, 86d, 86e, 86g, 87a, 87b, 87c, 87d, 87e, 87f, 88b, 88c, 88d, 88e, 88f, 89a, 91a1, 93a1, 94a2, 94b2, 95a1, 95b2, 96a1, 97a1, 98a1, 101a1, 102a1, 105a1, 106b1, 107a1, 108a1, 108b1, 109a1, 110a1, 110b1, 114a1, 116a1, 118a1, 118b1, 119b1, 120a1, 124a1, 125a1, 126a1, 126b1, 128a1, 128b1, 130a1, 131a1, 132a1, 134a1, 137a1, 138a1, 140a1, 142a1, 143a1, 144a1, 149a1, 151b1, 154a1, 158a1, 158b1, 159a1, 159a2, 160a3, 161b1, 162b1, 163a1, 164a2, 165a2, 166a1, 166a2, 167a1, 167a2, 168a2, 168b1, 171a1, 172b1, 173a1, 174a1, 175a1, 177a1, 178a1, 179a1, 180a1, 183a1, 184b2, 185a2, 185b1, 188a1, 188b1, 189a1, 189b1, 190b1, 192a1, 193a1, 195a1, 196a1, 197a1, 198a1, 199a1, 199b1, 200a2, 200b2, 201a1, 202a1, 204a1, 205a1, 206a1, 206b1, 207a1, 207b1, 208a1, 209a1, 211b1, 214a1, 214b1, 217a1, 217b1, 219a1, 220a1, 221b1, 222a1, 223a1, 224a1, 225a1, 227a1, 229a1, 230b1, 231a1, 232a1, 234a1, 235a1, 239a1, 240a1, 241a1, 242a1, 243a1, 244a1, 245b1, 246a1, 246b1, 247a1, 249a1, 252a1, 253a1, 254a2, 255a2, 255b1, 257a1, 258a1, 259a1, 260a1, 262a1, 263a1, 264a1, 267a1, 271a1, 271b1, 272a1, 273a1, 274b1, 276a1, 276b1, 277b1, 278a1, 278b1, 279a1, 279b2, 282a1, 284a1, 285a1, 287a1, 288a1, 290a1, 291a1, 292a1, 293a1, 294a1, 295a1, 296a1, 298a1, 299a1, 301a1, 302a1, 303a1, 304a1, 305a1, 306a1, 307a1, 311a1, 312a1, 314a1, 315a1, 316a1, 320a1

Anna B. Gardner: 12a, 13a, 13b, 15b, 17b, 17c, 17d, 18a, 19a, 19b, 20a, 20b, 20c, 20d, 22b, 93a2, 93a4, 94a1, 94a3, 94b1, 95b3, 96a2, 97a2, 97a3, 97a4, 97b2, 98a4, 101a2, 101a4, 102a4, 102a5, 105a3, 106b2, 106b3, 107a2, 107a3, 107b1, 107b2, 107b3, 107b4, 108a2, 108a3, 108b3, 109a3, 110a2, 110a3, 110b3, 114a2, 114a4, 116a2, 116a3, 116a4, 116a6, 118a2, 118a3, 118a4, 118b3, 119b2, 119b3, 119b4, 120a2, 120a4, 120a5, 124a2, 124a3, 125a3, 125a4, 126a2, 126a3, 127a1, 127a2, 127a4, 128a4, 128b2, 128b3, 128b4, 128b5, 130a2, 130a3, 130a4, 131a2, 131a3, 132a2, 132a3, 132a5, 134a2, 134a3, 134a5, 137a2, 138a2, 138a3, 138a4, 140a3, 142a2, 142a3, 143a2, 143a3, 143a4, 144a2, 147a2, 147a4, 149a2, 149a3, 149a5, 149a6, 151b2, 151b3, 154a4, 154a5, 157a1, 157a2, 157a3, 157a4, 158a2, 158a3, 158b2, 159a3, 159a4, 159a5, 160a1, 160a2, 161b2, 162b2, 162b3, 163a2, 163a4, 164a1,

164a3, 165a1, 165a3, 165a4, 166a3, 166a4, 167a3, 167a4, 168a1, 168a3, 168b2, 168b3, 169a2, 170a1, 171a2, 172b2, 172b3, 172b4, 173a2, 173a3, 174a2, 174a3, 175a2, 177a2, 177a3, 177a4, 178a2, 178a3, 179a3, 180a2, 180a3, 180a4, 183a2, 183a3, 183a5, 184b3, 185a1, 185a3, 185b4, 188a2, 188a3, 188b2, 188b3, 189a2, 189a3, 189a4, 189b3, 189b4, 190b2, 190b3, 190b4, 192a3, 192a4, 192a5, 193a2, 193a3, 193a4, 193a5, 195a3, 195a5, 196a2, 196a4, 196a5, 197a2, 197a3, 198a2, 198a4, 199a2, 199a4, 199b2, 199b3, 199b4, 199b5, 199b6, 200a1, 200a3, 200b3, 201a3, 202a2, 202a3, 202a4, 202a5, 204a2, 205a2, 205a3, 205a4, 205a5, 206a2, 206a3, 207b2, 207b3, 208a2, 208a3, 209a4, 209a7, 211b2, 211b4, 211b5, 214a3, 214a4, 214a5, 214b2, 214b3, 217a4, 217a5, 217b2, 217b4, 218a1, 219a2, 219a3, 220a2, 220a3, 221b2, 221b3, 221b4, 224a3, 225a2, 225a3, 225a6, 227a3, 227a4, 227a5, 229a3, 229a4, 229a5, 230b2, 230b3, 230b4, 231a2, 231a3, 232a2, 232a3, 234a2, 234a3, 235a2, 235a3, 235a5, 236a1, 239a2, 239a3, 240a2, 240a3, 240a4, 241a3, 242a2, 243a2, 243a3, 243a4, 243a5, 244a2, 244a3, 244a4, 244a5, 245b3, 245b4, 246a2, 246a3, 246b2, 246b3, 246b4, 247a2, 247a3, 247a4, 247a5, 249a3, 249a4, 249a5, 252a2, 252a3, 253a2, 253a3, 253a4, 254a1, 254a3, 255a1, 255a3, 255b2, 255b3, 255b4, 258a2, 258a4, 258a6, 259a2, 260a3, 260a5, 260a7, 263a3, 263a5, 263a6, 264a4, 267a3, 271a2, 271a3, 271b2, 271b3, 272a2, 272a5, 273a2, 273a3, 274b3, 276a2, 276a3, 276b2, 277b2, 277b4, 277b6, 278a2, 278b2, 279a2, 279a3, 279b1, 279b3, 282a2, 284a2, 285a4, 286a1, 287a2, 287a3, 288a2, 288a4, 288a5, 290a2, 290a4, 290a5, 292a2, 292a3, 292a4, 292a7, 293a2, 293a3, 293a4, 293a5, 294a2, 294a3, 294a4, 295a2, 296a3, 298a2, 299a3, 299a5, 301a2, 302a2, 302a3, 303a2, 303a3, 304a2, 304a3, 304a4, 305a2, 306a2, 306a3, 306a5, 307a2, 307a3, 307a4, 307b2, 307b3, 307b5, 311a2, 311a3, 311a4, 312a2, 314a2, 314a4, 320a4

Steve C. Garske, Great Lakes Indian Fish and Wildlife Commission: 103a1, 103b3, 112a2, 112a3, 112a5, 113a1, 122a2, 211a4, 233a2, 233a4, 233a5, 237a2, 237a4, 237a5, 264a5, 274b2, 275a1, 275a2

A. S. Hitchcock and A. Chase, *Manual of the Grasses of the United States*: 72c, 72f, 75d, 76e, 77b, 77e, 77f, 77g, 77h, 78b, 79f, 80a, 80f, 81c, 81d, 81f, 82f, 83d, 85e, 86f, 87g, 88a, 89b, 97b1, 103b1, 104a1, 106a1, 112a1, 131b1, 141a1, 147a1, 148a1, 151a1, 153a1, 162a1, 186b1, 198b2, 233a1, 237a1, 245a1, 253b1, 266a1, 273b1, 280a1, 281a1, 307b1, 308a1, 313a1

Emmet J. Judziewicz: 9c, 10a, 9c, 26b, 27b, 29a, 34a, 36b, 39b, 41a, 41b, 43a, 43b, 46b, 47a, 48a, 49b, 53a, 53b, 54a, 55a, 56a, 59a, 59b, 60a, 60c, 60d, 91a3, 95b1, 98a2, 99a2, 101a3, 102a3, 103b2, 104a2, 104a3, 108b2, 109a2, 109a4, 111a2, 111a3, 112a4, 118a5, 119a1, 122a1, 124a4, 125a2, 131b2, 136a1, 137a3, 139b2, 139b3, 140a2, 142a4, 145a1, 145a2, 146a1, 146a2, 148a2, 149a4, 150a1, 151a2, 153a2, 153a3, 153a4, 154a3, 159b1, 160a5, 162a2, 163a3, 163a5, 171a3, 172a1, 172b5, 172b6, 174a4, 175a3, 175a4, 177a5, 179a2, 179a4, 180a5, 183a4, 183a6, 183a7, 184a1, 184b1, 186b2, 189b2, 190b5, 195a2, 195a4, 195a6, 196a3, 196a6, 198b1, 199a3, 201a2, 204b2, 206b2, 207a2, 211a2, 213b1, 213b2, 214a2, 216a2, 216a3, 217a3,

217b3, 218a2, 223a2, 223a3, 225a5, 228a1, 228a2, 228a3, 228a4, 229a2, 233a3, 234a4, 237a3, 239a4, 241a2, 241a4, 244a6, 245a2, 245a3, 245a4, 245b2, 247b2, 248a1, 248a2, 249a2, 253b2, 257a2, 258a3, 258a5, 259a3, 259a4, 262a2, 266a2, 267a2, 272a4, 273b2, 273b3, 276b3, 277a1, 278a3, 278b3, 280a2, 281a2, 282a3, 282a4, 283a2, 283a3, 285a2, 290a3, 290a6, 305a3, 307b4, 312a3, 312a4, 313a2, 313a3, 313a4, 313a5, 318a2, 130b1

Kitty Kohout, Wisconsin Department of Natural Resources: 205a6, 224a2, 224a4, 301a3

Robert R. Kowal: 217a2

Matt Lavin, Montana State University: 315a2

Barney Lipscomb: 27a

Arthur Meeks: 225a4, 226a1, 226a2

Scott A. Milburn: 114a3

Christopher Noll: 11a, 18b, 102a2, 120a3, 121a1, 170a3, 170a4, 185b3, 192a2, 192a6, 202a6, 227a2, 250a2, 250a4, 250a5, 263a2, 284a3, 285a3, 286a2, 306a4

Suzanne Sanders, National Park Service: 151b4, 152a1, 276b4

Paul Skawinski: 209a2, 209a3, 209a5, 264a2, 264a3, 265a1, 310a2

Wisconsin Historical Society: 26a

John Zaborsky, University of Wisconsin–Madison: 198a3

Taxonomic Index

This index includes all scientific and common names of grasses (Poaceae). References to associated non-grass species are not included in this list. Bold-faced type indicates pages on which a taxon is described in detail.

Achnatherum
calamagrostis (only in cultivation), 56
extremorientalis (only in cultivation), 56
Agropyron, **91**
cristatum, **91**
dasystachyum (see *Elymus lanceolatus*)
magellanicum (see *Elymus magellanicus*)
repens (see *Elymus repens*)
smithii (see *Pascopyrum smithii*)
trachycaulum (see *Elymus trachycaulus*)
Agrostis, 18, **92–95**, 143
capillaris, 31, 92
gigantea, 14, 18, 53, **93**
hyemalis, 65, **94**
perennans, 43, 49, 64, **94**
scabra, 50, 52, 53, 65, 94, **95**
stolonifera, 51, 63, **95**
Alkali grass, 281
European, 282
Nuttall's, 283
Alopecurus, **96–98**
aequalis, 41, **96**, 98
carolinianus, **97**
geniculatus, **97**
pratensis, 53, 56, 64, **98**
Ammophila breviligulata, 51, 58, **99**, 138, 228
Andropogon, **100–101**
gerardii, 12, 34, 37–38, 58, 64, **101**
hallii, 31, 100
saccharoides (only in cultivation), 56
scoparius (see *Schizachyrium scoparium*)
virginicus, 32, 100, 285
Anthoxanthum, **102–3**
hirtum subsp. *arctica*, 38, 65, **102–3**
odoratum, 30, 53, **103**
Apera interrupta, 31, 66, **104**
Aristida, 55, 66, **104–8**
basiramea, 35, **105**
basiramea var. *curtissii* (see *Aristida dichotoma*)
desmantha, **106**
dichotoma, 34, **106**
intermedia (see *Aristida longespica*)
longespica, 34, **107**
necopina (see *Aristida longespica*)
oligantha, **107**
purpurascens, 35, 66, **108**
tuberculosa, 35–36, **108**

Aristideae, 24
Aristidoideae, 24
Arrhenatherum elatius, 56, **109**
Arundineae, 24
Arundinoideae, 24
Avena, 109, **110**
  fatua, 66, **110**
  sativa, 9, 15–16, 66, **110**
Avenella flexuosa, 51, 52, 63, **111**, 152
Avenula pubescens, 31, **112–13**

Bamboo
  dwarf, 58
  golden, 58, 60
  red, 57
  running, 56
  striped dwarf, 58
Banner grass, silver, 57
Barley, 220, 222
  little, 222
  squirrel-tail, 57, 220
Barnyard grass, 176
  American, 178
  bottlebrush, 179
  European, 177
Beach grass, 99
  American, 58, 99
  blue, 57
Bead grass, 258
Beakgrain, 154
Beckmannia syzigachne, 67, **114**
Bent grass, 92
  autumn, 94
  creeping, 95
  Rhode Island, 92
  silky, 104
Bermuda grass, 147
Big bluestem, 59, 100–101
Black-seeded rice grass, 259
Bluegrass, 267
  annual, 271
  bog, 276
  bulblet, 272
  Canada, 273
  forest, 278
  glaucous, 273
  grove, 270
  Kentucky, 277
  languid, 274
  marsh, 276
  plains, 271
  rough, 279
  Wolf's, 279
  wood, 274
  woodland, 278
Bluejoint, 133
  Canada, 134
Bluestem
  big, 58, 100–101
  little, 59, 100, 285–86
  sand, 100
  silver, 56
Bottlebrush grass, 59, 185, 186
Bottle grass, 154
Bouteloua, **115–19**
  curtipendula, 22, 34, 38, 58, **116–17**
  dactyloides, 21, 30, 56, 115, **118**
  gracilis, 56, **118–19**
  hirsuta, 34, 58, **119**
Brachyelytreae, 25
Brachyelytrum, 15, 23, **120–21**
  aristosum, 30, 47, 48, 64, **120**
  erectum, 44, 64, **120–21**
Bristle grass, barbed, 293
Briza media, 56, **122**
Brome, 123, 229
  Canadian, 130
  corn, 131
  ear-leaved, 128
  erect, 125
  European, 130
  field, 124
  fringed, 125
  prairie, 128

rye, 131
satin, 129
smooth, 126
soft, 126
Bromeae, 25
Bromus, 12, 17, 67, **123–32**, 203, 233
altissimus (see *Bromus latiglumis*)
arvensis, **124**
ciliatus, 39, 40, **125**
commutatus (see *Bromus racemosus*)
erectus, **125**
hordeaceus, **126**
inermis, 53, 63, 67, **126–27**, 263
japonicus (see *Bromus arvensis*)
kalmii, 17, 44, 50, **128**
latiglumis, 12, 13, 43–45, 47, 64–66, **128–29**
mollis (see *Bromus hordeaceus*)
nottowayanus, 43, 64–65, **129**
pubescens, 45, 47, 64, 128–29, **130**
pumpellianus, 126
purgans (see *Bromus latiglumis*)
racemosus, **130**
secalinus, **131**
squarrosus, 31, **131**
tectorum, 64, 67, **132**
Brook grass, 139
Broomcorn, 296
Broomsedge, 100
Buchloe dactyloides (see *Bouteloua dactyloides*)
Buffalo grass, 56, 118

Calamagrostis, **133–37**, 138
brachytricha (only in cultivation), 57
canadensis, 18, 19, 38–40, 49, 52, 133, **134–35**
epigeios, **136**
inexpansa (see *Calamagrostis stricta*)
neglecta (see *Calamagrostis stricta*)
stricta, 39, 52, **137**
× *acutiflora* (only in cultivation), 57, 59, 133
Calamovilfa longifolia, 35, 51, **138–39**
Canary grass, 260
common, 261
reed, 262
Catabrosa aquatica, 40, **139**
Catchfly grass, 224
Cenchrus, 35, 57, 59–60, **140**
longispinus, 35, 66, **140**
orientalis (see *Cenchrus purpurascens*)
purpurascens (only in cultivation), 57, 60, 294
spicatus (only in cultivation), 57, 59
Chasmantheae, 24
Chasmanthium latifolium, 23, 31, 57, **141**
Cheat grass, 132
Chlorideae, 24
Chloridoideae, 24
Chloris verticillata, 31, 65, **142**
Cinna, **143–44**
arundinacea, 45, 64, **143**
latifolia, 47, 49, 64, **144**
Coix lacryma-jobi (only in cultivation), 57
Coleataenia rigidula, **144–45**, 251
Columbus grass, 296
Cordgrass, prairie, 59, 299
Corn, 315–16
Cortaderia selloana (only in cultivation), 57
Crabgrass, 171
hairy, 173
slender, 171
smooth, 172
Crested wheat grass, 91
Crypsis schoenoides, 55, 67, **146**
Cup grass, 201
Chinese, 202
Prairie, 202
Cut-grass, rice, 225
Cymbopogon citratus, 57
Cynodon dactylon, 31, 67, **147**, 180
Cynosurus cristatus, **148**

Dactylis glomerata, 53, 63, 66, **149–50**
Danthonia, **150–52**
compressa, 30, 48, **151**
spicata, 44, 48, 50, 53, 112, **151–52**, 171
Danthonieae, 24
Danthonioideae, 24
Darnel, 230
Deer-tongue grass, 57, 159
Deschampsia, **152–53**
cespitosa, 40, 41, 52, 58, 63, **153**
flexuosa (see *Avenella flexuosa*)
Deyeuxia brachytricha (see *Calamagrostis brachytricha*)
Diarrhena, **154**
americana (see *Diarrhena obovata*)
obovata, 43, 45, 65, **154**
Diarrheneae, 25
Dichanthelium, 66, **155–70**, 251, 267
acuminatum, 157, 166, 170
acuminatum var. *fasciculatum*, 53, **157**
acuminatum var. *lindheimeri*, **158**
boreale, 157, **158**
clandestinum, 31, 57, 64, **159**
columbianum, 54, **159**, 161, 167
commonsianum var. *euchlamydeum*, 35, **160–61**
depauperatum, 44, 48, **162**
dichotomum, **162**
latifolium, 13, 43, 44, 65, **163**
leibergii, 35, 37, **163**
linearifolium, 44, 50, **164**
meridionale, 35, 54, **165**
oligosanthes var. *scribnerianum*, 23, 37, 157, **166**, 168
ovale subsp. *praecocius* (see *Dichanthelium villosissimum* var. *praecocius*)
ovale subsp. *pseudopubescens* (see *Dichanthelium commonsianum* var. *euchlamydeum*)
perlongum, 34, 37, **167**
villosissimum var. *praecocius*, 44, 166, **168**
wilcoxianum, 34, **168–69**
xanthophysum, 44, 50, 157, **170**
Digitaria, 19, 54, 55, **171–73**, 180
cognata, 35, 50, 54, 65, 66, **171–72**, 200, 252
filiformis, 32
ischaemum, 20, 63, 67, **172**
sanguinalis, 67, **173**
Diplachne, **174**
fascicularis (see *Diplachne fusca* subsp. *fascicularis*)
fusca subsp. *fascicularis*, 10, 55, 56, 67, **174**
Distichlis spicata, 21, 30, 55, **175**
Dog's-tail, 148
Dropseed, 302
alkali, 58
hidden, 303
meadow, 304
poverty, 307
prairie, 59, 306
sand, 305
small, 307

Echinochloa, 6, 19, 114, **176–79**
crus-galli, 54, 67, **177**, 178
esculenta, 67, **177**
muricata, 20, 45, 66, **178**
pungens (see *Echinochloa muricata*)
walteri, 46, 66, **179**
Ehrhartoideae, 24
Elbow grass, 253
Eleusine indica, 12, 67, **180**
Elymus, 64, 91, 179, **181–93**, 228
canadensis, 34, 37, 38, 51, 59, 64, 181, **182–83**, 192
curvatus, **184**
diversiglumis, **185**
hystrix, 43, 59, 65, 181, **185–86**, 192
interruptus (see *Elymus diversiglumis*)

lanceolatus, 51, 64, **186–87**
macgregorii, **188**
magellanicus (only in cultivation)
repens, 11, 17, 18, 54, 64, 148, 181, **188**, 228
riparius, 46, 64, **189**
trachycaulus, 50, 64, 181, **189–90**
villosus, 44, 59, **190–91**
virginicus, 46, 66, 181, 183, 185, **192**
virginicus var. *submuticus* (see *Elymus curvatus*)
wiegandii, 46, **193**
× *ebingeri*, 185, 192
× *maltei*, 183, 192
Eragrostideae, 24
Eragrostis, 17, 67, **194–201**
capillaris, 55, 65, 67, **195**
cilianensis, 55, 67, **196**
frankii, 46, 66, **197**
hypnoides, 46, 66, **198**
mexicana, 67, **198**
minor, 67, **199**
multicaulis (see *Eragrostis pilosa*)
neomexicana (see *Eragrostis mexicana*)
pectinacea, 55, 67, **199**
pilosa, **200**
poaeoides (see *Eragrostis minor*)
reptans, 32, 198
spectabilis, 17, 35, 54, 59, 65, 171, **200**, 252
trichodes, 31, 57, **201**
Erianthus ravennae (see *Tripidium ravennae*)
Eriochloa, **201–2**, 251
contracta, 32, 202
villosa, 30, 67, **202**
Eriocoma hymenoides, 32, 258
Eulalia, 57, 60, 235

Fairgrounds grass, 287
Fall witch grass, 171
False melic grass, 284
Fargesia rufa (only in cultivation), 57
Feather grass, 58
horsetail, 58
Ukrainian, 58
Feather reed grass, Korean, 57
Feathertop, 57
Fescue, 203
blue, 57
clustered, 206
clustered meadow, 231
gray, 57
hard, 208
large blue, 57
meadow, 228–29
nodding, 207
rat-tail, 204
red, 206
Rocky Mountain, 207
sheep, 57, 208
six-weeks, 205
tufted, 57
western, 204
Festuca, **203–8**, 229
amethystina (only in cultivation)
arundinacea (see *Lolium arundinaceum*)
diffusa (see *Festuca rubra*)
elatior (see *Lolium arundinaceum*)
glauca (only in cultivation)
myuros, 67, **204**
occidentalis, 51, **204**
octoflora, 34, 64, 66, 67, **205**
ovina (see *Festuca trachyphylla*)
paradoxa, 46, **206**
pratensis (see *Lolium pratense*)
rubra, 54, **206**
saximontana, 50, 51, **207**
subverticillata, 43, 65, **207**
trachyphylla, 54, **208**
Finger grass, 142
Fountain grass, 57, 60
Fowl meadow grass, 276

Foxtail, 289
bristly, 293
Carolina, 97
green, 294
marsh, 97
meadow, 56, 96, 98
nodding, 290
short-awned, 96
yellow, 292
Frost grass, 58

Gama grass, 58
Glyceria, 17, 40, **208–15**, 236, 281, 308
borealis, 39, 41, 42, **209–10**, 214
canadensis, 39, 40, **211**
fernaldii (see *Torreyochloa pallida*)
grandis, 17, 39, 41, 65, 209, **211–12**
maxima, 31, 57, **213**
pallida (see *Torreyochloa pallida*)
septentrionalis, 41, 209, **214**
striata, 39, 46, 49, **214–15**, 278
Goose grass, 180
Grama grass, 115
blue, 56, 118
hairy, 58, 119
side-oats, 58, 116
Graphephorum melicoides, **215–16**

Hackonechloa macra (only in cultivation), 57
Hackone grass, 57
Hairgrass, 152
crested, 59
tufted, 58, 153
wavy, 58, 111
Hard grass, 287
Hare's-tail, 57
Heleochloa schoenoides (see *Crypsis schoenoides*)
Helictotrichon sempervirens (only in cultivation), 57, 60
Hesperostipa, **216–18**, 249, 258
comata, **217**
spartea, 34, 37, **217–18**
Hierochloe odorata (see *Anthoxanthum hirtum* subsp. *arctica*)
Holcus, **219**
lanatus, 31, **219**
mollis (only in cultivation), 57
Hordeae, 25
Hordeum, 67, 91, **220–22**
jubatum, 54, 57, **220–21**
pusillum, 31, **221**
vulgare, 66, **222**
Hystrix patula (see *Elymus hystrix*)

Ice cream grass, 201
Indian grass, 59, 258, 295

Job's tears, 57
Johnson grass, 298
June grass, 59, 223

Koeleria, **222–23**
cristata (see *Koeleria macrantha*)
macrantha, 34, 36, 50, 59, **223**

Lace grass, 195
Lagurus ovatus (only in cultivation), 57
Lawngrass, Korean, 58
Leersia, 14, 15, **224–27**
lenticularis, 46, 65, **224**
oryzoides, 41, 46, 64, 140, **225–26**
virginica, 43, 46, 65, **227**
Lemon grass, 57
Leptochloa fusca (see *Diplachne fusca* subsp. *fascicularis*)
Leptoloma cognata (see *Digitaria cognata*)
Leymus arenarius, 30, 51, 52, 57, 64, 91, **227–28**
Little bluestem, 59, 100, 285–86
Lolium, 13, **228–31**
arundinaceum, 54, 64, 66, 203, **229–30**

multiflorum (see *Lolium perenne*)
perenne, 54, **230**
pratense, 54, 64, 66, 203, 229, **231**
tementulum (see *Lolium perenne*)
Love grass, 194
creeping, 198
Indian, 200
little, 199
Mexican, 198
purple, 59, 200
sand, 57, 201
sandbar, 197
tufted, 199
Lyme grass, 57, 227–28

Magellan grass, blue, 57
Maize, 315
Manna grass, 208
eastern, 214
Fernald's false, 308
fowl, 214
northern, 209
pale false, 310
rattlesnake, 211
reed, 57, 211
tall, 213
Meadow grass, fowl, 276
Melica, **231–33**
altissima (only in cultivation), 57
ciliata (only in cultivation), 57
mutica, 32, 231
nitens, 34, 231, **232**
smithii, 30, 47, 65, 231, **233**, 284
Meliceae, 25
Melic grass, 231
false, 284
hairy, 57
twinflower, 231
purple Siberian, 57
silky, 57
Smith's, 233
tall, 232
Melic-oats, 215
Milium effusum, 43, 47, 59, 64, 65, **234**
Millet, 57, 59, 177, 234, 254, 291
broomcorn, 254
foxtail, 291
Japanese, 177
pearl, 57, 59
proso, 254
wood, 59, 234
Miscanthus, **235–36**
floridulus (only in cultivation), 57
sacchariflorus, 30, 57, **235–36**
sinensis (only in cultivation), 57, 60, 235
Molinia, **236–37**
caerulea, 31, 58, **237**
littoralis (only in cultivation), 58
Moor grass, 58, 237
autumn, 58
blue, 58
green, 58
purple, 58, 237
tall, 58
Mountain rice grass, 267
Canada, 266
Muhlenbergia, 65, **238–48**
asperifolia, 55, 65, **239**
bushii, 32, 241
cuspidata, 35, **240**
frondosa, 46, **241**
glomerata, 39, 40, **242**, 244
mexicana, 54, **243**
racemosa, 37, 38, 50, 242, **244**
reverchonii (only in cultivation), 58
richardsonis, 30, 40, **245**
schreberi, 36, **245**
sobolifera, 35, **246**
sylvatica, 35, **246**
tenuiflora, 36, **247**
uniflora, 40, **247–48**
Muhly grass, 238
autumn embers, 58

Muhly grass (*continued*)
  bog, 247
  creeping, 246
  forest, 246
  mat, 245
  Mexican, 243
  nodding, 241
  plains, 240
  slender, 247
  wire-stemmed, 241
Munro grass, 145

Nassella viridula, 216, **249**, 258
Needle-and-thread, 217
Needle grass, 216, 217
  green, 249
Nimble-will, 245

Oatgrass, 50, 56, 57, 109, 112, 151
  blue, 57, 50
  downy alpine, 112
  flattened, 151
  poverty, 151
  tall, 109
  variegated, 56
Oats, 110
  wild, 110, 141
Orchard grass, 149
Oryzeae, 24
Oryzopsis, **250**, 258
  asperifolia, 47, 48, 64–66, **250**
  canadensis (see *Piptatheropsis canadensis*)
  pungens (see *Piptatheropsis pungens*)
  racemosa (see *Patis racemosa*)

Pampas grass, 57
  hardy, 58
Paniceae, 24
Panic grass, 144, 155, 251
  broad-leaved, 163
  fall, 253
  fen, 253
  forked, 162
  hairy, 158
  Leiberg's, 163
  Lindheimer's, 158
  linear-leaved, 164
  long-stalked, 167
  northern, 158
  pale, 170
  Philadelphia, 255
  poverty, 161
  prairie, 168
  puberulent, 159
  red-dot, 166
  red-topped, 144
  Scribner's, 166
  Shinners', 160
  slender, 165
  slender-leaved, 164
  slender-rosette, 170
  starved, 161
  twice-blooming, 155, 157
  Wilcox's, 168
Panicoideae, 24
Panicum, 144, **251–56**, 267
  boreale (see *Dichanthelium boreale*)
  capillare, 13, 54, 65, 66, 171, 200, 251, **252**
  columbianum (see *Dichanthelium columbianum*)
  commonsianum var. *euchlamydeum* (see *Dichanthelium commonsianum* var. *euchlamydeum*)
  depauperatum (see *Dichanthelium depauperatum*)
  dichotomiflorum, 23, 54, 67, 251, **253**
  dichotomum (see *Dichanthelium dichotomum*)
  flexile, 40, 66, **253**
  latifolium (see *Dichanthelium latifolium*)
  leibergii (see *Dichanthelium leibergii*)
  lindheimeri (see *Dichanthelium acuminatum* var. *lindheimeri*)

linearifolium (see *Dichanthelium linearifolium*)
meridionale (see *Dichanthelium meridionale*)
miliaceum, 67, **254**
oliganthes var. *scribnerianum* (see *Dichanthelium oligosanthes* var. *scribnerianum*)
perlongum (see *Dichanthelium perlongum*)
philadelphicum, 66, **255**
praecocius (see *Dichanthelium villosissimum* var. *praecocius*)
rigidulum (see *Coleataenia rigidula*)
subvillosum (see *Dichanthelium acuminatum* var. *fasciculatum*)
tuckermanii (see *Panicum philadelphicum*)
virgatum, 12, 19, 20, 36–38, 59, 64, 236, 251, **255–56**
xanthophysum (see *Dichanthelium xanthophysum*)
Pascopyrum smithii, 64, 91, **256–57**
Paspalum, 251, **257–58**
ciliatifolium (see *Paspalum setaceum*)
setaceum, 36, **258**
Patis racemosa, 23, 43, 44, 65, 66, 250, **258–59**
Pennisetum
alopecuroides (see *Cenchrus purpurascens*)
glaucum (see *Cenchrus spicatus*)
Phalaris, 138, **260–62**
arundinacea, 11, 18, 19, 38–40, 54, 58, 63, 134, 135, **260–61**
canariensis, **262**
Phleum pratense, 18, 54, 63, 96, **262–63**
Phragmites, **263–65**
australis, 39, 40, 63, **264–65**
australis subsp. *americanus*, 264
communis (see *Phragmites australis*)
Phyllostachys aurea (only in cultivation), 58, 60
Piptatheropsis, 250, 258, **266–67**
canadensis, 50, **266**
pungens, 48, 50, 65, **267**
Piptatherum canadense (see *Piptatheropsis canadensis*)
Piptatherum pungens (see *Piptatheropsis pungens*)
Pleioblastus, 58
pygmeus (only in cultivation), 58
viridistriatus (only in cultivation), 58
Plume grass, 58
Poa, 12, 14, 17, 215, **267–79**, 281, 308
alsodes, 47, 64, 65, **270**
annua, 64, 67, **271**, 287
arida, 31, 55, 64, **271**
bulbosa, 31, 54, 267, **272**
chapmaniana, 32, 265
compressa, 12, 50–52, 54, 55, 63, 263, 267, **273**
glauca, 52, 53, 64, **273**
interior, 32, 268
languida, 47, 65, **274**, 278
nemoralis, 47, 65, 267, **274–75**
paludigena, 40, 65, **276**
palustris, 39, 40, 49, **276–77**
pratensis, 11, 14, 50, 54, 63, 64, 267, **277**
saltuensis, 47, 65, 274, **278**
sylvestris, 43, 65, **278**
trivialis, **279**
wolfii, 47, 65, **279**
Poeae, 25
Polypogon, **280–81**
interruptus, 67, **280**
monspeliensis, 31, 67, **281**
Polypogon, ditch, 280
Pooideae, 25
Prickle grass, 146
Puccinellia, 208, **281–83**, 308
distans, 55, **282**

Puccinellia (*continued*)
fernaldii (see *Torreyochloa fernaldii*)
nuttalliana, 55, **283**
pallida (see *Torreyochloa pallida*)
Purple moor grass, 58, 237
Purple-top, 58, 311

Quackgrass, 188
Quaking grass, 56, 122

Rabbit's-foot grass, 281
Rattlesnake grass, 211
Ravenna grass, 58
Redtop, 93
Red-topped panic grass, 144
Reed, 263
common, 263–65
sand, 138
Reed canary grass, 260–61
Reed grass, 57, 59, 136–37
Chee, 57, 136
feather, 57, 59
Korean feather, 57
slim-stemmed, 137
Ribbon grass, 58
Rice, 317
northern wild, 318
southern wild, 320
Rice grass, 250, 259, 266–67
black-seeded, 259
Canada mountain, 266
Indian, 259
mountain, 267
rough-leaved, 250
Rough-leaved rice grass, 250
Rye, 288
Ryegrass, 228
perennial, 230
sand, 57

Sacaton grass, 58
Sacchareae, 25
Saccharum ravennae (see *Tripidium ravennae*)
Salt grass, 175
Sandbur, field, 140
Sand grass, purple, 311–12
Sand-reed, 138
Satin grass, 300
Schedonorus
arundinaceus (see *Lolium arundinaceum*)
pratensis (see *Lolium pratense*)
Schizachne purpurascens, 47, 48, 64, 65, 233, **284**
Schizachyrium scoparium, 34, 36, 37, 50, 59, 64, 100, **285–86**
Sclerochloa dura, 31, 55, 67, **287**
Scratch grass, 239
Sea oats, northern, 57
Secale cereale, 67, 91, **288**
Sesleria, 58
autumnalis (only in cultivation), 58
caerulea (only in cultivation), 58
heufleriana (only in cultivation), 58
Setaria, 6, 19, 54, 67, 280, **289–93**
faberi, 30, **290**
glauca (see *Setaria pumila*)
italica, **291**
lutescens (see *Setaria pumila*)
pumila, 20, 55, 63, **292**
verticillata, **293**
verticilliformis, 30, **293**
viridis, **294**
Shingle grass, 57, 141
Shorthusk, 120
northern, 120
southern, 120
Shorthusked grass, 120
Side-oats grama, 116
Silver grass, 57, 235
Amur, 57, 235
Chinese, 57, 60
giant Chinese, 57, 235

Slough grass, 114
Sorghastrum nutans, 34, 36, 37, 59, **295**
Sorghum, 296
Sorghum, **296–98**
bicolor, 67, **296**
halepense, **298**
× *almum*, **296**
Spangle grass, 57
Spartina pectinata, 38, 39, 41, 42, 59, 66, **299**
Spear grass, silver, 56
Sphenopholis, 143, 223, **300–302**
intermedia, 32, 39, 40, 52, 66, **301**
nitida, 300
obtusata, 44, **302**
Spodiopogon sibiricus (only in cultivation), 58
Sporobolus, 22, 64, **302–7**
airoides (only in cultivation), 58
asper (see *Sporobolus compositus*)
clandestinus, **303**
compositus, 34, **304**
cryptandrus, 36, **305**
heterolepis, 11, 13, 34, 37, 59, 64, **306**
neglectus, 55, 66, **307**
vaginiflorus, 55, 66, **307**
wrightii (only in cultivation), 58
Sprangle-top, bearded, 174
Squirrel-tail barley, 57
Stink grass, 196
Stipa, 58
comata (see *Hesperostipa comata*)
spartea (see *Hesperostipa spartea*)
tirsa (only in cultivation), 58
ucrainica (only in cultivation), 58
viridula (see *Nassella viridula*)
Stipeae, 25
Sweet grass, northern, 102
Sweet vernal grass, 103
Switch grass, 59, 256

Tall oatgrass, 109
Thinopyrum ponticum, 32, 181
Three-awned grass, 104–8
arrow-feathered, 108
curly, 106
Curtiss's, 106
dune, 108
fork-tipped, 105
Kearney's, 107
old field, 107
Tickle grass, 92
northern, 95
southern, 94
Timothy, 263
marsh wild, 242
upland wild, 243
Torreyochloa, 208, 215, 281, **308–10**
fernaldii, 40, 42, **308**
pallida, 40, 42, **310**
Tridens, 310
Tridens flavus, 30, 36, 58, **310–11**
Tripidium ravennae (only in cultivation), 58
Triplasis purpurea, 36, 64, 66, **311–12**
Tripsacum dactyloides (only in cultivation), 58
Trisetum, 313
Trisetum, **313**
melicoides (see *Graphephorum melicoides*)
spicatum, **313**
Triticum aestivum, 67, 91, **314**
Twice-blooming panic grass, 155

Velvet grass, 219
Ventenata dubia, 31, 67, **315**
Ventenata grass, 315
Vernal grass, sweet, 103
Vulpia
myuros (see *Festuca myuros*)
octoflora (see *Festuca octoflora*)

Wedge grass, 300
prairie, 302

Wedge grass (*continued*)
satin, 301
shiny, 300
Wheat, 314
Wheat grass, 91, 181, 186, 189–90, 256–57
crested, 91
slender, 189–90
tall, 181
thick-spiked, 186
western, 256–57
White grass, 227
Wild oats, 141
woodland, 57
Wild rice, 317–20
Wild-rye, 181
awnless, 184
Canada, 59, 183
downy, 59, 190–91
Ebinger's, 192
Macgregor's, 188
Malte's, 192
Minnesota, 185
riverbank, 189
silky, 59
Virginia, 192
Wiegand's, 193
Windmill grass, 142
Witch grass
common, 252
fall, 171
Wood millet, 59
Wood-reed, 143
common, 143
drooping, 144

Yorkshire fog, 219

Zea mays, 12, 20, 21, 67, **315–16**
Zizania, 14, 15, 20, 21, **317–20**
aquatica, 21, 42, 66, **318–19**
aquatica var. *angustifolia* (see *Zizania palustris*)
aquatica var. *interior* (see *Zizania palustris*)
palustris, 42, 66, **320**
Zoysia japonica, 58
Zoysieae, 24